CRC Series in **ANALYTICAL BIOTECHNOLOGY**

NEW METHODS IN PEPTIDE MAPPING FOR THE CHARACTERIZATION OF PROTEINS

CRC Series in
ANALYTICAL BIOTECHNOLOGY

Series Editor
William S. Hancock
Hewlett-Packard, Inc.

Advisory Editors

Barry Karger
Northeastern University

Csaba Horvath
Yale University

Fred Regnier
Purdue University

Donald B. Wetlaufer
University of Delaware

New and Forthcoming Titles

Deamidation and Isoaspartate Formation in Peptides and Proteins, *Dana Aswad*

New Methods in Peptide Mapping for the Characterization of Proteins, *William Hancock*

NMR in Drug Design in Biotechnology, *David J. Craik*

Capillary Electrophoresis in the Biotechnology Industry, *Pier Giorgio Righetti*

High Performance Ion-Exchange Chromatography of Proteins, *Istvan Mazsaroff*

Oxidation Reactions in Peptides and Proteins, *Eleanor Canova-Davis*

Methods for Assessing Genetic Stability, *Mickey Williams*

Methods for the Analysis of Antisense Oligonucleotides, *Aharon Cohen*

QC Methods in Biotechnology, *Alan Herman*

Impurity Analysis of rDNA Derived Proteins, *Vince Anicetti*

Chromatographic Separations of Peptide and Protein Samples, *Benny Welinder*

CRC Series in **ANALYTICAL BIOTECHNOLOGY**

NEW METHODS IN PEPTIDE MAPPING FOR THE CHARACTERIZATION OF PROTEINS

Edited by
William S. Hancock, Ph.D., D.Sc.
Hewlett-Packard Laboratories
Palo Alto, California

CRC Press
Boca Raton   New York   London   Tokyo

Library of Congress Cataloging-in-Publication Data

New methods in peptide mapping for the characterization of proteins/
edited by William S. Hancock.
    p.   cm. — (Analytical biotechnology)
    Includes bibliographical references and index.
    ISBN 0-8493-7822-2 (alk. paper)
    1. Proteins—Analysis. I. Hancock, William S. II. Series: CRC
series in analytical biotechnology.
QP551.N4936 1995
547.7'5046—dc20   95-34181
                                                                                                                              CIP

    This book contains information obtained from authentic and highly regarded sources. Reprinted material is quoted with permission, and sources are indicated. A wide variety of references are listed. Reasonable efforts have been made to publish reliable data and information, but the author and the publisher cannot assume responsibility for the validity of all materials or for the consequences of their use.

    Neither this book nor any part may be reproduced or transmitted in any form or by any means, electronic or mechanical, including photocopying, microfilming, and recording, or by any information storage or retrieval system, without prior permission in writing from the publisher.

    All rights reserved. Authorization to photocopy items for internal or personal use, or the personal or internal use of specific clients, may be granted by CRC Press, Inc., provided that $.50 per page photocopied is paid directly to Copyright Clearance Center, 27 Congress Street, Salem, MA 01970 USA. The fee code for users of the Transactional Reporting Service is ISBN 0-8493-7822-2/96/$0.00+$.50. The fee is subject to change without notice. For organizations that have been granted a photocopy license by the CCC, a separate system of payment has been arranged.

    CRC Press, Inc.'s consent does not extend to copying for general distribution, for promotion, for creating new works, or for resale. Specific permission must be obtained in writing from CRC Press for such copying.

    Direct all inquiries to CRC Press, Inc., 2000 Corporate Blvd., N.W., Boca Raton, Florida 33431.

© 1996 by CRC Press, Inc.

No claim to original U.S. Government works
International Standard Book Number 0-8493-7822-2
Library of Congress Card Number 95-34181
Printed in the United States of America 1 2 3 4 5 6 7 8 9 0
Printed on acid-free paper

SERIES PREFACE

Analytical protein chemistry has played a key role in the approval of new protein pharmaceuticals produced by recombinant DNA-technology as well as in the development of small molecule mimetics. Therefore, recent advances in this field are of strong interest to the industry, as well as to academic and government scientists who are contibuting to biotechnology developments. The scope of this series will include important topics in analytical chemistry such as capillary electrophoresis and mass spectrometry. The reviews will also include advances in organic chemistry, protein and nucleic acid chemistry, and relevant sections of biochemistry that contribute to an understanding of the properties of the biopolymers, such as mechanisms of deamidation and oxidation.

THE EDITOR

Dr. Hancock joined Hewlett-Packard Laboratories at the beginning of 1994 to develop a new generation of bio-instrumentation and to continue the characterization of complex bimolecules by hyphenated liquid-phase analysis and different forms of mass spectrometry.

Prior to joining Hewlett-Packard, Dr. Hancock spent eight years at Genentech Inc., where he developed novel analytical methods that were used for the FDA approval of some of the first biotechnology products such as human growth hormone and tissue plasminogen activator. He established the Analytical Chemistry Department at Genentech and later served as Director of Pharmacology and studied the application of analytical methods to protein metabolism.

In 1984, he was a Visting Scientist at Bureau of Drugs (FDA) and established the HPLC assay for insulin certification within the agency.

Dr. Hancock is an Associate Editor of *Analytical Chemistry*, a member of the editorial board of the *Journal of Chromatography* and Editor-in-Chief, CRC Series in Analytical Biotechnology. He is President of the Californian Separation Science Society and Chairman-elect for the 20th International Symposium on Column Liquid Chromatography (1996) and the 7th International Symposium on Capillary Electrophoresis (1997).

Dr. Hancock has 7 patents and more than 120 publications. He is also the author of 4 books which include, the CRC handbooks of *HPLC for the Separation of Amino Acids, Peptides, and Proteins*.

Contributors

James H. Bourell
Protein Chemistry Department
Genentech, Inc.
South San Francisco, California

Eleanor Canova-Davis, Ph.D.
Medical and Analytical Chemistry Department
Genentech, Inc.
South San Francisco, California

Richard M. Caprioli, Ph.D.
The University of Texas Medical School
Department of Biochemistry and Molecular Biology
Analytical Chemistry Center
Houston, Texas

Marian L. Eng, M.Sc.
Analytical Chemistry Department
Genentech, Inc.
South San Francisco, California

Beth L. Gillece-Castro, Ph.D.
Protein Chemistry Department
Genentech, Inc.
South San Francisco, California

Andrew W. Guzzetta
Genentech, Inc.
South San Francisco, California

William S. Hancock, Ph.D.
Genentech, Inc.
South San Francisco, California

William J. Henzel
Protein Chemistry Department
Genentech, Inc.
South San Francisco, California

Rodney G. Keck
Genentech, Inc.
South San Francisco, California

Bruce A. Keyt, Ph.D.
Cardiovascular Research
Genentech, Inc.
South San Francisco, California

Paoli Lecchi, Ph.D.
Laboratory of Analytical Chemistry
National Institutes of Health (NIDDK)
Bethesda, Maryland

Paul-Jane Lee, M.S.
Department of Biochemistry
Rutgers University
Piscataway, New Jersey

Victor T. Ling, M.S.
Analytical Chemistry Department
Genentech, Inc.
South San Francisco, California

Lydia M. Nuwaysir, Ph.D.
Protein Chemistry Department
Genentech, Inc.
South San Francisco, California

Kathy L. O'Connell
Protein Chemistry Department
Genentech, Inc.
South San Francisco, California

Eugene C. Rickard, Ph.D.
Eli Lilly and Company
Lilly Research Laboratories
Indianapolis, Indiana

Hans-Jürgen P. Sievert, Ph.D.
Hewlett-Packard Company
Wilmington, Delaware

Richard D. Smith, Ph.D.
Chemical Sciences Department
Pacific Northwest Laboratory
Richland, Washington

Lloyd R. Snyder, Ph.D.
LC Resources Inc.
Orinda, California

John T. Stults, Ph.D.
Protein Chemistry Department
Genentech, Inc.
South San Francisco, California

John K. Towns, Ph.D.
Eli Lilly and Company
Lilly Research Laboratories
Indianapolis, Indiana

Harold R. Udseth, Ph.D.
Chemical Sciences Department
Pacific Northwest Laboratory
Richland, Washington

Jon H. Wahl, Ph.D.
Chemical Sciences Department
Pacific Northwest Laboratory
Richland, Washington

TABLE OF CONTENTS

Chapter 1
**Subtleties of Peptide Mapping in the
Analysis of Protein Pharmaceuticals** 1
*Victor T. Ling, Marian L. Eng, Paul-Jane Lee, Rodney G. Keck, Bruce A.
Keyt, and Eleanor Canova-Davis*

Chapter 2
Optimization of Peptide Mapping via Computer Simulation 31
Lloyd R. Snyder

Chapter 3
**The Use of UV-Spectral Information in the Evaluation and
Interpretation of Tryptic Maps** 57
Hans-Jürgen P. Sievert

Chapter 4
**The Use of Capillary Electrophoresis for Peptide Mapping of
Proteins** 97
Eugene C. Rickard and John K. Towns

Chapter 5
Packed Capillary HPLC-Electrospray Ionization Mass Spectrometry 119
*John T. Stults, Beth L. Gillece-Castro, William J. Henzel, James H.
Bourell, Kathy O'Connell, and Lydia M. Nuwaysir*

Chapter 6
Capillary Electrophoresis—Mass Spectrometry in Peptide Mapping 143
Jon H. Wahl, Harold R. Udseth, and Richard D. Smith

Chapter 7
**Analyzing Reversed Phase Peptide Maps of Recombinant
Human Glycoproteins Using LC/ES/MS** 181
Andrew W. Guzzetta and William S. Hancock

Chapter 8
**Matrix-Assisted Laser Desorption Mass Spectrometry for Peptide
Mapping** 219
Paolo Lecchi and Richard M. Caprioli

Index 241

INTRODUCTION

After a decade of successful approvals in the biotechnology industry, we can note the key role analytical chemistry has played in the regulatory acceptance of protein pharmaceuticals produced by recombinant DNA technology. Ultimately, the quality of a pharmaceutical can only be established through appropriately designed human clinical trials.[1] However, a comprehensive analytical chemical program for the characterization and control of manufacture of the recombinant product often provides an important level of assurance in the approval process.

Reversed-phase high-performance liquid chromatography (RP-HPLC) has been invaluable for the characterization of these pharmaceuticals.[2] This technique can be used to establish chromatographic identity between the recombinant and natural material, as well as demonstrating that the purification train has produced a high-purity product. Also, RP-HPLC has proven to be a powerful technique for the isolation and quantitation of variants of the target sequence. Such variants may arise from the inherent heterogeneity of the molecule, e.g., the presence of carbohydrate chains or because of degradative reactions such as deamidation.[3,4]

A major problem in the detection of such variants is the unpredictable nature of the chromatographic properties of closely related variants and thus the appearance of a single peak in a chromatographic analysis may not be a reliable estimate of product purity. An example of the strengths and weaknesses of RP-HPLC occurred with the low-level substitution of norleucine for methionine in some bacterial expression systems. This substitution was observed in both interleukin-2 and bovine growth hormone and was attributed to an insufficient supply of amino acids, in particular methionine and leucine, during the fermentation process. Using similar conditions to the published studies, a preparation of methionyl-human growth hormone (met-hGH) could be isolated which contained a 2% level of substitution of norleucine for methionine.[5]

The presence of norleucine could be confirmed by a combination of analytical techniques, such as amino acid analysis, N-terminal sequencing, FAB-mass spectrometry, and reversed phase HPLC (RP-HPLC). In fact, RP-HPLC provided the most sensitive and rapid assay for the norleucine variants. However, the low pH system of TFA/ACN was only able to separate one of the four substitutions (norleucine for methionine at position 125), while a neutral pH mobile phase showed a dramatic difference in selectivity. This study, as well as others,[2,3] illustrates the unpredictable nature of the ability of RP-HPLC to separate protein variants.

In the late 1970s, we and others adapted RP-HPLC to peptide mapping and applied the procedure to the confirmation of the linear sequence of proteins.[6] This procedure was particularly appropriate for the products of rDNA technology where the expected sequence could be predicted with some certainty from the cDNA used in the cloning step. In this situation, the bulk of the sequence could probably be assumed as being present, but the N- and C-terminal sequences needed confirmation. In addition, the possibility of more unlikely events such as deletion of segments of the protein, or mutational events, could be monitored. With experience, other changes were detected such as unexpected side reactions in the cleavage of fusion proteins and degradative processes that occurred during manufacture or formulation of the bulk drug substance. Chapter 1 by Canova-Davis et al. gives an excellent account of the experience derived from the production of several biotechnology products. It is impressive to note the consistent ability of the peptide map to detect the variety of chemical transformations possible for such complex macromolecules. It is therefore not surprising that both the industry and the regulatory agencies have adopted the tryptic map as the premier method for both the characterization and release of protein products.

However, the characterizations of high molecular weight proteins, and in particular glycoproteins, have exposed the limitations of tryptic mapping techniques, where a significant number of coelutions often can be detected by a second analytical technique. The degree of coelutions can be minimized by careful optimization of the HPLC conditions used in the peptide map, but such a process can be time-consuming and susceptible to differences between analysts or laboratories.

Chapter 2 by Snyder gives a useful discussion of approaches for the use of computer simulation for the rapid optimization of the peptide map. With the widespread availability of diode-array detectors, it is often possible to conveniently record the spectra of all peaks. It is underappreciated, however, that the spectral information is valuable both for the detection of coelutions and for identification of the nature of a given peak if a set of standards is available. Chapter 3 by Sievert gives a convincing demonstration of the power of spectra in the interpretation of peptide maps.

Another approach to the problem of coelutions is to take advantage of the resolving power of capillary electrophoresis (HPCE) as well as the orthogonal nature of the separation (charge instead of hydrophobicity). Chapter 4 by Rickard and Towns gives an excellent account of the application of HPCE to peptide mapping and gives practical tips on developing the most effective conditions. These authors have had extensive experience at Eli Lilly and Company for the application of HPCE to the analysis of biotechnology products such as human growth hormone, and this success is being repeated at other biotechnology companies.

For complex proteins such as glycosylated samples, or high molecular weight rDNA-derived products such as monoclonal antibodies, the extra dimension of mass spectrometry may be required to fully characterize a peptide map. With LC/MS the mass spectrometer allows on-line identification of peptides and coelutions are readily detected. Many electrospray MS systems must use very low flow rates, and thus one must resort to the use of capillary chromatography and/or stream splitters.

Chapter 5 by Stults et al. gives an excellent description of these approaches and shows the power of electrospray mass spectrometry (ESI-MS) as an on-line HPLC detector. In a similar manner, Chapter 6 by Smith et al. describes how ESI-MS can be successfully coupled to capillary electrophoresis. While this was initially a technically challenging interface, on-line detection is even more important in HPCE due to the very small sample loads and the lack of other approaches to monitor these separations. Recent advances in instrumentation, including some that are now available, have served to make the CE-MS based upon ESI-MS more routine.

The application of peptide mapping to glycoprotein characterization has been hindered by the difficulties of both separating and detecting families of closely related glycoforms. In the past, the use of mass spectrometric approaches such as FAB (fast atom bombardment) on-line with HPLC has suffered from the difficulty of ionizing polar samples, such as glycopeptides.[7] The arrival of electrospray mass spectrometry (ESI-MS) has represented a tremendous advance in that it is ideally suited for HPLC-MS. Chapter 7 by Guzzetta and Hancock describes the application of ESI-MS to analyze complex glycoproteins via on-line analysis of HPLC-tryptic maps. In this approach the carbohydrates are left attached to the peptide and are analyzed at the same time as the other peptides. Mapping of the glycopeptides is particularly useful for variants prepared by site-directed mutagenesis, as these samples are often available only in small amounts.[8]

Matrix-assisted laser desorption mass spectroscopy when combined with time of flight MS (known as MALDI-TOF) has become a promising new approach to characterize polypeptides without some of the disadvantages of ESI and with potentially higher sensitivity. Chapter 8 by Lecchi and Caprioli gives a review of the early experiments on monitoring peptide mapping by this new approach.

In conclusion, this text is a strong demonstration of the rapid developments that have occurred since the introduction of the HPLC-peptide map in 1978. The chapters have been designed to encourage those currently characterizing proteins to make full use of the power of the method, while at the same time being aware of its limitations. Analytical chemists now understand that the map cannot completely resolve complex mixtures of peptides or quantitate peptides much below the 5% level. Such observations, however, have acted as a stimulant to improve both the separation and detection areas and have assured a strong future for this important analytical method.

REFERENCES

1. Anicetti, V., Keyt, B.A., and Hancock, W.S., Purity analysis of protein pharmaceuticals produced by rDNA technology, *Trends Biochem. Sci.,* 7, 342, 1989.
2. Frenz, J., Hancock, W.S., Henzel, W., and Horvath, C., Reversed phase chromatography in analytical biotechnology of proteins, in *HPLC of Biological Macromolecules,* Marcel Dekker, New York, 1990, 145.
3. Hancock, W.S., Canova-Davis, E., Chloupek, R.C., Wu, S.L., Baldonado, I.P., Battersby, J.E., Spellman, M.W., Basa, L.J., and Chakel, J.A., *Therapeutic Peptides and Proteins: Assessing the New Technologies,* Banbury Report 29, Cold Spring Harbor Laboratory Press, Cold Spring Harbor, N.Y., 1988, 95.
4. Becker, G.W., Tackitt, P.M., Bromer, W.W., Lefeber, D.S., and Riggin, R.M., Isolation and characterisation of a sulfoxide and desamido derivative of human growth hormone, *Biotech. Appl. Biochem.,* 10, 326, 1987.
5. Canova-Davis, E., Teshima, G.M., Kessler, T.J., Lee, P.J., Guzzetta, A.W., and Hancock, W.S., Strategies for an analytical examination of biological pharmaceuticals, *ACS Symp.,* 434, 90, 1990.
6. Hancock, W.S., Bishop, C.A., and Hearn, M.T.W., The use of high pressure liquid chromatography for the peptide mapping of proteins, *Anal. Biochem.,* 89, 203, 1978.
7. Covey, T.R., Huang, E.C., and Henion, J.D., Structural characterisation of protein tryptic peptides via liquid chromatography/mass spectrometry and collision induced dissociation of their doubly charged molecular ions, *Anal. Chem.,* 63, 1193, 1991.
8. Ling, V., Guzzetta, A.W., Canova-Davis, E., Stults, J.T., and Hancock, W.S., Characterisation of the tryptic map of recombinant DNA derived tissue plasminogen activator by high-performance liquid chromatography-electrospray ionization mass spectrometry, *Anal. Chem.,* 63, 2909, 1991.

CHAPTER 1

SUBTLETIES OF PEPTIDE MAPPING IN THE ANALYSIS OF PROTEIN PHARMACEUTICALS

Victor T. Ling, Marian L. Eng, Paul-Jane Lee, Rodney G. Keck, Bruce A. Keyt, and Eleanor Canova-Davis

CONTENTS

1 Introduction 2

2 Peptide Maps as Protein Fingerprints 3
   2.1 Peptide Maps for Disulfide Assignments 3
       2.1.1 Prevention of Disulfide Exchanges 5
       2.1.2 Transpeptidation During Analytical Proteolysis 8
   2.2. Peptide Mapping for Variant Identification 9
       2.2.1 Internal Cleavages 9
       2.2.2 Amino- and Carboxyl-Termini Truncations 11
           2.2.2.1 Isolation Artifacts 11
           2.2.2.2 Chemical Degradations 15
       2.2.3 Isomerization of Aspartic Acid and Proline
            Residues 16
       2.2.4 Deamidation of Asparagine Residues 19
       2.2.5 Oxidation of Methionine Residues 20
       2.2.6 Amino Acid Substitutions 22
       2.2.7 Chemical Modifications 22
       2.2.8 Peptide Mapping Artifacts 23

3 Discussion 24

Acknowledgments 25

References 25

1 INTRODUCTION

Sanger et al.[1-4] determined the amino acid sequence of insulin in the early 1950s without the use of high-purity reagents, modern chemical procedures, or the sophisticated analytical instruments that are prevalent in protein chemistry today. Instead, they relied upon nonspecific acid hydrolysis and enzymatic digestions using pepsin or a crude pancreatic extract which yielded families of peptides with similar sequences. More than 150 short peptides were generated from insulin by using these methods. These peptides were purified by adsorption onto charcoal or ion-exchange resins with subsequent ionophoresis on filter paper followed by elution for analyses. This procedure was extremely labor-intensive and required large amounts of peptide.

A cardinal improvement was effected in later years by the purification of enzymes with high specificity for a particular type of peptide bond, resulting in a reduction in the number of fragments generated during digestion. Additionally, the enzymes chosen were capable of cleaving the desired bond in near-quantitative yield. It was concluded that chymotrypsin, pepsin, and papain were too nonspecific, and thrombin was too restricted.[5] The proteases with the greatest utility have been trypsin with specificity at arginine and lysine residues, clostripain with specificity at arginine residues,[6] endoproteinase Lys-C with specificity at lysine residues,[7] and *Staphylococcus aureus* V_8 (endoproteinase Glu-C) with specificity at glutamic and aspartic acid residues.[8] This development of controlled proteolytic digests was followed by the separation of the resultant peptides using two-dimensional ascending paper chromatography and electrophoresis.[9] Ingram[10] called the resulting chromatogram the "fingerprint" of the protein. In this manner he was able to identify the single amino acid substitution in hemoglobin S.[10]

Peptide separations with higher resolving power were obtained by Katz et al.[11] utilizing descending paper chromatography followed by high-voltage electrophoresis. This method was reported as being capable of comparing digestion mixtures and detecting differences as "slight" as the replacement of a single amino acid. The resulting patterns of up to 80 peptide spots were referred to as "peptide maps".

The next advance in peptide mapping was the use of ion-exchange chromatography in volatile organic developers[12] followed by adaptation to the automatic chromatographic equipment normally used for amino acid analyses.[13] This procedure was more reproducible and yielded purer peptides than those obtained from the paper methods.

It was several years before the natural progression to the use of reversed-phase high performance liquid chromatography (HPLC) from the reversed-phase partition chromatography of Howard and Martin[14] was investigated.[15] The technique quickly evolved from the initial use of phosphoric acid-containing mobile phases to the volatile trialkylammonium phosphate,[16] formic acid,[17] and trifluoroacetic acid (TFA) ion-pairing reagents.[18]

Reversed-phase HPLC separations are relatively fast and the resolution is extremely good. Separation conditions are easily modified by gradient programming to provide optimal separations. Trifluoroacetic acid solutions are excellent solvents in which to chromatograph peptides of all sizes with the following advantages: (1) volatility; (2) transparency in the 210 to 230 nm range so that detection of all peptides is sensitive and nondestructive; and (3) generally good recovery. Hence, the technique of peptide mapping matured over a score of years to a point where the routine analysis of macromolecules is possible. The ability to separate and recover peptides in nanomole amounts makes reversed-phase HPLC a particularly convenient means of preparing peptides for studies of protein structure.

2 PEPTIDE MAPS AS PROTEIN FINGERPRINTS

Peptide mapping has become a useful technique to establish the fidelity of protein translation in biological systems, especially in the case of recombinant DNA-derived proteins. In addition to providing definitive data on the primary sequence of a protein, peptide mapping techniques which involve the separation of relatively small peptides can also provide information concerning disulfide linkages and protein variants. A search is ordinarily conducted for oxidations and deamidations. In the case of proteins prepared using recombinant methods particular attention is paid to detecting signal sequences, proteolytically degraded species, or chemical modifications resulting from isolation procedures.

The number of peptide fragments generated by a particular enzyme can be predicted by considering the specificity of the protease involved. The fragments are conventionally numbered sequentially from the amino terminus through to the carboxyl terminus.

2.1 Peptide Maps for Disulfide Assignments

Since some proteins including relaxin, insulin-like growth factor (IGF-I), human growth hormone (hGH), recombinant human tissue factor, recombinant human immunodeficiency virus envelope glycoprotein (gp120), and a soluble form of human CD4 can be completely digested enzymatically without first reducing their disulfide linkages, it is possible to identify the peptides which are involved in each linkage.[19-26] Analyses of this type were used to verify the correct pairing of the cysteine residues involved in disulfide bridge formation during the combination reaction between the A- and B-chains of relaxin.[27] The relaxin molecule (Figure 1) is composed of two nonidentical peptide chains linked by two disulfide bridges with an additional intrachain disulfide linkage in the smaller A-chain of 24 amino

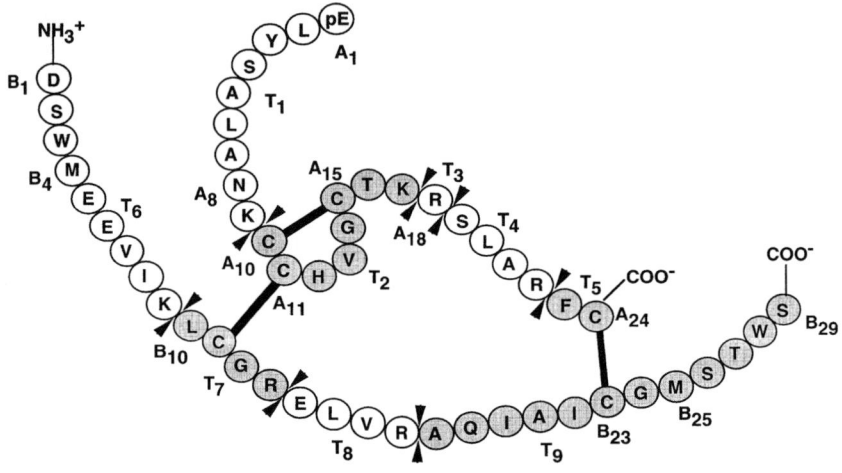

FIGURE 1
Primary sequence of human relaxin with disulfide bonds and tryptic cleavage sites indicated. The A-chain amino terminal residue is pyroglutamic acid (pE). The "T" denotes a tryptic fragment numbered consecutively from the amino termini beginning with the A-chain and continuing through the B-chain.

acids. The B-chain is 29 amino acids in length. It was possible to detect peptides which were a result of disulfide shuffling under the digest conditions (pH 8.2) and hence identify areas of the map to monitor for the detection of misfolded variants.

The identification of the disulfide linkages in hGH was unequivocal since each linkage was contained in separate tryptic fragments (Figure 2). The disulfide-linked pairs Cys^{53}-Cys^{165} and Cys^{182}-Cys^{189} could be unambiguously assigned.[23] The cystine-containing peptides can then be characterized by comparing reversed-phase behavior under reduced and nonreduced conditions.[28] This technique will lead to the disappearance of the disulfide-linked peptides and the appearance of the cysteine-containing peptides in the map of the reduced digest. It is thus possible to estimate the amount of free sulfhydryl groups in the nonreduced protein by searching for peaks in the positions where the cysteine-containing peptides were shown to elute. No cysteine-containing peptides were found.

Since both relaxin and IGF-I contain adjacent cysteine residues, two of the linkages required additional cleavages, using tandem mass spectrometry in the case of relaxin[20] and Edman degradation plus enzymatic digestion for IGF-I.[21,22] It was actually possible to isolate an improperly folded IGF-I molecule (Figure 3B). The V_8 protease map is shown in Figure 4. The misfolded V_2-V_7 peptide fragment elutes at an earlier retention time. This fragment was isolated and shown to be in a Cys^6-Cys^{47} and Cys^{48}-Cys^{52} configuration by Edman degradation and mass spectral analysis of the

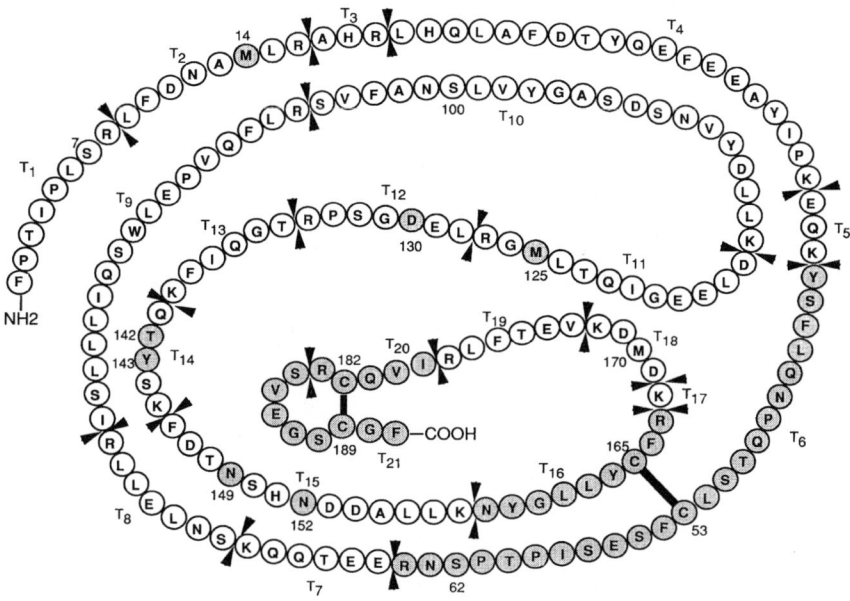

FIGURE 2
Primary sequence of human growth hormone with disulfide bonds, tryptic cleavage sites, and resulting peptides indicated.

resultant product.[21,22] The expected arrangement, from its homology to insulin,[29] was Cys^6-Cys^{48} and Cys^{47}-Cys^{52} (Figure 3A). In the course of assigning the disulfide linkages in the relaxin molecule it was necessary to chemically synthesize both T_2-T_7 forms (Figure 5), since attempts to fragment the peptides chemically between Cys^{A10} and Cys^{A11} (Figure 1) were unsuccessful. Figure 6 illustrates that the improperly folded form elutes at an earlier retention time, just as in the case for the improperly folded form of IGF-I.

2.1.1 Prevention of Disulfide Exchanges

When the tryptic digestion of relaxin was conducted at pH 8.2, all possible cystine-containing peptide pairs were seen.[19] This observation indicated that a disulfide exchange was occurring in the basic medium. Adjusting the buffer pH to 7.3 prevented this exchange reaction and allowed the proper disulfide assignments to be made: Cys^{A11}-Cys^{B11}, Cys^{A10}-Cys^{A15}, and Cys^{A24}-Cys^{B23} (Figure 1).[20] The addition of iodoacetic acid to the digestion reaction at pH 8.2 produced a profile similar to the one obtained when the digestion was conducted at pH 7.3. This suggested that disulfide exchanges were responsible for the extra peaks, since an alkylating reagent prevented the reaction.

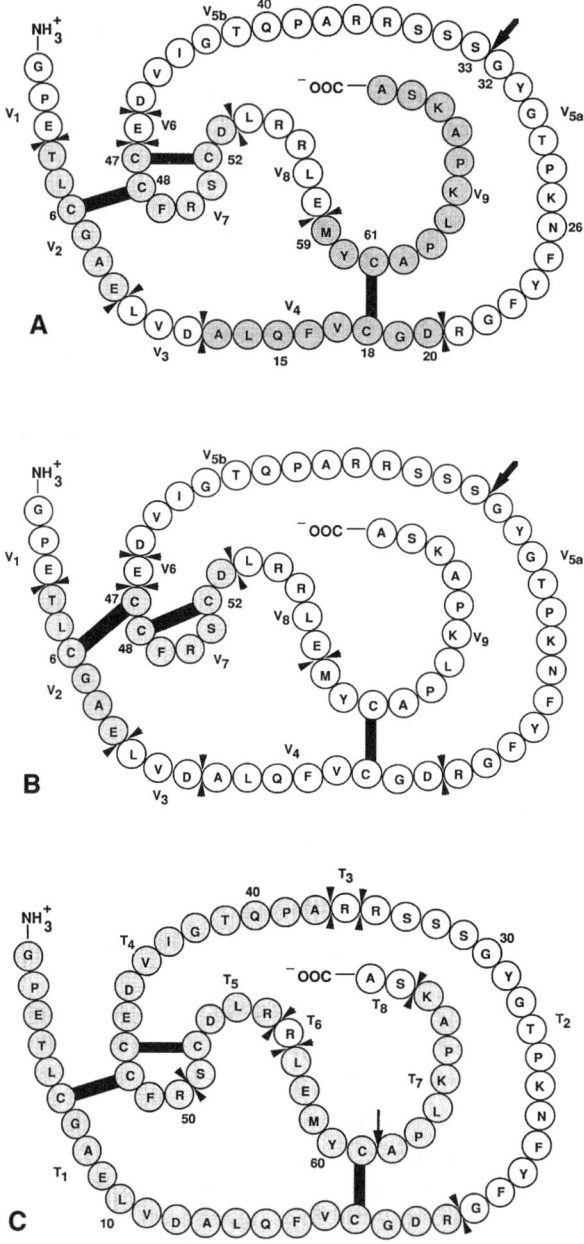

FIGURE 3
Primary sequence of human IGF-I with (A) disulfide bonds and V_8 protease cleavage sites, (B) improperly folded disulfide bonds and V_8 protease cleavage sites, and (C) disulfide bonds and tryptic cleavage sites indicated. The "V" denotes a V_8 protease fragment; the "T" denotes a tryptic fragment.

mAU
500
400
300
200
100
0
V1
V3
V8
V5b,6
V2 -V7
V5a
V4 -V9
10
20
30
40
Time (min.)
mAU
180
160
140
120
100
80
60
40
20
0
V1
V3
V8
V5b,6
V2 -V7
V5a
V4 -V9
10
20
30
40
Time (min.)

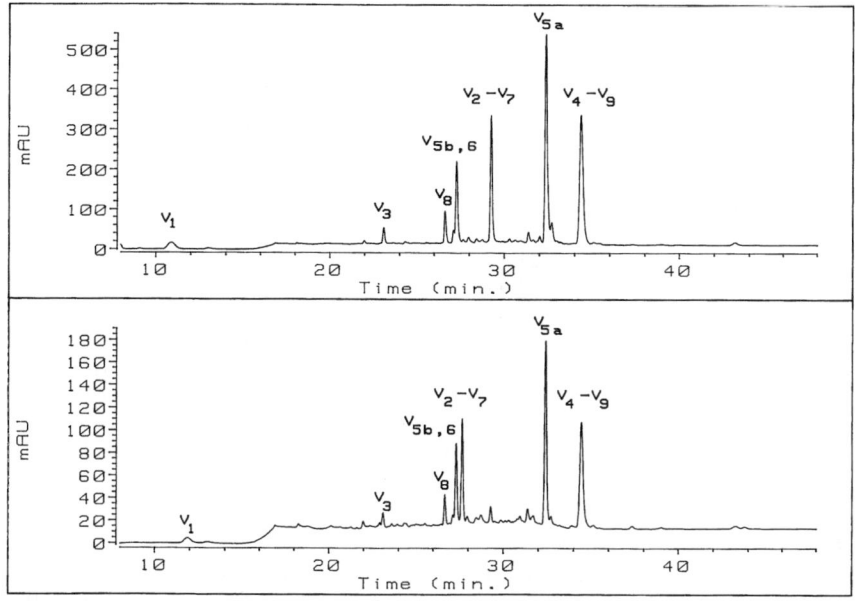

FIGURE 4
Reversed-phase HPLC of a V_8 protease digestion of human IGF-I (upper panel) and an improperly folded variant (lower panel). Peaks are numbered according to cleavages as described in Section 2 of the text and as given in Figure 3A. Human IGF-I was digested with V_8 protease in 100 m*M* phosphate buffer, pH 6.5, at an enzyme to substrate ratio of 1:3 (w/w) for 30 h at 37°C. The chromatogram was developed on a Vydac C_{18} reversed-phase column with a gradient of acetonitrile containing 0.1% TFA. Flow rate was 0.5 mL/min and column temperature was 40°C.

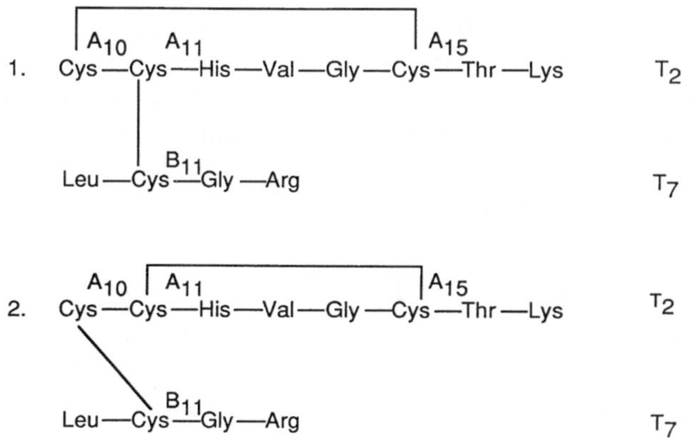

1.
A10
A11
A15
Cys—Cys—His—Val—Gly—Cys—Thr—Lys
T2
B11
Leu—Cys—Gly—Arg
T7
2.
A10
A11
A15
Cys—Cys—His—Val—Gly—Cys—Thr—Lys
T2
B11
Leu—Cys—Gly—Arg
T7

FIGURE 5
Structures of chemically synthesized relaxin T_2-T_7 peptides. Structure 2 is the improperly folded form.

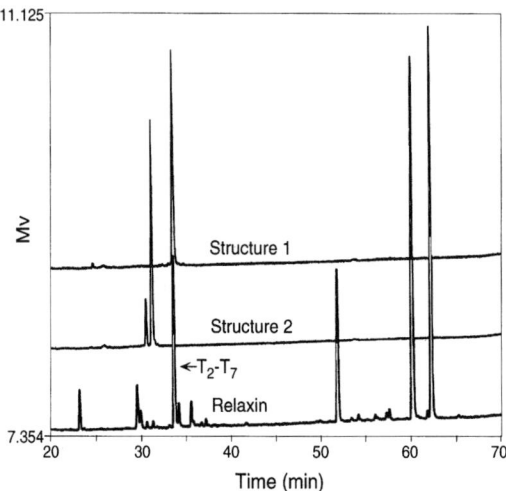

FIGURE 6
Reversed-phase HPLC of chemically synthesized $T_2.T_7$ disulfide-bonded isomers with structures as shown in Figure 5. A recombinant relaxin sample (375 µg) was dissolved in 750 µL of 10 mM Tris, pH 7.2, in 10 mM sodium acetate and 2 mM CaCl$_2$. Digestion was conducted with L-1-(tosylamino)-2-phenylethyl chloromethyl ketone-treated trypsin (Cooper Biomedical) at 37°C for 4 h. An aliquot of 1:100 enzyme to substrate by weight was added at zero time and another after 2 h of digestion. The reaction was terminated with 5 µL of 2 N HCl. The resulting peptide mixture was separated by reversed-phase HPLC and monitored at 214 nm. The column was packed with Nucleosil C_{18} resin (4.6 × 150 mm, 5 µm, 300 Å). Elution was effected with a linear gradient from 0 to 50% acetonitrile containing 0.1% TFA over 100 min. The flow rate was 1 mL/min, and the column temperature was controlled at 35°C.

Digestion of IGF-I with trypsin will not yield peptide fragments conducive to disulfide assignments (Figure 3C). Hence, another protease, the V_8 protease from *Staphylococcus aureus*, was chosen. Again, digestion at the optimal pH for V_8 protease, approximately pH 8, led to disulfide exchanges. Figure 7 is a reversed-phase HPLC profile of a digest of IGF-I using V_8 protease at pH 7.4. Disulfide exchange was still occurring as shown by the presence of V_2-V_9 and V_7-V_9 disulfide-linked peptides. An adjustment of the buffer to pH 6.5 was necessary to prevent these exchanges.[30]

2.1.2 Transpeptidation During Analytical Proteolysis

In many instances, peptides are generated by enzymatic digestion in neutral or acidic buffers to avoid disulfide interchanges as observed during the digestion of relaxin[19] and IGF-I. This usually constitutes less than optimal pH conditions for enzymatic hydrolysis of peptide bonds. Conversely, it has been reported that the pH optimum for trypsin-catalyzed peptide bond synthesis is pH 6.5.[31] As a consequence of these lower pH conditions, minor amounts of transpeptidation products have been found in the digests obtained after

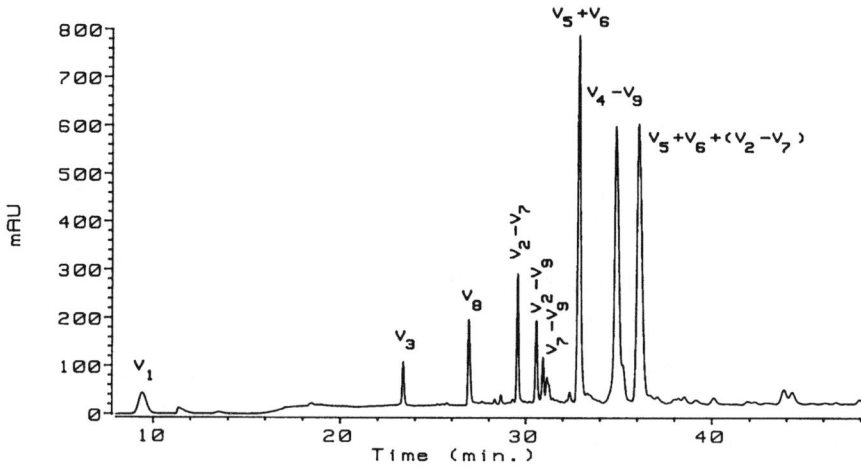

FIGURE 7
Reversed-phase HPLC of a pH 7.4, V_8 protease digestion of human IGF-I. Conditions were similar to those as described in the legend for Figure 4. Peak labels refer to the peptides identified in Figure 3A.

treatment with such enzymes as trypsin or *Staphylococcus aureus* V_8 protease. Hence, if peptide mapping is to be used as a probe for detecting variants in a protein preparation, it is imperative that the investigator considers the extreme pH sensitivity of such transpeptidation reactions. Slight losses in pH control might greatly affect the amounts of the minor peaks observed.

Digestion of relaxin with trypsin at pH 7.3 yields two peptides: $T_{2,3}$ (A10-18) and T_7 (B10-13) linked together by a disulfide bond (Figure 1). An unexpected component at a 10% level was identified to be the T_2-T_7 peptide pair where T_3 (Arg^{A18}) has formed a peptide bond with the amino-terminal Leu^{B10} of the T_7 peptide. It was also observed that the digestion of IGF-I at pH 6.4 with V_8 protease normally yields two peptides — V_4 (13-20) and V_9 (59-70) — linked by a disulfide bridge (Figure 3A). A minor peak at a 1 to 2% level was identified to be a single polypeptide resulting from the formation of a peptide bond between the amino-terminal Met^{59} of V_9 and the carboxyl-terminal Asp^{20} of V_4, with the disulfide bond intact. It is noteworthy that in both cases the transpeptidation occurred between peptides that were held in close proximity by disulfide linkages, increasing their effective concentration.[30]

2.2 PEPTIDE MAPPING FOR VARIANT IDENTIFICATION

2.2.1 *Internal Cleavages*

Human growth hormone, a polypeptide of 191 amino acids (Figure 2), was first expressed in *Escherichia coli* using recombinant DNA techniques that resulted in production of the protein in the cytoplasm as a methionyl analog.[32] Development of an *E. coli* secretion system allowed the production of rhGH lacking the amino-terminal methionine. However, a certain portion of the molecules was cleaved to a two-chain form by a cellular enzyme that may be located in the cell membrane.[33] This clipped variant was isolated by ion-exchange chromatography and characterized by tryptic mapping which located the cleavage site between Thr^{142} and Tyr^{143} in the T_{14} tryptic fragment QTYSK (residues 141-145). The T_{14} peptide was missing from the map and the tripeptide YSK was detected eluting at an earlier retention time. The dipeptide Gln^{141}Thr^{142} was shown by amino acid analysis to elute in the void volume.[37] The recovery of the adjacent peptide T_{13} (residues 135–140) was unaffected, suggesting no further proteolysis. An amino-terminal sequence analysis revealed the presence of two amino-terminal sequences in equimolar amounts and determined the new sequence to be YSKFDTNSHN-, confirming the tryptic mapping results: the variant was a two-chain form of rhGH.

In contrast to the relative simplicity of hGH, recombinant tissue plasminogen activator (r-tPA) is a single-chain glycosylated protein of 527 amino acids (Figure 8) with an apparent molecular weight of 65 kDa on reduced SDS-PAGE. A reconstituted solution of r-tPA was titrated to pH 5.5 and stored for 10 months at 30°C to generate degradation products. Amino-terminal sequence analysis indicated that the r-tPA molecule had been cleaved at several sites. Minor sequences were observed beginning at residues 28, 276, and 463. These products were separated by reversed-phase HPLC (Figure 9) after reduction with tributylphosphine. Fractions were collected and analyzed by SDS-PAGE and amino acid composition. The peaks eluting between 75 and 90 min were identified as glycosylated forms of residues 1-275. The two major peaks at 117 and 122 min were identified as 1-527 and 276-527, respectively. The peak eluting at 65 min was identified as 1-27 by amino acid composition and by coelution of a chemically synthesized peptide in which the Cys^6 residue was carboxylmethylated. While the tryptic map of hGH can be generated from the intact molecule, retaining disulfide linkages, it is necessary to first reduce and alkylate r-tPA to obtain digestion of the protein into its expected tryptic fragments.[34] Even such a complex map as that of r-tPA (Figure 10) was useful for the definitive identification of a cleavage occurring in the protease domain (276-527) upon incubation at elevated temperature in an acidic buffer. Peak 7 material from the reversed-phase HPLC separation of the reduced protein (Figure 9) was subjected to digestion with trypsin (Figure 10). It can be seen that the tryptic fragments comprising residues 463-527 (T_{47}-T_{51}) are absent, and that T_{46} is present in the treated material placing the cleavage reaction at position Arg^{462}-Ser^{463} of the r-tPA molecule (Figure 8) corroborating the amino-terminal sequence analysis result.

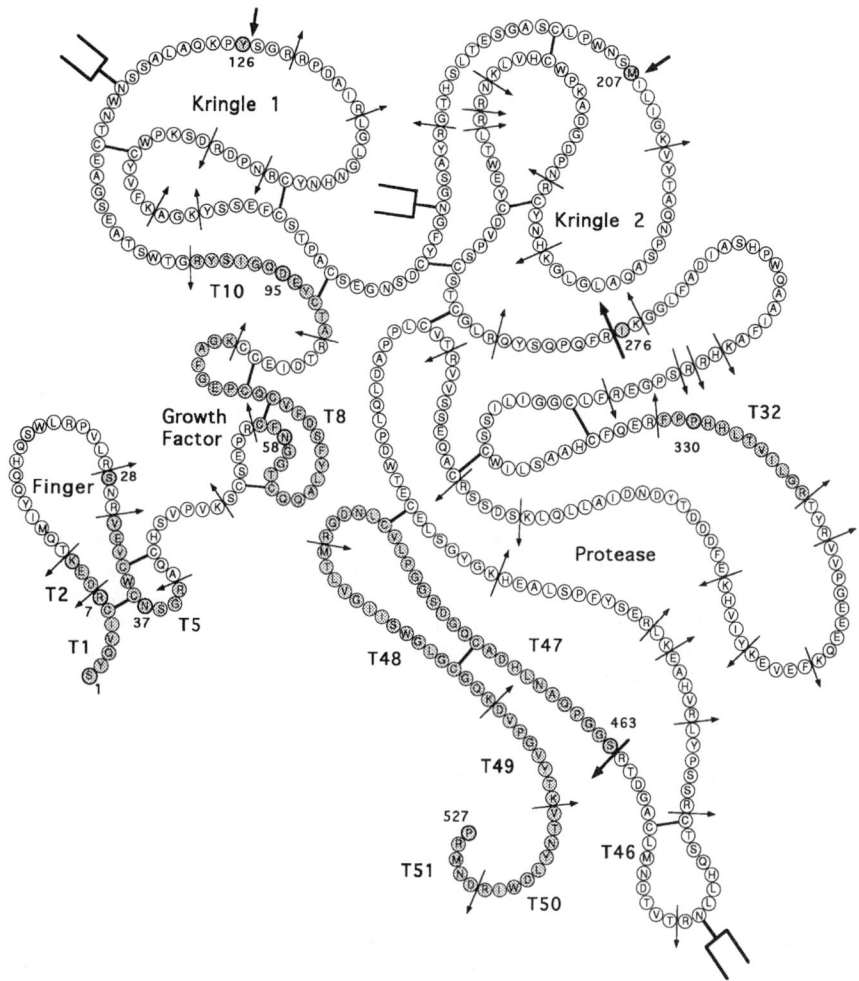

FIGURE 8
Primary sequence of r-tPA. The linear sequence of r-tPA with disulfide bonds indicated as dark lines; tryptic sites as light arrows; non-tryptic cleavage sites as dark arrows; modified residues as dark circles with sequence numbers indicated; and glycosylation sites as "tuning forks". Specific tryptic peptides (numbered T_1, T_2, etc.) that are discussed in the text are indicated as shaded peptides. The protein is organized in five modules; finger, growth factor, kringle 1, kringle 2, and protease; based on homology with other plasma proteins.

2.2.2 Amino- and Carboxyl-Termini Truncations

2.2.2.1 Isolation Artifacts

In addition to the vigilance extended for the detection of endopeptidase action on recombinant DNA-derived proteins, care must be taken to ensure

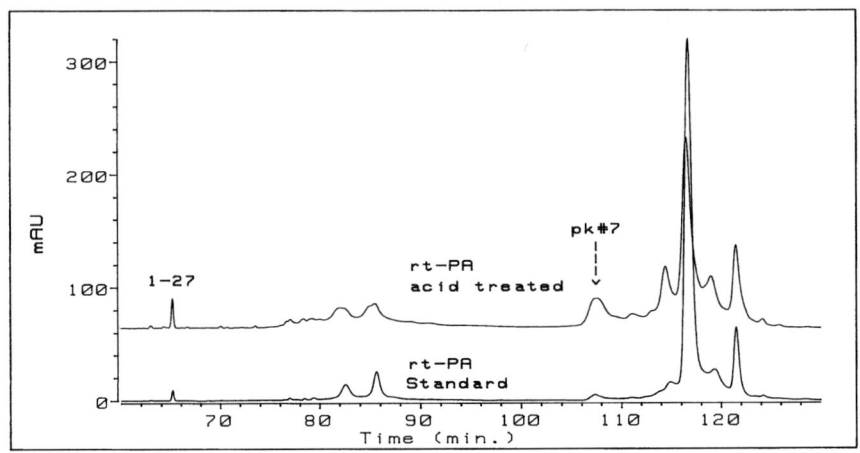

FIGURE 9
Comparison of the reversed-phase HPLC profile of a reduced r-tPA control and the acid-treated r-tPA preparation. The samples (100 µg) were reduced with 2.0 mM tributylphosphine in 0.25 M Tris-HCl, pH 8.3, containing 3.5 M guanidine. The chromatography was performed on a Brownlee RP-300 column thermostatted at 40°C with a 0.5 mL/min flow rate and monitored at 214 nm. Elution was effected with an acetonitrile gradient containing 0.1% TFA as follows: from 30% acetonitrile to 35% in 70 min and held for 5 min; to 37% in 10 min; to 50% in 35 min; and ramping to 70% in 8 min.

that the secreted protein contains the proper amino-terminal amino acid. It is possible that processing of the protein will lead to truncated molecules due to a nonspecificity of processing proteases. One such variant was detected in recombinant hGH, *des*-Phe¹hGH,[35,36] and one in recombinant hIGF-I, *des*-Gly¹hIGF-I. Analysis of secreted recombinant hGH using hydrophobic interaction chromatography led to the discovery of a minor component. Tryptic mapping revealed that the T_1 fragment was missing and a new peak appeared at an earlier retention time. Analysis of the new peptide by fast atom bombardment mass spectrometry identified the *des*-Phe¹ modification.[35]

The discovery of an improperly processed hIGF-I variant was also due to the development of a more discriminating reversed-phase HPLC analysis. When a sodium phosphate pH 4.0 buffer was substituted for the standard 0.1% TFA usually added to the running buffer, a hIGF-I variant was uncovered (Figure 11). In this case V_8 protease mapping was useful in locating the modification at the amino terminus (Figure 12). It can be seen that the V_1 peptide is missing, suggesting a modification at the amino terminus. Analysis by electrospray mass spectrometry on the intact variant indicated the loss of a glycine residue, with the observed mass of 7592.5 being 57 units less then the expected mass of 7649.7. This identified the nonspecific cleavage as occurring between Gly¹ and Pro². An amino-terminal sequence analysis corroborated this finding; i.e., only proline was detected in cycle 1.

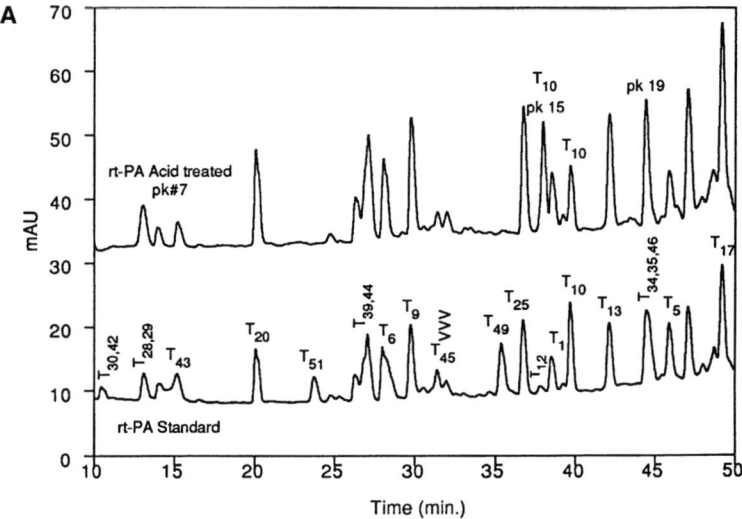

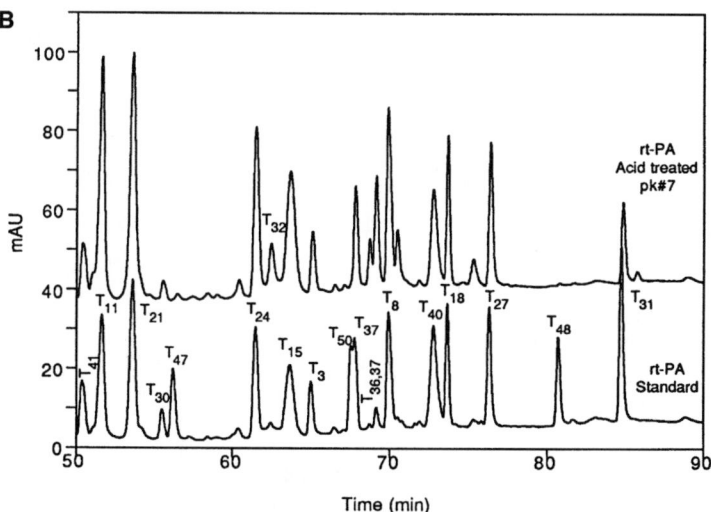

FIGURE 10
Comparison of the tryptic maps of reduced (A) and S-carboxylmethylated (B) r-tPA control and degradation product isolated from the acid-treated sample (Figure 9). The chromatography was performed on a Nova-pak C_{18} column thermostatted at 40°C with a 1.0 mL/min flow rate, and monitored at 214 nm. Elution was effected with an acetonitrile gradient containing 0.1% TFA as follows: from 0 to 30% acetonitrile in 60 min after a 5 min hold with 0.1% aqueous TFA, and then to 60% in 30 min. The samples were reduced with 1.0 mM tributylphosphine in 0.25 M Tris-HCl, pH 8.5, containing 4.0 M guanidine for 40 min at 37°C. Alkylation was effected with 5.0 mM iodoacetate at room temperature shielded from light for 1 h. The resultant protein was buffer-exchanged into 0.1 M ammonium bicarbonate, using a G-25 column (PD-10, Pharmacia) and treated with TPCK-trypsin to give a final enzyme:substrate ratio of 1:100 (w/w). The digestion proceeded for 2 h at 37°C, followed by a second aliquot of enzyme, and analyzed after a total of 4 h.

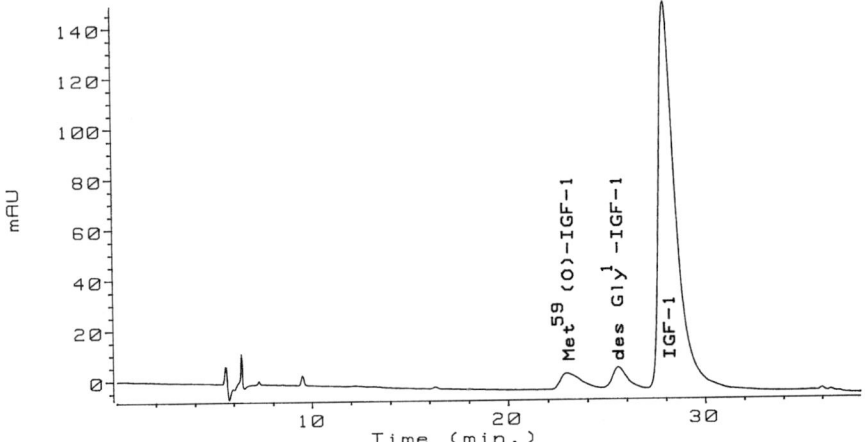

FIGURE 11
Reversed-phase HPLC of a preparation of recombinant human IGF-I. Separation was achieved isocratically on a Vydac C_{18} column using a mobile phase of 28% acetonitrile in water, buffered at pH 4.0 with 100 mM potassium phosphate. Flow rate was 0.5 mL/min at a column temperature of 60°C.

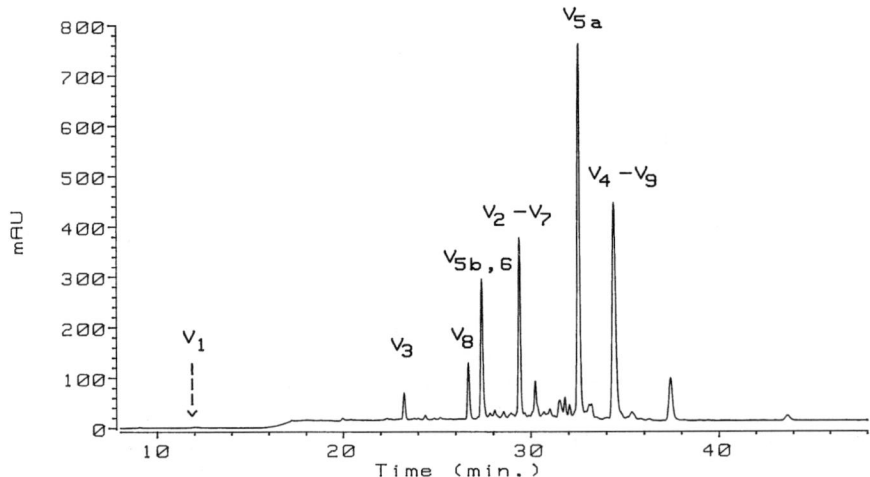

FIGURE 12
V_8 protease map of *des*-Gly[1]IGF-I fraction from Figure 11. Conditions were as described in the legend for Figure 4.

When recombinant hGH was treated with carboxypeptidase A and then digested with trypsin, the carboxyl-terminal T_{20}-T_{21} disulfide-linked tryptic fragment disappeared and two new peaks were observed eluting at earlier retention times. These new peaks were analyzed and found to be the peptides corresponding to the loss of Phe^{191} and Gly^{190}Phe^{191}.[37] These earlier eluting peaks were not present in the maps of isolated recombinant hGH preparations, indicating that carboxyl-terminal processing does not occur above the 2% sensitivity limit.[37]

Tryptic mapping serves as an excellent method for detection of amino-terminal proteolysis during manufacture, even in a protein as large as r-tPA. A chemically synthesized *des*-Ser^{1}T$_1$ (2-7) tryptic fragment (YQVICR, S-carboxylmethylated) is well resolved from the intact T$_1$ tryptic peptide and elutes in a clear region of the map.[37] This peak was not observed in r-tPA preparations.

2.2.2.2 Chemical Degradations

Determination of shelf life is an important aspect in protein pharmaceutical development. A standard approach involves the use of accelerated stability studies to identify probable degradation products. These studies often use elevated temperature or exposure to light as stress-inducing treatments to simulate sample aging. Reversed-phase HPLC is commonly used as a stability indicating assay as this technique is capable of detecting many changes in the protein, very often at levels of 0.1% of the total sample.

Preparations of recombinant hGH and hIGF-I were incubated at elevated temperature for various periods of time and then analyzed by reversed-phase HPLC. Figure 13 shows the profile of a recombinant hIGF-I preparation treated at 50°C for 3 months. The IGF-I V$_8$ protease map of the material eluting at about 16 min is shown in Figure 14. The V$_1$ peptide (GPE) is missing and the Glu^3 residue was found to be uncleaved by the protease, resulting in a Glu^{3}V$_2$-V$_7$ (ETLCGAE/CCFRSCD, disulfide-linked) fragment at 30 min as identified by mass spectrometry. Similarly, the tryptic map of hGH revealed the disappearance of the T$_1$ peptide (FPTIPLSR) and the appearance of a new peptide which was identified as *des*-Phe^{1}Pro^{2}T$_1$ by mass spectrometry.[38] A comparison of the sequences of these proteins resulted in an explanation for such a reaction. The second residue in both proteins is proline, noted for its participation in diketopiperazine formation (Figure 15) with subsequent cleavage of a dipeptide.[39,40] A comprehensive study of this reaction has been conducted using hGH.[38] Similar conclusions apply to IGF-I; i.e., diketopiperazine formation is a likely mechanism for the cleavage.

The B-chain amino terminal sequence Asp^{1}Ser^2 of relaxin is particularly acid labile with cleavage occurring after the Asp residue.[41] Hence, the reported lack of an amino-terminal Asp^{B1}, as detected by tryptic mapping,[19] may

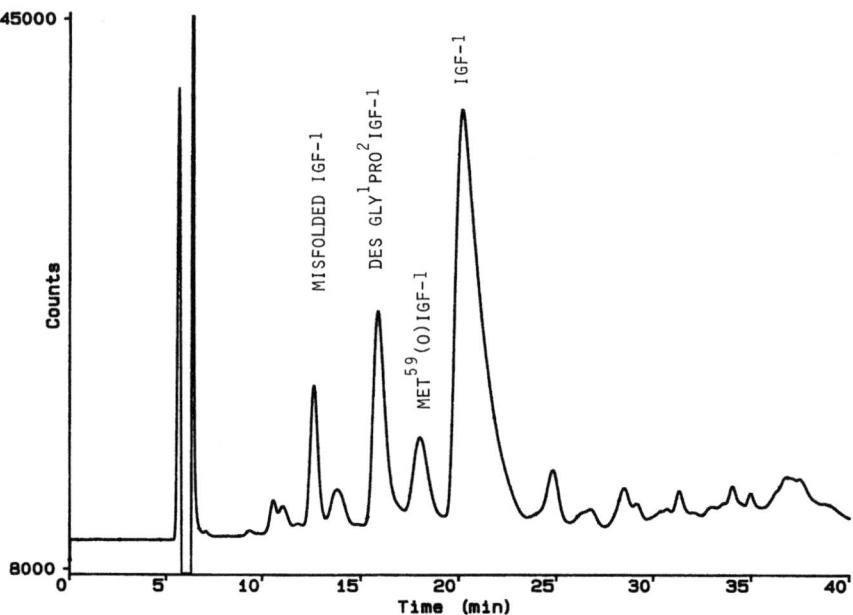

FIGURE 13
Reversed-phase HPLC of recombinant human IGF-I stored at 50°C for 3 months. Separation was achieved isocratically on a Vydac C_{18} column using a mobile phase of 28% aqueous acetonitrile containing 0.1% TFA, at a flow rate of 0.5 mL/min and column temperature of 60°C.

be due to spontaneous chemical cleavage during purification of the protein utilizing reversed-phase HPLC in acidic mobile phases.

2.2.3 Isomerization of Aspartic Acid and Proline Residues

Isomerization of aspartic acid side chains is one type of common spontaneous chemical reaction known to occur in proteins. The generally accepted mechanism involves two steps.[42] In the first step, regarded as the rate limiting step, an intramolecular cyclic imide intermediate is formed. The imide is then readily hydrolyzed, resulting in the formation of either the original α– or the β-aspartyl (isoaspartyl) form (Figure 16), depending upon which bond to the nitrogen atom in the imide ring is broken. This intermediate form was isolated from a preparation of recombinant hGH and characterized by tryptic mapping.[43] The imide was located at Asp^{130} (Figure 2) which is followed by Gly^{131} in the T_{12} peptide LEDGSPR (128-134): a sequence reported to allow such isomerizations to occur.[44] The order of elution of the T_{12} peptides found was β > α > imide. The imide was positively identified by tandem mass spectrometry. A similar situation exists in porcine growth hormone.[45] The sequence $Asp^{129}Gly^{130}$ forms the imide, with subsequent hydrolysis to the

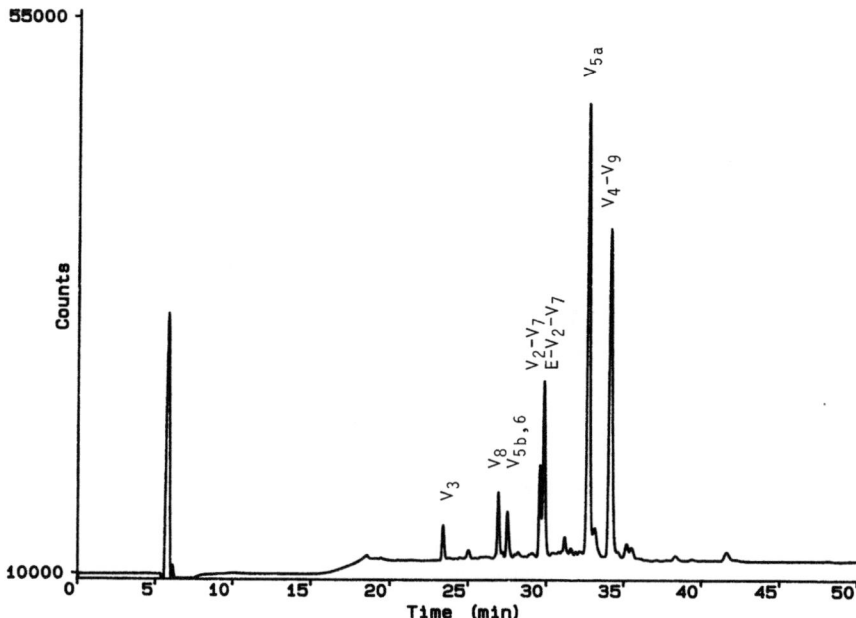

FIGURE 14
V_8 protease map of peak 4 material from Figure 13. Conditions were as described in the legend for Figure 4.

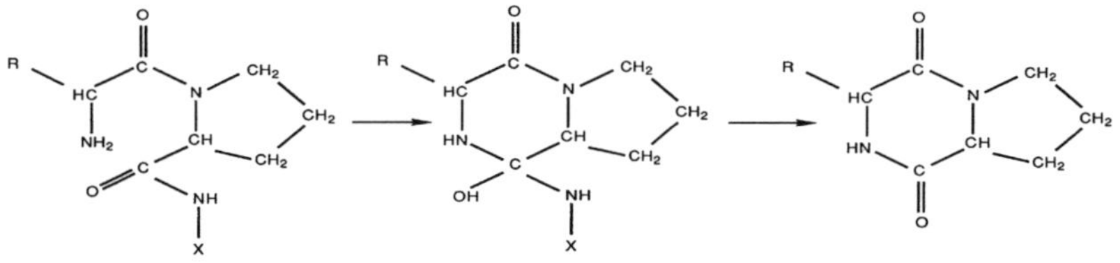

FIGURE 15
Decomposition of peptides via diketopiperazine formation.

β-form. Again, tryptic mapping determined the order of elution as above (β > α > imide).

The r-tPA tryptic fragment T_{10} ATCYEDQGISYR (90-101) from the digest of peak 7 material (Figure 9) was found to elute at two different retention times (Figure 10). The earlier eluting fragment (Figure 10, peak 15) was analyzed by amino-terminal Edman degradation for sequence information. The main sequence (75%) stopped at Glu^{94}, which is followed by Asp^{95}, suggesting that the acidic conditions had catalyzed isomerization to β-Asp^{95}.

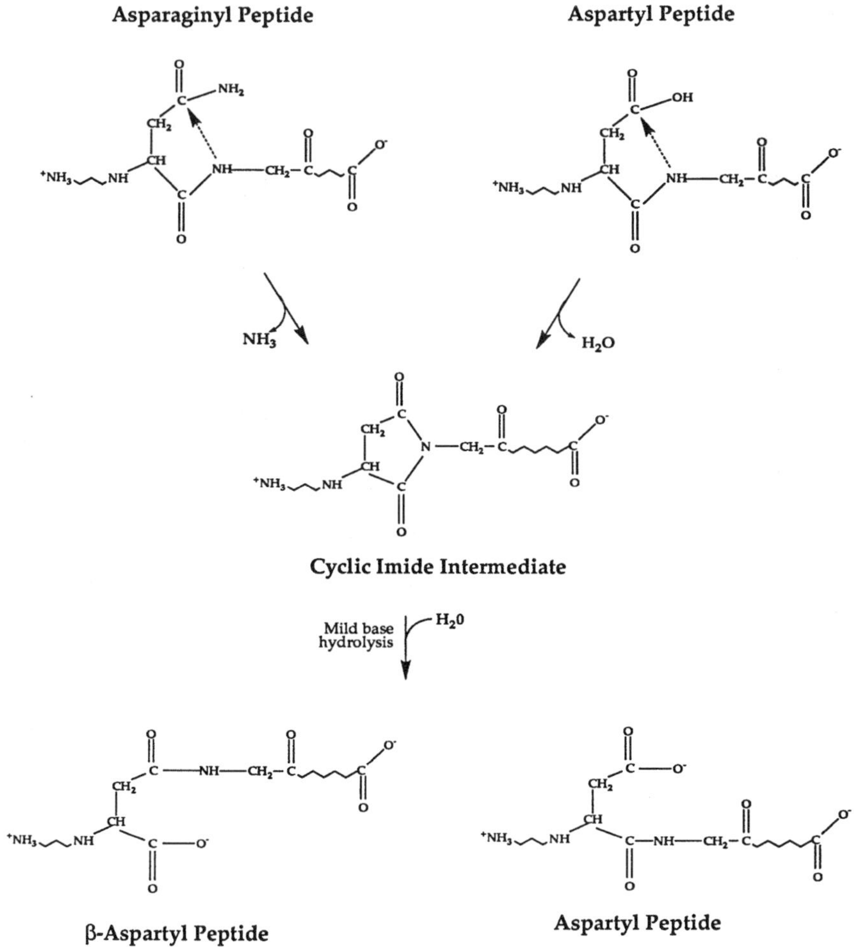

FIGURE 16
Mechanisms for the formation of β-aspartyl linkages upon isomerization of aspartate residues or deamidation of asparagine residues.

The secondary sequence, found at a 25% level, was the incompletely cleaved $T_{1,2}$ sequence (SYQVICRDEK).

Tryptic mapping conventionally utilizes reversed-phase HPLC separation with UV-absorbance detection. Mass spectrometric detection was shown to provide an extra dimension of information (mass-to-charge ratio) that is invaluable for identifying the eluting peptides.[46] Through this improvement, the proline isomers of a tryptic peptide generated from r-tPA were observed to elute over a broad range of retention times. It had previously been demonstrated that peptide T_{32} (FPPHHLTVILGR) eluted as a multiplet.[34] Proline conformers have been separated by reversed-phase HPLC[47,48] and, as this peptide contains two proline residues at positions 329 and 330 (Figure 8), the presence of cis-trans isomers was proposed. A portion of the total ion current yielded a continuum of ion signal with a mass of 1385.4, which corresponded to a broad peak in the UV profile. Approximately four peaks were discernible, consistent with the four possible isomers expected from two proline residues in sequence.[46]

2.2.4 Deamidation of Asparagine Residues

Deamidation of asparagine residues is a potential degradation pathway in the aging of a protein. In general, deamidation of an asparagine residue proceeds through a cyclic imide intermediate, which can hydrolyze to form either an α- or a β-aspartic acid residue (Figure 16). Tryptic mapping has been quite effective in monitoring the deamidation of recombinant hGH.[49-51] The tryptic map analysis of a chemically deamidated preparation of recombinant hGH showed loss of peptide T_{15} which contains Asn at positions 149 and 152 (Figure 2). In addition, three new peaks with later retention times were detected and shown to be deamidated forms of T_{15} by mass spectral and amino-terminal sequence analyses. By this procedure, it is possible to identify deamidated peptides. Deamidated peptides usually elute slightly later than the native peptide, giving rise to doublet or triplet peak patterns signifying the α, β, and imide forms of aspartic acid residues.[43] In the case of recombinant hGH, the three peaks were the α and β form of $Asp^{149}T_{15}$[51] and $Asp^{149,152}T_{15}$.[79]

No evidence for the deamidation of relaxin or IGF-I was apparent using this technique. Since deamidation of asparagine side chains is commonly seen in proteins, Asn^{A8} in the relaxin A-chain (Figure 1) was of interest. The T_1 peptide was chemically synthesized with a β-aspartic acid residue at that position so as to determine its elution time in the tryptic map of relaxin. It eluted in a position completely devoid of any absorbing peptides in the map of a chemically synthesized relaxin molecule containing 33 amino acids in the B-chain (Figure 17), indicating that this asparagine residue is not particularly susceptible to deamidation. This is not surprising since it is followed by a lysine — an amino acid that does not favor the formation of the cyclic imide which is an intermediate in the deamidation reaction.[44] A similar situation exists with IGF-I; Asn^{26}, which is also followed by a lysine residue, is not deamidated (Figure 3).

Owing to the complexity of the tryptic map of r-tPA (with at least 60 chromatographic peaks), it is difficult to identify all of the sites of deamidation in this manner. Two sites, one in T_5 containing an $Asn^{37}Ser^{38}$ sequence, and another in T_8 containing an $Asn^{58}Gly^{59}$ sequence (Figure 8) were identified by tryptic mapping and confirmed by mass spectral analyses.[28]

In recent years much attention has been directed to the cell surface glycoprotein CD4 found on peripheral T-cells after discovery of its involvement in the onset of AIDS-related symptoms in HIV-infected patients.[52,53]

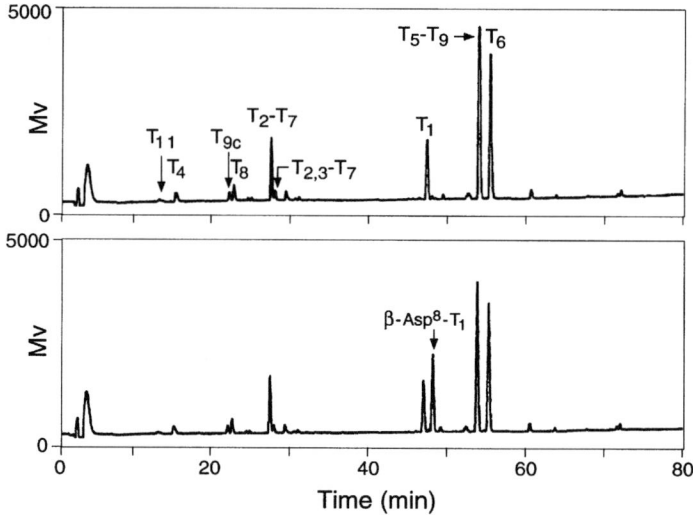

FIGURE 17
Tryptic maps of chemically synthesized human relaxin illustrating the position of elution of the β-Asp⁸T₁ peptide. Chromatography was conducted as described in the legend to Figure 6 except that the linear gradient was effected from 3 to 30% acetonitrile containing 0.08% TFA over 54 min then to 70% acetonitrile in 40 min.

Teshima et al.[54] found that deamidation of a soluble CD4 construction lacking the transmembrane and cytoplasmic domains resulted in a reduced binding capacity for the HIV-1 envelope glycoprotein gp120. Isolation of a charge variant of the recombinant CD4 and analysis by tryptic mapping located the deamidation at Asn^{52}, reputed to be in the binding region.[55,56] The sequence in this region is Leu^{51}Asn^{52}Asp^{53}; not the usual sequence of a neighboring serine or glycine which are conducive to succinimide formation. Methylation of rCD4 by protein carboxyl methyltransferase using radiolabeled methyl donor S-adenosyl-L-methionine suggested that most of the aspartate was in the isoaspartyl form.[54]

2.2.5 Oxidation of Methionine Residues

Oxidation of methionyl residues is another common degradation pathway in the aging of proteins (Figure 18). The sulfoxide form is the most prevalent oxidation product of methionine. It is also, however, the most difficult to detect by conventional means such as amino acid composition or amino-terminal sequence analyses. The separation power of reversed-phase HPLC tryptic mapping offers the best opportunity for detecting and purifying low levels of oxidized peptides for mass spectral analysis.

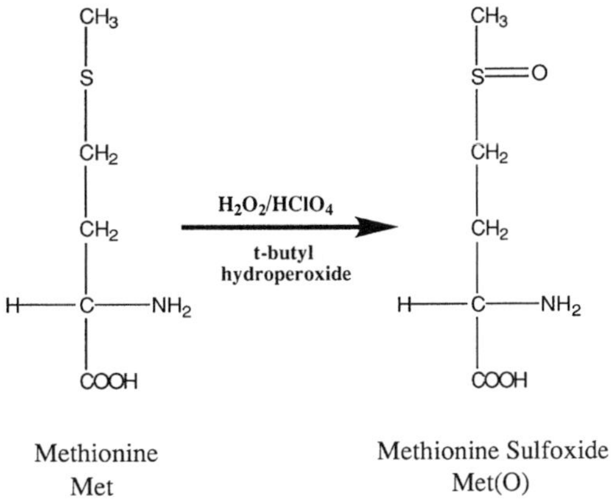

FIGURE 18
Product from the oxidation of methionine. A number of chemical reagents can lead to the oxidation of methionine, namely, hydrogen peroxide (H_2O_2), perchloric acid ($HClO_4$), and t-butyl hydroperoxide. Their use in the study of methionine oxidation is common.

Human growth hormone contains three potential sites for methionine oxidation; Met^{14}, Met^{125}, and Met^{170} (Figure 2). Oxidation has been detected at both Met^{14} and Met^{125}.[50,57-60] Met^{170} is not oxidized in the native state,[61] consistent with its location in the interior of the molecule according to the recently published hGH-receptor crystal structure.[62]

It has been shown that Met^{14} and Met^{125} have very different solvent accessibility under reversed-phase chromatographic conditions.[60] Therefore, the Met^{14}-oxidized variant is better resolved from native hGH at neutral pH, whereas the Met^{125}-oxidized material is better resolved from hGH at low pH. Hence, the tryptic map of hGH in which the $Met^{14}T_2$ peptide LFDNAMLR (9-16) and the Met^{125} T_{11} peptide DLEEGIQTLMGR (116-127) are well resolved has been instrumental in identifying the sites of oxidation in hGH impurity peaks. Tryptic mapping revealed a change in the retention for the methionine-containing fragments to earlier retention times, consistent with the more hydrophilic character of the methionyl sulfoxide residue. Mass spectral analyses of the modified peptides showed 16 mass units greater than expected, suggesting the addition of an oxygen atom.

Tryptic mapping of relaxin revealed low levels of methionine oxidations in the B-chain at both methionine residues: positions 4 and 25.[19] A collisionally activated dissociation mass spectral analysis of the T_9 peptide AQIAICGMSTWSK (B18-30), which contains both methionine and tryptophan residues revealed the presence of methionine sulfoxide at position 25.[27,63] It was subsequently reported that the major route of degradation of human relaxin exposed to 3600 foot-candles of light was through oxidation of the methionine residues, with Met^{B25} being the most susceptible as revealed by tryptic mapping.[63]

Similarly, IGF-I with structural homology to relaxin is easily oxidized at its only methionine residue at position 59.[64,65] In this instance, peptide mapping with V_8 protease was utilized since a tryptic digest would lead to the methionine residue being incorporated into a large disulfide-linked fragment (Figure 3C). It is interesting to note that insulin, the "structural grandparent" of molecules like relaxin and IGF-I, does not contain methionine residues.[4]

2.2.6 Amino Acid Substitutions

A common amino acid substitution, as the result of an error in the translation step, is the incorporation of norleucine in place of methionine. This was first reported by Munier and Cohen in 1956.[66] A decade later Anfinsen and Corley[67] isolated a norleucine variant of *Staphylococcal* nuclease and utilized the then established technique of peptide mapping by two-dimensional paper chromatography/high voltage electrophoresis to localize the norleucine incorporation. Two decades later the norleucine replacement for methionine was detected in recombinant DNA-derived proteins biosynthesized in *E. coli*.[68,69] These workers had the luxury of the presently used technique of reversed-phase HPLC tryptic mapping to aid in obtaining and analyzing the norleucine-containing peptides.

Similarly, in the present decade Forsberg et al.[64] analyzed an IGF-I variant with a 30% reduced affinity for its receptor after digestion with pepsin and confirmed that the lone methionine in IFG-I at position 59 was replaced by norleucine (Figure 3). Amino-terminal sequence and amino acid composition analyses can confirm the presence of norleucine. A mass spectral analysis can suggest that a norleucine substitution for methionine has occurred since it results in a mass 18 units less than expected. Amino acid analyses are definitive in the quantitation of norleucine incorporation, but are not as sensitive for low levels (less than 0.1% of total protein) as an optimized reversed-phase HPLC analysis of the intact protein.

A tyrosine to glutamine substitution has been uncovered by tryptic mapping of an antibody directed against the HER-2 receptor biosynthesized in Chinese hamster ovary cells. Although it has been reported that the Gln t-RNA synthetase is receptive to mischarging of the sup^F amber suppressor t-RNA^{Tyr},[70] it was determined that this substitution was a result of a mutation after transfection.[71]

The substitution of Arg^{275} in r-tPA (Figure 8) by Glu^{275} by site-directed mutagenesis results in a molecule wherein a tryptic cleavage site is eliminated.[72] Hence, the expected tryptic fragments T_{25} QYSQPQFR (268-275) and T_{26} IK (276-277) are replaced with a single peptide (268-277). This new peptide is resolved from the 49 other expected tryptic fragments and can be detected in the map.[37]

2.2.7 Chemical Modifications

Initially, small polypeptide hormones were biosynthesized by recombinant DNA-derived technology as fusion proteins owing to their intracellular degradation by endogenous proteolytic enzymes.[73] Since IGF-I has an amino-terminal glycine residue, a hydroxylamine cleavage was considered reasonable. Hence, a fragment of protein A, chosen for its usefulness during purification by affinity chromatography, was coupled through an asparagine residue to IGF-I, thereby creating the necessary Asn-Gly hydroxylamine-sensitive linkage.[74] A variant produced by this cleaving reagent was reported early in 1992.[65] The usual procedure of peptide mapping using V_8 protease followed by mass spectral analyses uncovered V_8 protease fragments that were 16 mass units higher than expected. Oxidation was suspected; however, these peptides did not contain the usual oxidation-susceptible amino acids methionine and tryptophan (Figure 3A): V_4 ALQFVCGD (13-20), V_{5a} RGFYFNKPTGYG (21-32), and $V_{5b,6}$ SSSRRAPQTGIVDE (33-46). A tandem mass spectral analysis of V_4 implicated a glutamine residue as the site of modification. It then became clear that the well-known reaction of hydroxylamine with amide-containing amino acids was involved,[75] yielding hydroxamic acids at the Gln[15], Asn[26], and Gln[40] positions; i.e., the amide group –C(O)NH$_2$ was converted to a hydroxamate group –C(O)NOH.

2.2.8 Peptide Mapping Artifacts

In addition to confirming the integrity of the primary structure of recombinant hGH, tryptic mapping revealed a variety of nontryptic-like cleavages. Also noted was the cyclization of a newly released amino-terminal glutamine to pyroglutamic acid in the T_{14} tryptic fragment QTYSK.[23,76] The atypical tryptic cleavages occurred between Leu[6] and Ser[7], Ser[62] and Asn[63], and Asn[99] and Ser[100] in hGH. A similar atypical tryptic cleavage was observed in r-tPA between Tyr[126] and Ser[127].[46] Each of these, interestingly, involves a serine residue at one of the positions around the cleavage site. Although these cleavages are atypical they usually are consistent and therefore do not present a problem once they are identified.

Atypical cleavages by trypsin have also been seen at methionine residues: in the T_9 peptide fragment (18-29) in the B-chain of relaxin between Met[B25] and Ser[B26],[19,20] in calmodulin between Met[145] and Thr[146],[77] and in r-tPA between Met[207] and Ile[208].[34,46]

It was necessary to develop a V_8 protease map for IGF-I since trypsin will cleave in such a manner as to result in eight fragments, four of which are held together by disulfide bonds (Figure 3C). This protease is not as stable as trypsin; that is, different preparations tend to have varying cleavage patterns. A sample digest from a less discriminating enzyme preparation is shown in Figure 19. The peaks "a" through "d" were collected and analyzed by mass spectrometry (Table 1) and amino-terminal Edman degradation. In addition to the Gly[32]Ser[33] cleavage usually seen,[65,78] this particular preparation not only cleaved nonspecifically at the Gly[30]Tyr[31] and Cys[61]Ala[62] positions, but also did not completely cleave at the expected Glu[46]Cys[47] position. This observation illustrates the need to include the digest of a reference sample for each experiment.

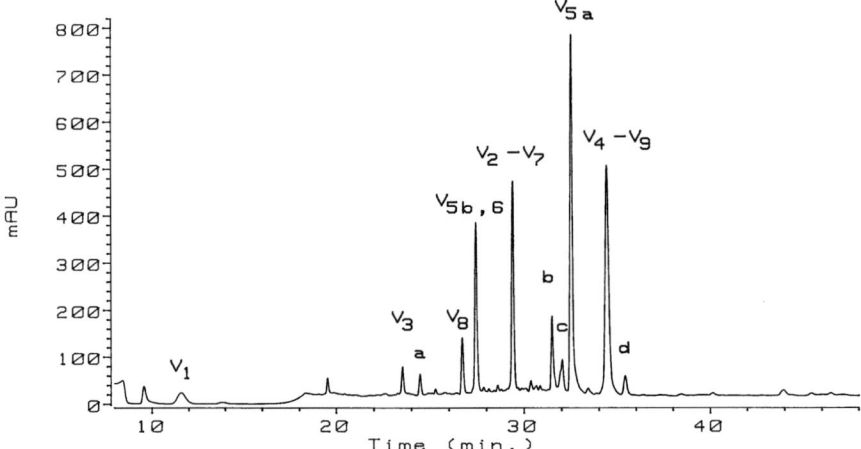

FIGURE 19
Peptide map of human IGF-I obtained with a suboptimal V_8 protease enzyme preparation. Conditions were as described in the legend for Figure 4.

TABLE 1

Mass Spectral Analysis of V_8 Protease Fragments Generated From IGF-I

Peak (From Fig. 19)	Observed mass	Assignments[a]
a	882.5	A^{62}PLKPAKSA^{70}
b	1187.2	R^{21}GFYFNKPTG^{30}
c	2907.7	T^4LCGAE^9/S^{33}SSRRAPQ TGIVDECCFRSCD^{53}
d	1265.5	A^{13}LQFVCGD^{20}/M^{59}YC^{61}

[a] The assignments of peaks a, b, and c were confirmed by amino-terminal sequence analysis. Peaks c and d contained disulfide-linked peptides.

3 DISCUSSION

Peptide mapping by HPLC is a particularly useful technique for defining the primary sequence of proteins. It is hence one of the best means of showing consistency of protein structure for samples obtained from different lots of material. Modifications of any residues, or additions, substitutions, or deletions of residues generally manifest themselves as new peaks in the chromatogram and/or changes in peak intensities. However, care must be taken in choosing an appropriate protease; for example, a tryptic map would not detect any proteolysis by an enzyme with a similar specificity, such as plasmin.[37]

Peptide mapping has been shown to be a powerful tool for the detection of the proteolysis of recombinant proteins by cellular peptidases. The analyses of the two-chain and des-Phe[1]rhGH species are cases in point. Notably, peptide mapping is ordinarily the first line of attack in solving the problem of disulfide assignments. On occasion, it has also been the first indicator that a mistranslation or mutation has occurred, as demonstrated by the norleucine substitutions found in recombinant interleukin-2[68] and bovine growth hormone,[69] and the more recent finding of a tyrosine to glutamine substitution in the antibody engineered with specificity for the HER-2 receptor.[71] These observations are reminiscent of the very early use of two-dimensional peptide mapping by Ingram[10] in the identification of a hemoglobin mutant. Since peptide mapping can easily detect degradation products derived from deamidation and oxidation, it is an important technique for determining the shelf life of a product during stability programs.

The finding of transpeptidation products in enzyme digests warrants that caution is advised in relying upon these mapping techniques to detect the presence of minor variants in recombinant-derived proteins. It is imperative that these artifacts of the analytical technique be recognized as such to avoid concerns raised with respect to purity, and hence to the safety and efficacy of a protein pharmaceutical.

ACKNOWLEDGMENTS

The authors would like to thank Ida Baldonado for the reversed-phase HPLC of the chemically synthesized β-Asp^{8}T$_1$ relaxin peptide, Jeremy Kessler for the reversed-phase HPLC of the chemically synthesized T$_2$-T$_7$ disulfide-bonded isomers of relaxin, and Glen Teshima for his helpful scientific insights.

REFERENCES

1. Sanger, F. and Tuppy, H., The amino-acid sequence in the phenylalanyl chain of insulin. I. The identification of lower peptides from partial hydrolysates *Biochem. J.*, 49, 463, 1951.
2. Sanger, F. and Tuppy, H., The amino acid sequence in the phenylalanyl chain of insulin. II. The investigation of peptides from enzymic hydrolysates, *Biochem. J.*, 49, 481, 1951.
3. Sanger, F. and Thompson, E.O.P., The amino acid sequence in the glycyl chain of insulin. I. The identification of lower peptides from partial hydrolysates, *Biochem. J.*, 53, 353, 1953.

4. Sanger, F. and Thompson, E.O.P. The amino acid sequence in the glycyl chain of insulin. II. The investigation of peptides from enzymic hydrolysates, *Biochem. J.,* 53, 366, 1953.
5. Graf, L., Barat, E., Borvendeg, J., Hermann, I., and Patthy, A., Action of thrombin on ovine, bovine, and human pituitary growth hormones, *Eur. J. Biochem.,* 64, 333, 1976.
6. Mitchell, W.M. and Harrington, W.F., Purification and properties of Clostridiopeptidase B (Clostripain), *J. Biol. Chem.,* 243, 4683, 1968.
7. Masaki, T., Nakamura, K., Isono, M., and Soejima, M., A new proteolytic enzyme from *Achromobacter lyticus* M497-1, *Agric. Biol. Chem.,* 42, 1443, 1978.
8. Drapeau, G.R., Boily, Y., and Houmard, J., Purification and properties of an extracellular protease of *Staphylococcus aureus, J. Biol. Chem.,* 247, 6720, 1972.
9. Ingram, V.M., A specific chemical difference between the globins of normal human and sickle cell anaemia haemoglobin, *Nature,* 178, 792, 1956.
10. Ingram, V.M., Abnormal human haemoglobins. I. The comparison of normal human and sickle cell haemoglobins by "fingerprinting", *Biochim. Biophys. Acta,* 28, 539, 1958.
11. Katz, A.M., Dreyer, W.J., and Anfinsen, C.B., Peptide separation by two-dimensional chromatography and electrophoresis, *J. Biol. Chem.,* 234, 2897, 1959.
12. Schroeder, W.A., Jones, R.T., Cormick, J., and McCalla, K., Chromatographic separation of peptides on ion exchange resins. Separation of peptides from enzymatic hydrolyzates of the α, β, and γ chains of human hemoglobins, *Anal. Chem.,* 34, 1570, 1962.
13. Jones, R.T., Structural studies of aminoethylated hemoglobins by automatic peptide chromatography, *Cold Spring Harbor Symp. Quant. Biol.,* 29, 297, 1964.
14. Howard, G.A. and Martin, A.J.P., The separation of the C_{12}-C_{18} fatty acids by reversed-phase partition chromatography, *Biochem. J.,* 46, 532, 1950.
15. Hancock, W.S., Bishop, C.A., Prestidge, R.L., and Hearn, M.T.W., The use of high pressure liquid chromatography (hplc) for peptide mapping of proteins IV, *Anal. Biochem.,* 89, 203, 1978.
16. Rivier, J.E., Use of trialkyl ammonium phosphate (TAAP) buffers in reverse phase HPLC for high resolution and high recovery of peptides and proteins, *J. Liq. Chrom.,* 1, 343, 1978.
17. Gerber, G.E., Anderegg, R.J., Herlihy, W.C., Gray, C.P., Biemann, K., and Khorana, H.G., Partial primary structure of bacteriorhodopsin. Sequencing methods for membrane proteins, *Proc. Natl. Acad. Sci. U.S.A.,* 76, 227, 1979.
18. Mahoney, W.C. and Hermodson, M.A., Separation of large denatured peptides by reverse phase high performance liquid chromatography. Trifluoroacetic acid as a peptide solvent, *J. Biol. Chem.,* 255, 11199, 1980.
19. Canova-Davis, E., Baldonado, I.P., and Teshima, G.M., Characterization of chemically synthesized human relaxin by high-performance liquid chromatography, *J. Chromatogr.,* 508, 81, 1990.
20. Canova-Davis, E., Kessler, T.J., Lee, P.-J., Fei, D.T.W., Griffin, P., Stults, J.T., Wade, J.D., and Rinderknecht, E., Use of recombinant DNA derived human relaxin to probe the structure of the native protein, *Biochemistry,* 30, 6006, 1991.
21. Raschdorf, F., Dahinden, R., Maerki, W., Richter, W.J., and Merryweather, J.P., Location of disulphide bonds in human insulin-like growth factors (IGFs) synthesized by recombinant DNA technology, *Biomed. Environ. Mass Spectrom.,* 16, 3, 1988.
22. Axelsson, K., Johansson, S., Eketorp, G., Zazzi, H., Hemmendorf, B., and Gellerfors, P., Disulfide arrangement of human insulin-like growth factor I derived from yeast and plasma, *Eur. J. Biochem.,* 206, 987, 1992.
23. Kohr, W.J., Keck, R., and Harkins, R.N., Characterization of intact and trypsin-digested biosynthetic human growth hormone by high-pressure liquid chromatography, *Anal. Biochem.,* 122, 348, 1982.

24. Paborsky, L.R. and Harris, R.J., Post-translational modifications of recombinant human tissue factor, *Thrombosis Res.*, 60, 367, 1990.
25. Leonard, C.K., Spellman, M.W., Riddle, L., Harris, R.J., Thomas, J.N., and Gregory, T.J., Assignment of intrachain disulfide bonds and characterization of potential glycosylation sites of the Type 1 recombinant human immunodeficiency virus envelope glycoprotein (gp120) expressed in Chinese hamster ovary cells, *J. Biol. Chem.*, 265, 10373, 1990.
26. Harris, R.J., Chamow, S.M., Gregory, T.J., and Spellman, M.W., Characterization of a soluble form of human CD4. Peptide analyses confirm the expected amino acid sequence, identify glycosylation sites and demonstrate the presence of three disulfide bonds, *Eur. J. Biochem.*, 188, 291, 1990.
27. Vandlen, R.L., Winslow, J.W., Kohr, W.J., Bourell, J.H., Stults, J.T., and Canova-Davis, E., Instrumentation needs for the biotechnology laboratory, in *Biological Mass Spectrometry*, Burlingame, A.L. and McCloskey, J.A., Eds., Elsevier, Amsterdam, 1990, 579.
28. Canova-Davis, E., Teshima G.M., Kessler, T.J., Lee, P.-J., Guzzetta, A.W., and Hancock, W.S., Strategies for an analytical examination of biological pharmaceuticals, in *Analytical Biochemistry ACS Symposium Series 434*, Horvath, C. and Nikelly, J.G., Eds., American Chemical Society, Washington, D.C., 1990, 90.
29. Ryle, A.P., Sanger, F., Smith, L.F., and Kitai, R., The disulphide bonds of insulin, *Biochem. J.*, 60, 541, 1955.
30. Canova-Davis, E., Kessler, T.J., and Ling, V.T., Transpeptidation during the analytical proteolysis of proteins, *Anal. Biochem.*, 196, 39, 1991.
31. Homandberg, G.A. and Chaiken, I.M., Trypsin-catalyzed conversion of Staphylococcal nuclease-T fragment complexes to covalent forms, *J. Biol. Chem.*, 255, 4903, 1980.
32. Goeddel, D.V., Heyneker, H.L., Hozumi, T., Arentzen, R., Itakura, K., Yansura, D.G., Ross, M.J., Miozzari, G., Crea, R., and Seeburg, P.H., Direct expression in *Escherichia coli* of a DNA sequence coding for human growth hormone, *Nature*, 281, 544, 1979.
33. Canova-Davis, E., Baldonado, I.P., Moore, J.A., Rudman, C.G., Bennett, W.F., and Hancock, W.S., Properties of a cleaved two-chain form of recombinant human growth hormone, *Int. J. Peptide Protein Res.*, 35, 17, 1990.
34. Chloupek, R.C., Harris, R.J., Leonard, C.K., Keck, R.G., Keyt, B.A., Spellman, M.W., Jones, A.J.S., and Hancock, W.S., Study of the primary structure of recombinant tissue plasminogen activator by reversed-phase high-performance liquid chromatographic tryptic mapping, *J. Chromatogr.*, 463, 375, 1989.
35. Wu, S.-L., Hancock, W.S., Pavlu, B., and Gellerfors, P., Application of high-performance hydrophobic-interaction chromatography to the characterization of recombinant DNA-derived human growth hormone, *J. Chromatogr.*, 500, 595, 1990.
36. Chloupek, R.C., Battersby, J.E., and Hancock, W.S., Practical considerations for assessing product quality of biosynthetic proteins by HPLC, in *High-Performance Liquid Chromatography of Peptides and Proteins: Separation, Analysis, and Conformation*, Mant, C.T. and Hodges, R.S., Eds., CRC Press, Boca Raton, FL, 1991, 825.
37. Hancock W.S., Canova-Davis, E., Battersby, J., and Chloupek, R., The use and limitations of reversed phase HPLC for the analysis of recombinant proteins and their tryptic digests, in *Biotechnologically Derived Medical Agents: The Scientific Basis of Their Regulation*, Gueriguian, J.L., Fattorusso, V., and Poggiolini, D., Eds., Raven Press, New York, 1988, 29.
38. Battersby, J.E., Hancock, W.S., Canova-Davis, E., Oeswein, J., and O'Connor, B., Diketopiperazine formation and N-terminal degradation in recombinant human growth hormone, *Int. J. Peptide Protein Res.*, 44, 215, 1994.

39. Gisin, B.F. and Merrifield, R.B., Carboxyl-catalyzed intramolecular aminolysis. A side reaction in solid-phase peptide synthesis, *J. Am. Chem. Soc.*, 94, 3102, 1972.
40. Khosla, M.C., Smeby, R.R., and Bumpus, F.M., Failure sequence in solid-phase peptide synthesis due to the presence of an N-alkylamino acid, *J. Am. Chem. Soc.*, 94, 4721, 1972.
41. Tsung, C.M. and Fraenkel-Conrat, H., Preferential release of aspartic acid by dilute acid treatment of tryptic peptides, *Biochemistry*, 4, 793, 1965.
42. Geiger, T. and Clarke, S., Deamidation, isomerization, and racemization at asparaginyl and aspartyl residues in peptides, *J. Biol. Chem.*, 262, 785, 1987.
43. Teshima, G., Stults, J.T., Ling, V., and Canova-Davis, E., Isolation and characterization of a succinimide variant of methionyl human growth hormone, *J. Biol. Chem.*, 266, 13544, 1991.
44. Bodanszky, M. and Kwei, J.Z., Side reactions in peptide synthesis. VII. Sequence dependence in the formation of aminosuccinyl derivatives from β-benzyl-aspartyl peptides, *Int. J. Peptide Protein Res.*, 12, 69, 1978.
45. Violand, B.N., Schlittler, M.R., Kolodziej, E.W., Toren, P.C., Cabonce, M.A., Siegel, N.R., Duffin, K.L., Zobel, J.F., Smith, C.E., and Tou, J.S., Isolation and characterization of porcine somatotropin containing a succinimide residue in place of aspartate, *Protein Sci.*, 1, 1634, 1992.
46. Ling, V., Guzzetta, A.W., Canova-Davis, E., Stults, J.T., Hancock, W.S., Covey, T.R., and Shushan, B.I., Characterization of the tryptic map of recombinant DNA derived tissue plasminogen activator by high-performance liquid chromatography-electrospray ionization mass spectrometry, *Anal. Chem.*, 63, 2909, 1991.
47. Jacobson, J., Melander, W., Vaisnys, G., and Horvath, C., Kinetic study on *cis-trans* proline isomerization by high-performance liquid chromatography, *J. Phys. Chem.*, 88, 4536, 1984.
48. Gesquire, J.C., Diesis, E., Cung, M.T., and Tartar, A., Slow isomerization of some proline-containing peptides inducing peak splitting during reversed-phase high-performance liquid chromatography, *J. Chromatogr.*, 478, 121, 1989.
49. Hancock, W.S., Canova-Davis, E., Chloupek, R.C., Wu, S.-L., Baldonado, I.P., Battersby, J.E., Spellman, M.W., Basa, L.J., and Chakel, J.A., Banbury Report, Characterization of degradation products of recombinant human growth hormone, in *Therapeutic Peptides and Proteins, Assessing the New Technologies*, Cold Spring Harbor Laboratory Press, Cold Spring Harbor, N.Y., 1988, 95.
50. Becker, G.W., Tackitt, P.M., Bromer, W.W., Lefeber, D.S., and Riggin, R.M., Isolation and characterization of a sulfoxide and a desamido derivative of biosynthetic human growth hormone, *Biotechnol. Appl. Biochem.*, 10, 326, 1988.
51. Johnson, B.A., Shirokawa, J.M., Hancock, W.S., Spellman, M.W., Basa, L.J., and Aswad, D.W., Formation of isoaspartate at two distinct sites during *in vitro* aging of human growth hormone, *J. Biol. Chem.*, 264, 14262, 1989.
52. Curran, J.W., Morgan, W.M., Hardy, A.M., Jaffe, H.W., Darrow, W.W., and Dowdle, W.R., The epidemiology of AIDS. Current status and future prospects, *Science*, 229, 1352, 1985.
53. Weiss, R., *RNA Tumor Viruses: Molecular Biology of Tumor Viruses*, Weiss, R., Teich, N., Varmus, H., and Coffin, J., Eds., Cold Spring Harbor Press, Cold Spring Harbor, NY, 1985, 405.
54. Teshima, G., Porter, J., Yim, K., Ling, V., and Guzzetta, A., Deamidation of soluble CD4 at asparagine-52 results in reduced binding capacity for the HIV-1 envelope glycoprotein gp120, *Biochemistry*, 30, 3916, 1991.
55. Peterson, A. and Seed, B., Genetic analysis of monoclonal antibody and HIV binding sites on the human lymphocyte antigen CD4, *Cell*, 54, 65, 1988.

56. Ashkenazi, A., Presta, L.G., Marsters, S.A., Camerato, T.R., Rosenthal, K.A., Fendly, B.M., and Capon, D.J., Mapping the CD4 binding site for human immunodeficiency virus by alanine-scanning mutagenesis, *Proc. Natl. Acad. Sci. U.S.A.,* 87, 7150, 1990.
57. Teh, L.-C., Murphy, L.J., Huq, N.L., Surus, A.S., Friesen, H.G., Lazarus, L., and Chapman, G.E., Methionine oxidation in human growth hormone and human chorionic somatomammotropin. Effects on receptor binding and biological activities, *J. Biol. Chem.,* 262, 6472, 1987.
58. Canova-Davis, E., Chloupek, R.C., Baldonado, I.P., Battersby, J.E., Spellman, M.W., Basa, L.J., O'Connor, B., Pearlman, R., Quan, C., Chakel, J.A., Stults, J.T., and Hancock, W.S., Analysis by FAB-MS and LC of proteins produced by either biosynthetic or chemical techniques, *Am. Biotechnol. Lab.,* 5, 8, 1988.
59. Gellerfors, P., Pavlu, B., Axelsson, K., Nyhlen, C., and Johansson, S., Separation and identification of growth hormone variants with high performance liquid chromatography techniques, *Acta Paediatr. Scand. Suppl.,* 370, 93, 1990.
60. Teshima, G. and Canova-Davis, E., Separation of oxidized human growth hormone variants by reversed-phase high-performance liquid chromatography. Effect of mobile phase pH and organic modifier, *J. Chromatogr.,* 625, 207, 1992.
61. Houghten, R.A., Glaser, C.B., and Li, C.H., Human somatotropin. Reaction with hydrogen peroxide, *Arch. Biochem. Biophys.,* 178, 350, 1977.
62. DeVos, A.M., Ultsch, M., and Kossiakoff, A.A., Human growth hormone and extracellular domain of its receptor. Crystal structure of the complex, *Science,* 255, 306, 1992.
63. Cipolla D.C. and Shire, S.J., Analysis of oxidized human relaxin by reverse phase HPLC, mass spectrometry, and bioassays, in *Techniques in Protein Chemistry II,* Villafranca, J.J., Ed., Academic Press, San Diego, 1991, 543.
64. Forsberg, G., Palm, G., Ekebacke, A., Josephson, S., and Hartmanis, M., Separation and characterization of modified variants of recombinant human insulin-like growth factor I derived from a fusion protein secreted from *Escherichia coli, Biochem. J.,* 271, 357, 1990.
65. Canova-Davis, E., Eng, M., Mukku, V., Reifsnyder, D.H., Olson, C.V., and Ling, V.T., Chemical heterogeneity as a result of hydroxylamine cleavage of a fusion protein of human insulin-like growth factor I, *Biochem. J.,* 285, 207, 1992.
66. Munier, R. and Cohen, G.N., Incorporation d'analogues structuraux d'aminoacides dans les proteines bacteriennes, *Biochim. Biophys. Acta,* 21, 592, 1956.
67. Anfinsen, C.B. and Corley, L.G., An active variant of Staphylococcal nuclease containing norleucine in place of methionine, *J. Biol. Chem.,* 244, 5149, 1969.
68. Lu, H.S., Tsai, L.B., Kenney, W.C., and Lai, P.-H., Identification of unusual replacement of methionine by norleucine in recombinant interleukin-2 produced by *E. coli, Biochem. Biophys. Res. Commun.,* 156, 807, 1988.
69. Bogosian, G., Violand, B.N., Dorward-King, E.J., Workman, W.E., Jung, P.E., and Kane, J.F., Biosynthesis and incorporation into protein of norleucine by *Escherichia coli, J. Biol. Chem.,* 264, 531, 1989.
70. Swanson, R., Hoben, P., Sumner-Smith, M., Uemura, H., Watson, L., and Soll, D., Accuracy of *in vivo* aminoacylation requires proper balance of tRNA and aminoacyl-tRNA synthetase, *Science,* 241, 1548, 1988.
71. Harris, R., Murnane, A.A., Utter, S.L., Wagner, K.L., Cox, E.T., Polastri, G.D., Helder, J.C., and Sliwkowski, M.B., Peptide mapping for assessment of genetic heterogeneity of cell lines, detection of a low level Tyr to Gln sequence variant in a recombinant antibody, *Bio/Technology,* 11, 1293, 1993.

72. Tate, K.M., Higgins, D.L., Holmes, W.E., Winkler, M.E., Heyneker, H.L., and Vehar, G.A., Functional role of proteolytic cleavage at arginine-275 of human tissue plasminogen activator as assessed by site-directed mutagenesis, *Biochemistry*, 26, 338, 1987.
73. Itakura, K., Hirose, T., Crea, R., Riggs, A.D., Heyneker, H.L., Bolivar, F., and Boyer, H.W., Expression in *Escherichia coli* of a chemically synthesized gene for the hormone somatostatin, *Science*, 198, 1056, 1977.
74. Moks, T., Abrahmsen, L., Osterlof, B., Josephson, S., Ostling, M., Enfors, S.-O., Persson, I., Nilsson, B., and Uhlen, M., Large-scale affinity purification of human insulin-like growth factor I from culture medium of *Escherichia coli, Biotechnology,* 5, 379, 1987.
75. Ramachandran, L.K. and Narita, K., Reactions involving the amide and carboxyl groups of tobacco mosaic virus (TMV) protein, *Biochim. Biophys. Acta,* 30, 616, 1958.
76. Bennett, W.F., Chloupek, R., Harris, R., Canova-Davis, E., Keck, R., Chakel, J., Hancock, W.S., Gellefors, P., and Pavlu, B., Characterization of natural-sequence recombinant human growth hormone, in *Advances in Growth Hormone and Growth Factor Research,* Muller, E.E., Cocchi, D., and Locatelli, V., Eds., Pythagora Press, Roma, 1989, 29.
77. Watterson, D.M., Sharief, F., and Vanaman, T.C., The complete amino acid sequence of the Ca^{2+}-dependent modulator protein (calmodulin) of bovine brain, *J. Biol. Chem.,* 255, 962, 1980.
78. Allen, G., *Laboratory Techniques in Biochemistry and Molecular Biology,* Work, T.S. and Burdon, R.H., Eds., Elsevier/North Holland, Amsterdam, 1981, 58.
79. Ling, V.T., Eng, M.L., Lee, P.-J., Keck, R.G., Keyt, B.A., and Canova-Davis, E., Unpublished observations.

CHAPTER 2

OPTIMIZATION OF PEPTIDE MAPPING VIA COMPUTER SIMULATION

L.R. Snyder

CONTENTS

1	Introduction	31
2	Theory: DryLab Computer Simulation	34
3	Some Examples of Computer Simulation for Peptide Samples	36
	3.1 Mixture of 23 Synthetic Peptides	36
	3.2 Recombinant Human Growth Hormone (rhGH) Tryptic Digest	37
	3.2.1 Effects of a Change in the Column	39
	3.3 Recombinant Tissue Plasminogen Activator (r-tPA) Tryptic Digest	41
4	Discussion	45
	4.1 Simultaneous Change in Gradient Steepness and Column Temperature	46
	4.2 Potential Problems in the Use of Computer Simulation for the Separation of Protein Digests	51
	4.2.1 Predictive Accuracy	51
	4.2.2 Convenience of Experimental vs. Simulated Separations	53
5	Conclusions	54
References		54

1 INTRODUCTION

Other chapters in this volume have discussed the importance of peptide mapping for various applications; high-performance liquid chromatography (HPLC) is now widely used for this purpose.[1,2] Typically, an enzymatic digest of the protein is prepared followed by reversed-phase HPLC separation using an acetonitrile/water gradient (with added trifluoroacetic acid [TFA]). The usual challenge at this point is to achieve the complete (baseline) separation of the resulting mixture of "primary" peptides, i.e., those resulting from the expected cleavage of the protein. This may prove relatively easy for a digest which contains only a small number (<20) of primary peptides, as in the example of Figure 1A for a tryptic digest of calmodulin. The similar separation of a digest containing 30 or more primary peptides will be more difficult, e.g., the tryptic digest of recombinant tissue plasminogen activator (r-tPA, 54 primary peptides) whose separation is shown in Figure 1B.

The traditional approach to improving separations as in Figure 1 is to decrease gradient steepness so as to improve overall resolution, and to use longer, smaller-particle columns of greater efficiency.[4] However, statistical considerations[5] suggest that this approach (by itself) cannot succeed for samples represented by Figure 1B by using HPLC columns that are presently available, i.e., with plate number N < 30,000. When the crowding of peaks within the chromatogram exceeds some limit, e.g., ten peaks in a typical peptide-digest chromatogram,* there is a reasonable probability that two or more peaks will overlap to some extent — assuming a random distribution of peak retention times. In view of this fundamental restriction, the complete separation of typical protein digest samples requires more than just longer gradients and better columns.

Workers who must separate and analyze protein digests are aware of the foregoing problem. A change in separation *selectivity* is one means of resolving overlapping peaks found in these complex reversed-phase chromatograms, e.g., by means of a change in pH,[8] ion-pairing,[8,9] column type,[10] buffer or buffer concentration,[11] temperature,[3] or gradient steepness.[3,12-14] A change in selectivity may also be combined with two or more sequential separations.[11] Each of the foregoing options usually leads to a considerable increase in experimental effort. When a change in selectivity is attempted *without* the use of multiple, sequential separations, interpretation of the resulting chromatograms is often quite difficult. The initial complexity of chromatograms as in Figure 1B is then compounded; changes in selectivity lead to changes in relative retention, the resolution of peak pairs that were originally overlapped, and the overlapping of new peak pairs. Keeping track

---

* See the discussion in References 5 and 6. Typical peptide-digest chromatograms will have a peak capacity of 100 to 200,[4,7] which suggests that there is a 50% probability for samples containing 9 to 13 components that one or more peak pairs will overlap with $R_s < 1$.

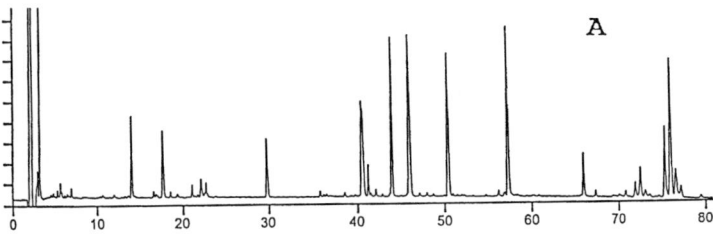
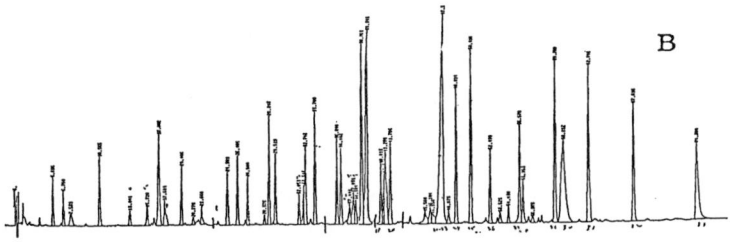
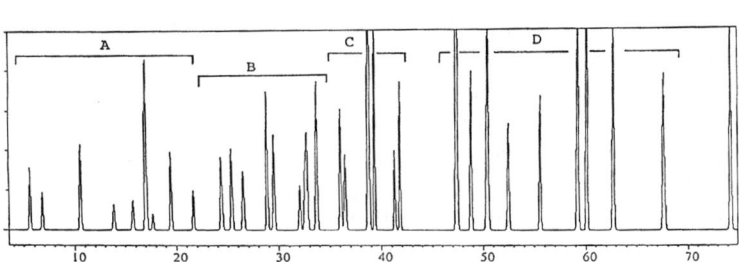

FIGURE 1
Separation of typical protein digests by reversed-phase HPLC using acetonitrile/water (TFA) gradients; Zorbax 80 SB-18 column. (A) Tryptic digest of calmodulin; 15 × 0.46 cm, 5 to 60% B in 90 min; 0.75 mL/min; 35°C; (B) tryptic digest of r-tPA; 0 to 60% B in 120 min; 1.0 mL/min; 40°C; (C) computer simulation of separation in (B). (From Chloupek, R.C., Hancock, W.S., Marchylo, B.A., Kirkland, J.J., Boyes, B.E., and Snyder, L.R., *J. Chromatogr.*, 686, 45, 1994. With permission.)

of these changes from one chromatogram to another — and assessing their significance — then becomes a formidable task.

The use of *computer simulation* as an aid in optimizing the separation of protein digests can provide better separations, requiring a smaller number of experimental runs and much less effort on the part of the chromatographer. Equally important, computer simulation can lead to a better understanding of the separation, i.e., how good a separation is possible, the shortest possible run time, etc. Two approaches to computer simulation for application to protein-digest separations are now commercially available. One approach (ProDigest-LC, Synthetic Peptides Inc., Alberta, Canada) is based on a knowledge of the primary structure of the protein and the resulting peptide products.[15,16] By means of basic chromatographic theory and some approximate relationships that relate the retention of a peptide to its structure and experimental conditions, it is possible to predict a chromatogram for a known protein digest in the absence of any experimental data. This computer program can be useful in the initial assessment of various approaches to a successful separation, but ProDigest-LC predictions are too approximate to allow the selection of conditions for an optimum peak spacing or selectivity.

A second software package (DryLab, LC Resources Inc., Walnut Creek, CA) is based on different principles. The separation of peptide digests (and other samples) can be predicted as a function of experimental conditions, but two experimental runs for the sample of interest must be carried out first — in order to "calibrate" the computer program for the separation characteristics of each sample component. For example, these initial experiments might involve a gradient from 0 to 60% acetonitrile/water in times of 30 and 90 min. These and other conditions are then entered into the program along with retention times and areas for the peaks in each chromatogram. At this point, various experimental conditions can be changed (initial and final %B, gradient time or steepness, gradient shape, column dimensions, particle size, and flowrate) and the resulting separation can be visualized either as a stimulated chromatogram (e.g., Figure 1C) or as a table of retention times.

Comparisons of predicted (DryLab) and actual separations of peptides and other compounds show generally good agreement, e.g., retention times within ±1% and resolution (R_s) within ±10%. However, the main benefit of this software in the separation of protein digests is that changes in gradient steepness — which can be simulated by DryLab — often lead to significant changes in peak spacing. This represents a simple and convenient way to optimize selectivity and separation for these samples.

This chapter will present a review of some previous applications of DryLab simulation for peptide samples, discuss how computer simulation can be used to benefit the practical chromatographer in different ways, and comment on some limitations of this approach and the need for further improvements.

2 THEORY: DRYLAB COMPUTER SIMULATION

Based on the starting experimental data (for two runs), the computer calculates values of k_w and S for each peptide[17] corresponding to the *isocratic* retention (k) of each compound as a function of mobile phase composition:

$$\log k = \log k_w - S \phi \qquad (1)$$

Here, k_w is the capacity factor value for water as the mobile phase, and ϕ is the volume-fraction of acetonitrile (B) in the mobile phase (equal to %B/100). The DryLab program next allows separation to be predicted[13,18,19] as a function of gradient conditions and other variables. Changes in gradient steepness provide simultaneous control over absolute retention (k or k*) and relative retention or selectivity. This can be expressed by the well-known relationship for resolution as a function of conditions:

$$R_s = (1/4)\ (\alpha - 1)\ N^{1/2}\ (k/[k+1]) \qquad (2)$$

In gradient elution, the average capacity factor k* is given as

$$k^* = 0.85/b \qquad (3)$$

where

$$b = V_m\ \Delta\phi\ S/(t_G\ F) \qquad (4)$$

Here, V_m is the column dead volume, $\Delta\phi$ is the change in ϕ (or %B) during the gradient, and F is the flowrate.

As gradient steepness b is decreased, k* and resolution increase for the chromatogram as a whole. However, changes in b can result in changes in relative retention and α for different peak pairs. The result is that the separation of a given peak pair may either increase or decrease as gradient steepness is decreased. This is illustrated for the later-eluting peaks of r-tPA in Figure 2, corresponding to the group of peaks labeled "D" in Figure 1C. With a 60-min gradient, peak X3-C is partly resolved from X3-A and -B, but all three peaks coelute for a 120-min gradient. Peaks T18 and T40, on the other hand, coelute in the 60-min gradient but are well separated in the 120-min gradient.

Since our goal for a given sample is usually the separation of *all* peaks of interest, there will usually be an optimum (intermediate) gradient steepness b for a given sample. The recognition of this optimum for samples as complex as that of Figure 1B will usually not be obvious. As we will see, however, computer simulation allows the selection of an optimum gradient steepness for a given sample by means of a *resolution map* (see Figure 7).

Because different parts of the chromatogram (as in Figure 2) will respond differently to changes in b, it will sometimes be advantageous to optimize gradient steepness for different parts of the chromatogram; i.e., via the use of multisegment gradients.[20] If an empirical trial-and-error approach is attempted, the optimization of a segmented gradient will usually require a large number of experimental runs. Computer simulation can achieve the same result with only two experimental runs (plus a few hours at the computer).

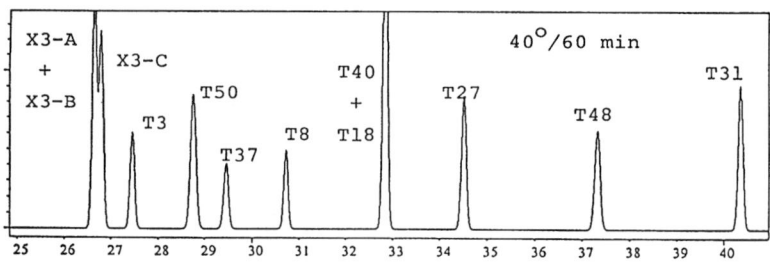

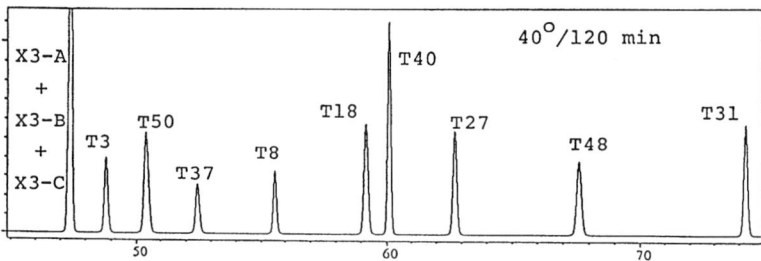

FIGURE 2
Separation of group D of Figure 1C (r-tPA digest) as a function of gradient time or steepness (other conditions the same). Adapted from Reference 3.

3 SOME EXAMPLES OF COMPUTER SIMULATION FOR PEPTIDE SAMPLES

3.1 Mixture of 23 Synthetic Peptides

The separation of this sample is summarized in Figures 3 and 4.[13] Figures 3A and B show the experimental input runs for 5 to 50% B gradients in 45 and 180 min; these data were used for computer simulation. The experimental conditions of Figure 3A were next used to predict the same separation; the computer simulation (45-min gradient) is shown in Figure 3C. There is good agreement between the two chromatograms (R_s values) for a best-fit column plate number N = 10,000. Finally, Figure 3D shows a resolution map for this separation, i.e., a plot of *critical* resolution (value of R_s for the poorest-resolved peak pair) as a function of gradient time.

The resolution map of Figure 3D shows that a maximum sample resolution is obtained (R_s = 1.1) for a gradient time of 85 min. The critical peak pairs (numbers in Figure 3D) for these conditions are peaks 11/12 and 7/8. The simulated chromatogram for this separation is shown in Figure 4A. It is not possible in this case to further improve *resolution* by use of a segmented gradient. However, the *gradient time* of 85 min in Figure 4A can be shortened to only 45 min by the use of the segmented gradient shown in Figure 4B. Finally, by decreasing the flowrate (from 1.0 to 0.4 mL/min) while holding b constant (see Equation 4), the optimized peak spacing of Figure 4B can be maintained, while the resolution for all peaks is increased (due to an increase in the plate number N). The resulting predicted separation of Figure 4C has a resolution $R_s = 1.4$ for the most overlapped peak pair (almost baseline separation), in a run time of 112 min.

An experimental run for the optimized conditions of Figure 4C was carried out and is shown in Figure 4D. A comparison of the two chromatograms (Figures 4C and D) shows good agreement; retention times match within ±1.4% (CV) and resolution is predicted with an average accuracy of ±8%. Returning to Figure 3D, each minimum in this plot corresponds to a peak reversal, i.e., change in relative retention. For gradient times between 25 and 500 min there are 7 such peak reversals, illustrating the important effect of gradient time or steepness on selectivity in the separation of this sample. Such selectivity effects are quite common in the separation of peptide samples, as is further illustrated in the chromatograms of Figure 5, other peptide separations reported in Reference 12, and the following examples.

3.2 Recombinant Human Growth Hormone (rhGH) Tryptic Digest

Prior to the application of computer simulation for the separation of this sample,[14] it was known that there are 20 major and 2 minor peptides that can be resolved in acetonitrile/water gradients. Separation of this sample was carried out initially using 30-min and 120-min gradients, as shown in Figures 6a and b. These two runs were used as input for DryLab computer simulation. However, before using computer simulation for predictive purposes, it is advisable to first verify the accuracy of these predictions. Errors in peak assignment (peak tracking, see later discussion) may have occurred, a wrong value of the system dwell volume[21] may have been assumed, etc. Figures 6c and d compare experimental and predicted separations for a run of intermediate gradient time (t_G = 60 min). In this case, experimental (c) and predicted (d) retention times agree within an average error of ±0.1 min (0.3%), so we can conclude that further computer simulations will be reliable.

Figure 7a shows a resolution map for the separations of Figure 6. The significance of this map is illustrated further in Figure 7b, which relates critical peak pairs (shaded) to different parts of the map. Peaks 9/10 are seen to be critical for gradient times around 50 min; peaks 14/15 become critical for gradient times of 60 to 74 min, and peaks 14/15 are critical when the gradient time exceeds 74 min. Figure 7a offers two choices for an optimum gradient time: 74 min or 4 hr. A resolution R_s = 1.3 is attainable with the 74-min gradient, while the 4-h gradient offers better resolution (R_s = 2.1) in a longer time. The shorter gradient was selected for further study. Figure 8a

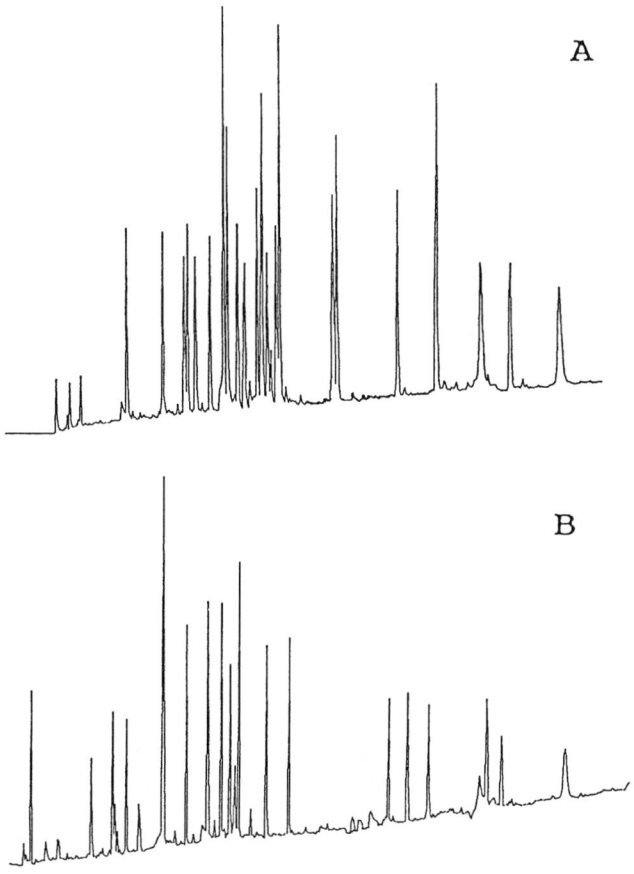

FIGURE 3
Separation of a sample containing 23 synthetic peptides. Column, 25 × 0.46 cm Zorbax Rx; 5 to 50% acetonitrile/water gradient (0.1% added TFA); 1.0 mL/min; 30°C. (A) Experimental 45-min gradient; (B) experimental 180-min gradient; (C) simulated 45-min gradient, N = 10,000; (D) resolution map based on 5 to 50% B gradient. (From Dolan, J.W., Lommen, D.C., and Snyder, L.R., *J. Chromatogr.*, 485, 91, 1989. With permission.)

shows experimental and predicted chromatograms for a 0 to 47% B gradient in a time of 74 min. Again, the experimental and predicted separations are in good agreement.

As in the case of Figure 4, the use of a segmented gradient does not allow a further increase in sample resolution for the rhGH sample, but run time can be shortened. This is shown in Figure 8b, where run time is reduced to 53 min — a saving of almost 30%. The rhGH separations of Figures 6 and 8 are similar in some respects to those of the synthetic peptide mixture (Figures 3 and 4). In each case, it was possible to obtain nearly baseline separation of the peptides of interest within a convenient run time. The

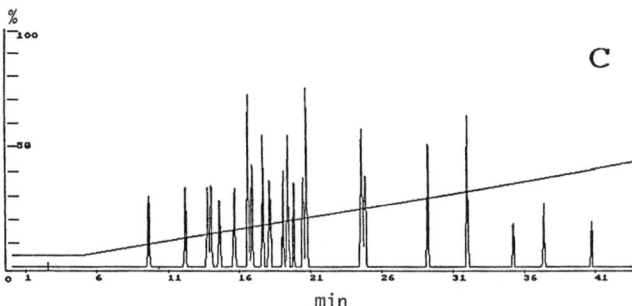

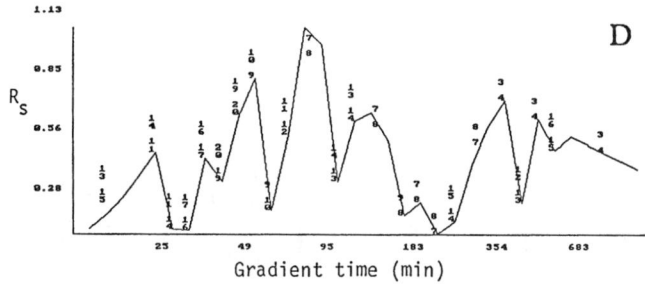

FIGURE 3 (continued)

tryptic digest sample (Figures 6 and 8), however, exhibits some additional complexity. A number of additional, minor peaks can be seen in these separations, especially in the latter part of the chromatogram. These peaks are assumed to arise from an incomplete or imperfect enzymatic hydrolysis of the starting protein; some of these peaks (e.g., #19 and #22) are about as large as some peaks of primary interest (e.g., #5 and #9). While these minor peaks do not greatly complicate the chromatograms of the rhGH digest, minor peaks can become a problem when the number of major peptide peaks in a protein digest ≫20.

3.2.1 Effect of a Change in the Column

The use of a given column for these peptide separations at low pH often results in changes in retention and separation over time. Even larger changes in retention may occur when a new column is used. This is illustrated for the present rhGH sample in Figures 9a and b. The critical peak region of Figure 8b is reproduced in Figure 9a; Figure 9b shows the same portion of the chromatogram run under identical conditions except for the use of a new column of the same type (Nucleosil C_{18}). The resolution of peaks 11/12 ($R_s = 0.9$) is seen to be poor in this separation on the second column. When

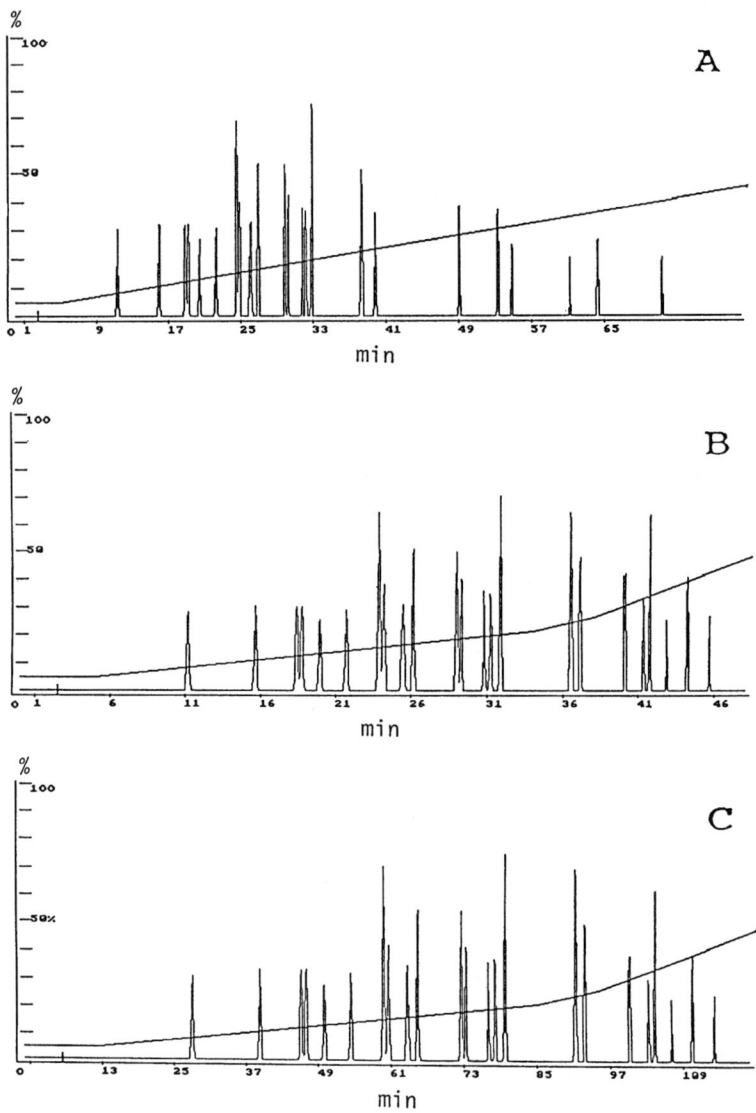

FIGURE 4
Optimized separations of synthetic peptide sample of Figure 3; conditions the same unless indicated otherwise. (A) Simulated separation for 0 to 50% B in 85 min (optimum gradient steepness); (B) simulated separation for multisegment gradient (5/21/26/50% B in 0/29/33/45 min); (C) simulated separation as in (B) except 0.4 mL/min with gradient times adjusted proportionately (0/72.5/82.5/112.5 min); (D) experimental run for conditions of (C).

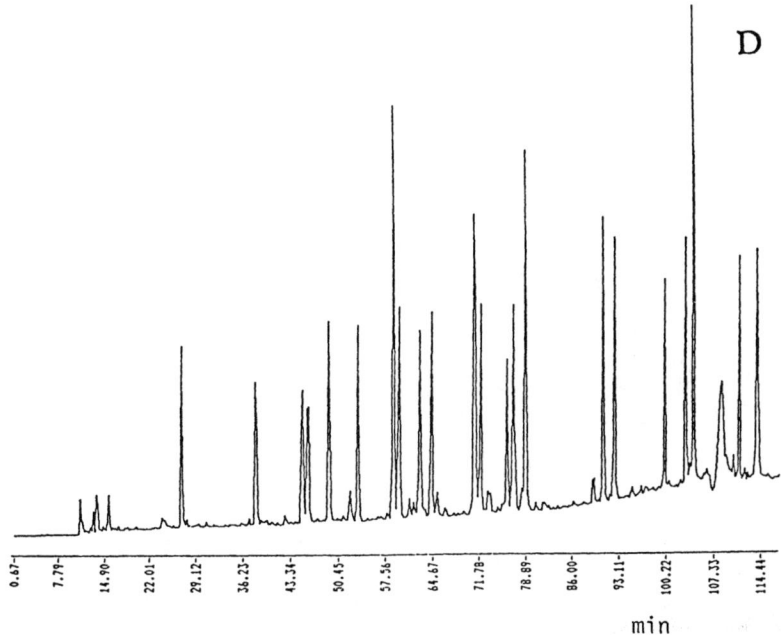

FIGURE 4 (continued)

a change in separation as in Figure 9b occurs, it is possible to reoptimize the separation by carrying out two experimental runs with the new column for subsequent computer simulation. When this was done for the latter column, it was possible to increase the resolution of peaks 11/12 for an acceptable separation.

The substitution of an entirely different column normally leads to even larger changes in retention and resolution, as seen in Figure 9c (StableBond C18; Rockland Technologies). Peaks #14/15 are now the critical pair, with a resolution R_s = 1.6, i.e., better than in Figures 9a or b. In the case of the latter column, a further increase in resolution is possible by means of an initially flatter gradient (see the resolution map of Figure 9e). For this flatter gradient (Figure 9d), the minimum resolution can be increased to R_s = 2.5.

3.3 Recombinant Tissue Plasminogen Activator (r-tPA) Tryptic Digest

This sample comprises 54 major peptides. As seen in the representative separation of Figure 1B, the resulting chromatogram is rather complex. In fact, for the conditions of Figure 1B (40°C), seven major peptides are

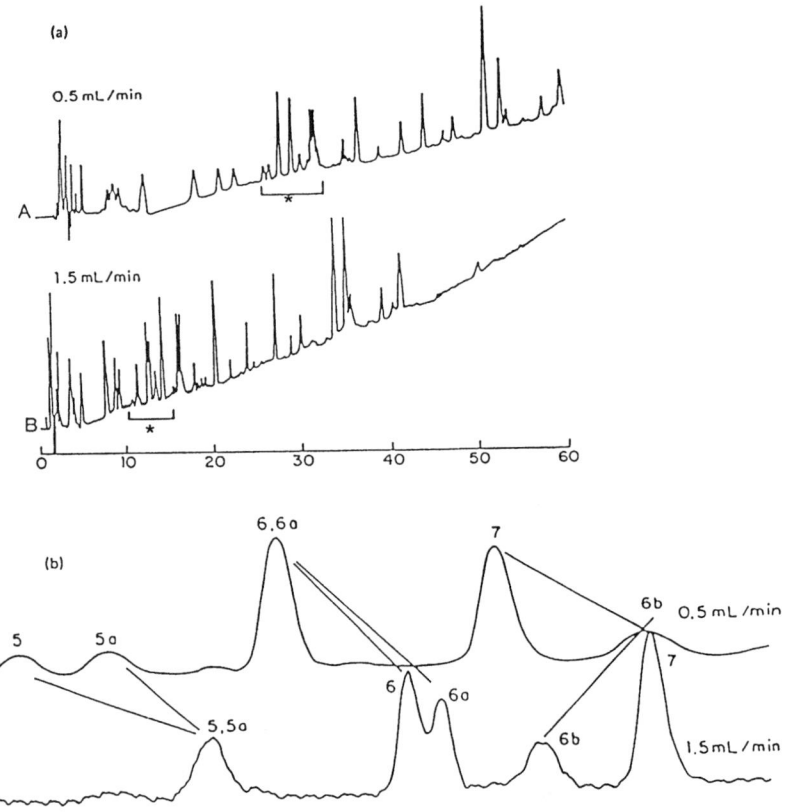

FIGURE 5
Separation of a myoglobin tryptic digest as a function of gradient steepness (flowrate varies). Conditions: 8 × 0.62 cm Zorbax PEP-RP1 column; 10 to 70%v acetonitrile/(0.1% trifluoroacetic acid/water) in 60 min. (a) Chromatograms for flow rates of 0.5 and 1.5 mL/min (b = 0.52 and 0.17, respectively); (b) expanded portions of chromatograms of (a) indicated by (*). Adapted from Reference 12.

completely unresolved, regardless of gradient steepness. This can be seen (Figures 10 to 13) for the separation of each of the four peptide groups of Figure 1C as a function of gradient time or steepness. At 60 °C, four *different* peptides are unresolved, regardless of gradient time; it is thus possible to separate any individual peptide from adjacent peaks by means of an optimum choice of gradient steepness and temperature. The use of computer simulation allows a quick survey of the different separation options, and makes clear which separations are possible via a single HPLC run — and which are not.

The application of computer simulation to samples as complex as the r-tPA digest is less obvious than in the case of the prior examples of Figures 3, 4, and 6 to 9. Thus, the complete separation of all r-tPA peptides of interest is not possible for any choice of gradient conditions or temperature.

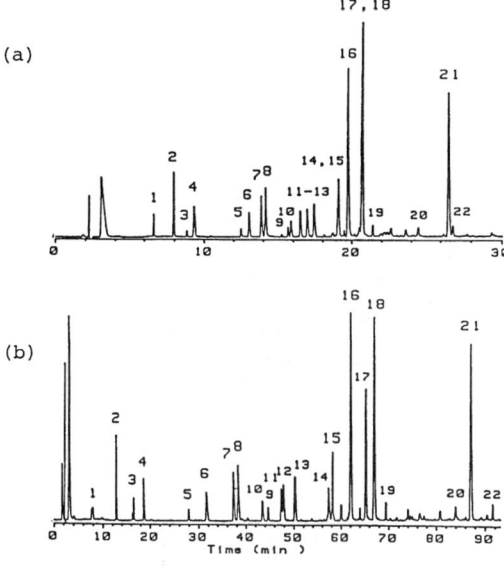

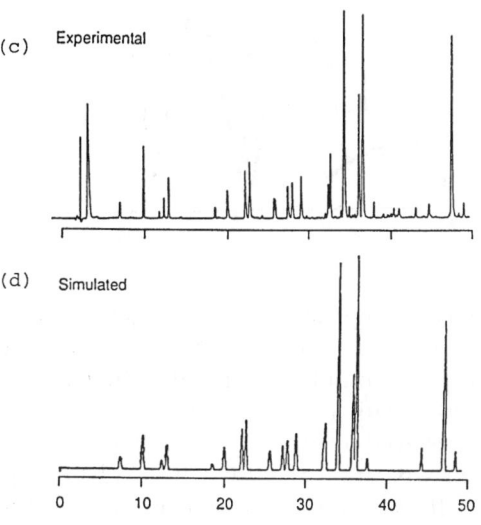

FIGURE 6
Separations of rhGH sample for input to computer simulation. Conditions: 15 × 0.46 cm Nucleosil C_{18} column; 0 to 60% B gradients; A solvent is 0.1% TFA-water; B solvent is 0.08% TFA-acetonitrile; 40°C; 1.0 mL/min; detection at 220 nm: (a) 30-min gradient time; (b) 120-min gradient time; (c) 60-min gradient time; (d) same as (c) except DryLab computer simulation.[14]

Furthermore, the great complexity of these samples, including the presence of a large number of minor peaks, renders the use of computer simulation more complicated and inconvenient (see the following section on "Peak Tracking"). Computer simulation can nevertheless be of value in exploring different possibilities and in keeping track of different peptides as experimental conditions are varied (e.g., Figures 10 to 13). Thus, assume we desire the

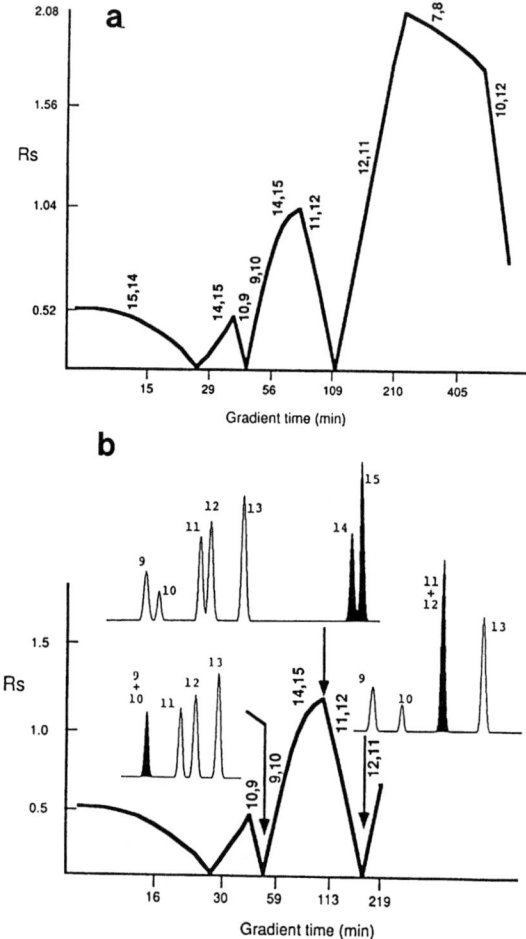

FIGURE 7
Resolution map for rhGH separation of Figure 6. (a) Map showing critical peak pair numbers; (b) same except critical peaks shown for different gradient times.[14]

complete separation of the major peptides in group A of Figure 1C. At a temperature of 40°C, the three peptides X1-A, -B, and -C are unresolved for any gradient time (Figure 10). These three peptides can be partially separated at 60°C with a 60-min gradient. Similarly, peptides T34-35 and T46 cannot be resolved at 60°C — regardless of gradient time (Figure 11). However, their separation at 40°C with a 120-min gradient is relatively easy.

If we want the best separation of the last four peptides in group B (Figure 11), it appears that a temperature of 40°C and an intermediate gradient time will be optimum. This is confirmed in Figure 14A, where the resolution map for group B at 40°C is shown. For an optimum gradient time of 95 min, a minimum resolution of $R_s = 1.2$ is possible (Figure 14B). The separation of peptide groups C and D of Figure 1C can be optimized in similar fashion by the simultaneous variation of temperature and gradient steepness (see

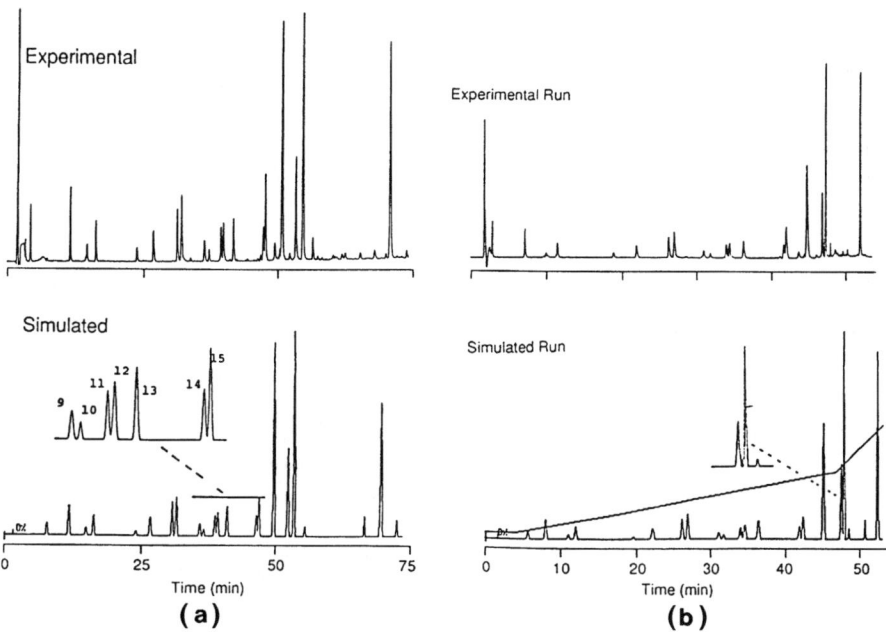

FIGURE 8
Optimized separation of rhGH sample. Conditions as in Figure 6 unless noted otherwise. (a) Linear gradient, 0 to 47% B in 74 min; (b) segmented gradient, 2/32/47% B in 0/48/53 min.[14]

Reference 3 for further discussion). While no single HPLC run results in the separation of all of these major peptides, any given peptide can be resolved from adjacent peaks by a suitable choice of experimental conditions. Similarly, two or more runs can be designed that together allow the separation and analysis of all the peptides of interest in samples of r-tPA. This composite procedure would combine data from each run for peaks that are adequately resolved.

4 DISCUSSION

Computer simulations require less than 1% of the time required for corresponding experimental runs. Consequently, more experiments can be carried out during method development, better (higher resolution, shorter run time) separation procedures can be developed, and the total cost and effort of method development can be greatly reduced. There are other possible advantages of computer simulation: a better understanding and control of the final separation, more rugged and reliable HPLC procedures, and a lower consumption of valuable samples during method development.

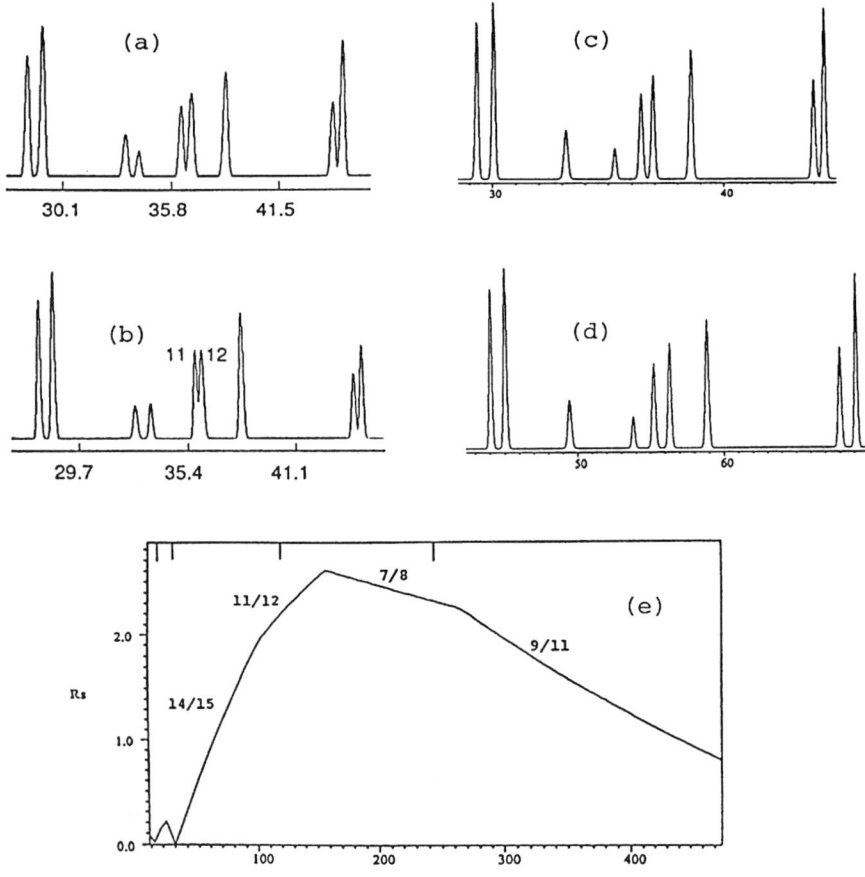

FIGURE 9
Effect of a new column on separation of the rhGH sample. Conditions as in Figure 8b unless noted otherwise. (a) Original Nucleosil C_{18} column; (b) new Nucleosil C_{18} column; (c) Zorbax 80 SB-18 column; (d) reoptimized separation for latter column (2/32/47% B in 0/80/86 min); (e) resolution map for Zorbax 80 SB-18 column.

4.1 Simultaneous Change in Gradient Steepness and Column Temperature

There is now a substantial body of data which confirms that useful changes in peptide selectivity (as in Figures 10 to 13) often result from the variation of either gradient steepness or temperature. It has been found that the effects of these two variables on selectivity are relatively independent of each other;[3] thus the simultaneous optimization of these two variables represents a powerful and convenient means of optimizing peak spacing and resolution. This approach is very much facilitated by the use of computer simulation, as described in this chapter. In fact, it is questionable whether a

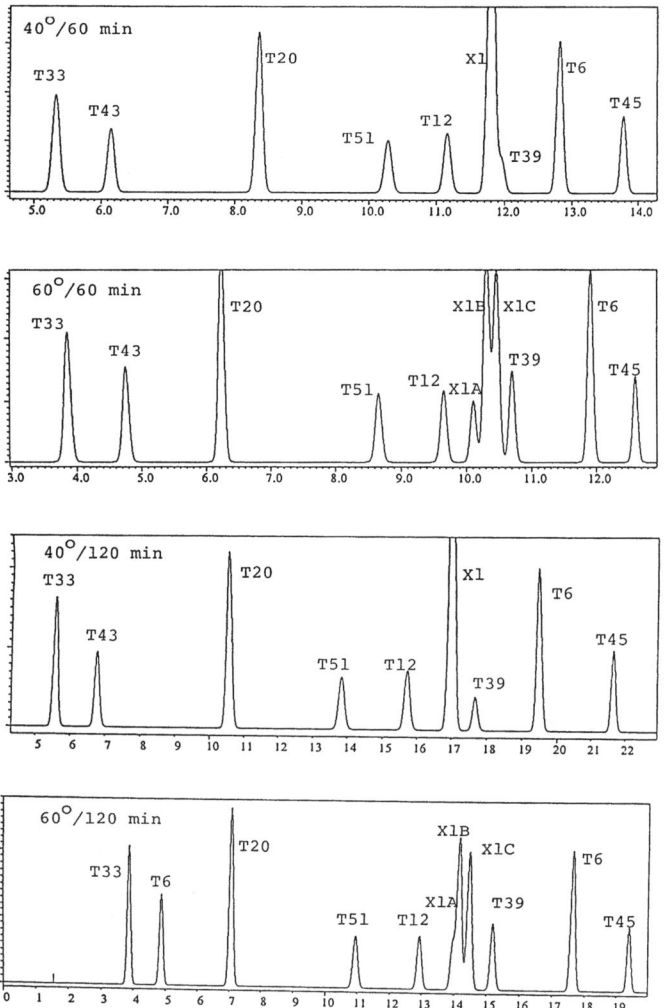

FIGURE 10
r-tPA trypsin digest showing effects of changes in temperature and gradient steepness on the separation of group A. Temperature and gradient times indicated in upper left of each figure. Computer simulations using DryLab software. (From Chloupek, R.C., Hancock, W.S., Marchylo, B.A., Kirkland, J.J., Boyes, B.E., and Snyder, L.R., *J. Chromatogr.*, 686, 31, 1994. With permission.)

trial-and-error approach based on the manual manipulation of experimental data is even feasible in most cases.

Only a small number of runs are required to optimize separation as a function of temperature and gradient steepness. As an example, consider the separation of the rhGH digest using the Zorbax 80 SB-18 column (as in Figure 9c). Two runs (30- and 120-min gradients) were carried out at

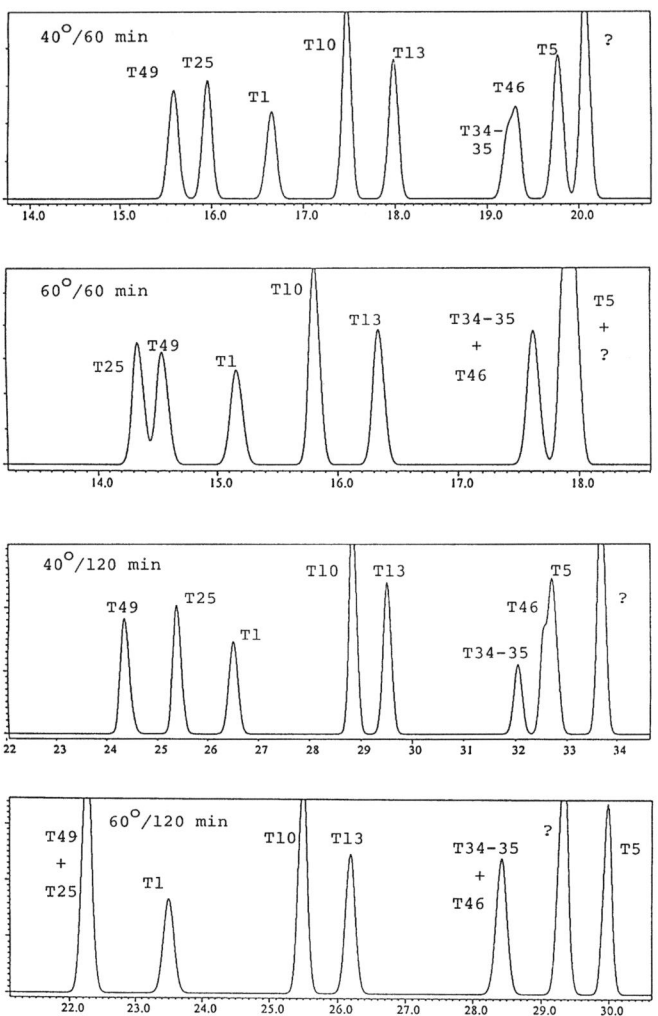

FIGURE 11
r-tPA trypsin digest. Effect of changes in temperature and gradient steepness on the separation of group B. Temperature and gradient times indicated in upper left of each figure. Computer simulations using DryLab software. (From Chloupek, R.C., Hancock, W.S., Marchylo, B.A., Kirkland, J.J., Boyes, B.E., and Snyder, L.R., *J. Chromatogr.*, 686, 31, 1994. With permission.)

temperatures of 20, 40, and 60°C, thus allowing computer simulation at each temperature. Resolution maps for this sample at each temperature are shown in Figure 15. Each resolution map is seen to have a quite different appearance; the optimum gradient time varies from 70 min for separation at 20°C to 250 min at 60°C, and the maximum attainable resolution varies from $R_s = 0.9$ (20°C, poor) to $R_s = 2.8$ (60°C, excellent). We normally desire baseline resolution ($R_s = 1.5$) for all peaks and the shortest possible run time. From

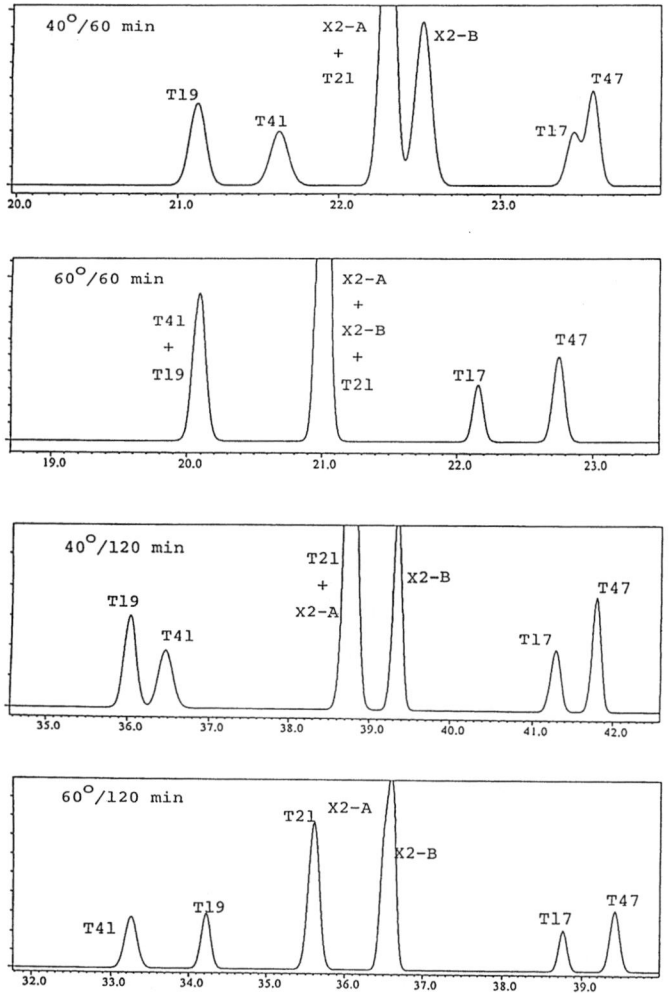

FIGURE 12
r-tPA trypsin digest. Effect of changes in temperature and gradient steepness on the separation of group C. Temperature and gradient times indicated in upper left of each figure. Computer simulations using DryLab software. (From Chloupek, R.C., Hancock, W.S., Marchylo, B.A., Kirkland, J.J., Boyes, B.E., and Snyder, L.R., *J. Chromatogr.*, 686, 31, 1994. With permission.)

the data of Figure 15, $R_s > 1.5$ can be achieved with an 85-min gradient at 40°C or a 150-min gradient at 60°C, but not at all at 20°C. Thus in this case, a temperature of 40°C is better than either 20°C or 60°C.

The latter approach to the separation and analysis of protein digests will not always permit the complete resolution of all peptides of interest in a single HPLC run. The complexity of some samples, e.g., r-tPA simply does not allow this. However, alternatives to the use of a single (optimized) run are

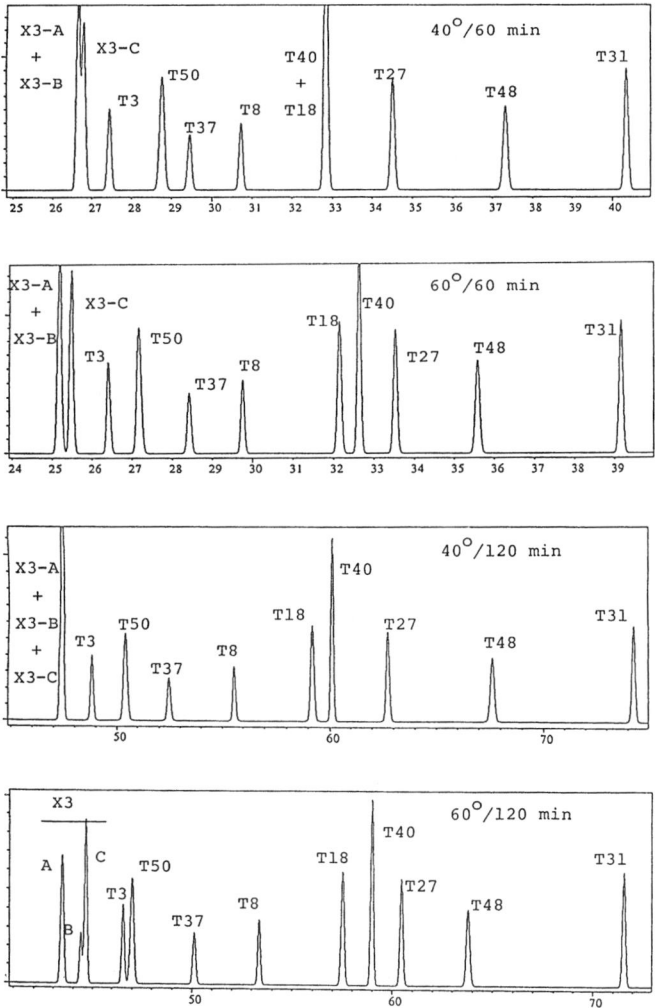

FIGURE 13
r-tPA trypsin digest. Effect of changes in temperature and gradient steepness on the separation of group D. Temperature and gradient times indicated in upper left of each figure. Computer simulations using DryLab software. (From Chloupek, R.C., Hancock, W.S., Marchylo, B.A., Kirkland, J.J., Boyes, B.E., and Snyder, L.R., *J. Chromatogr.*, 686, 31, 1994. With permission.)

possible: (a) composite runs as discussed above, (b) the use of multiple, sequential runs (e.g., 2-D separation, column switching, etc.), or (c) a further increase in the peak capacity of the chromatogram by an increase in column plate number or decrease in gradient steepness. Computer simulation is well suited for the exploration of these options, based on a small number of initial experimental runs.

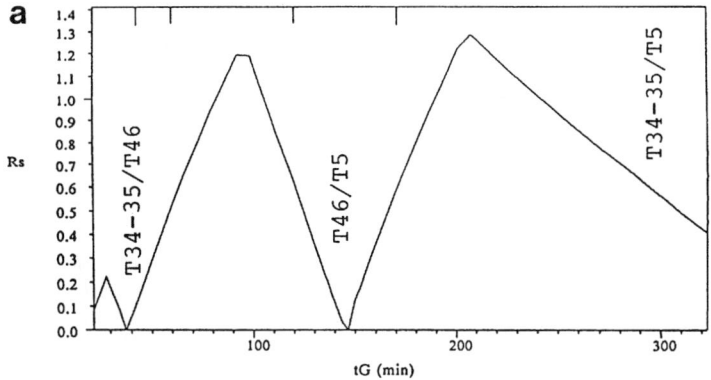

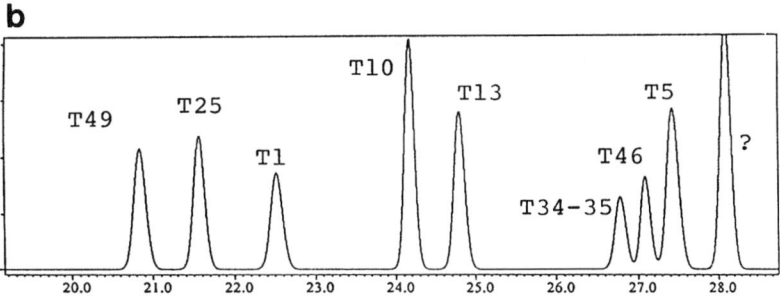

FIGURE 14
r-tPA trypsin digest. Resolution in map (a) and optimized separation (b) of group B at 40°C. Other conditions as in Figure 11. See text for details.

4.2 Potential Problems in the Use of Computer Simulation for the Separation of Protein Digests

The use of computer simulation in practical applications is potentially limited by two considerations: predictive accuracy and the convenience of using this technique. Once these two issues are resolved, computer simulation can be used in place of most of the experimental separations now required to develop HPLC methods for peptide or other samples.

4.2.1. Predictive Accuracy

Computer simulation as described here has been used for a large number of different sample types other than peptides. Deviations between experimental and predicted separations are generally small and of minor practical significance, as illustrated by the representative examples presented in this chapter. Possible causes of predictive inaccuracy have been discussed elsewhere.[13,17,21-25] Errors in simulated separations can arise from any of the

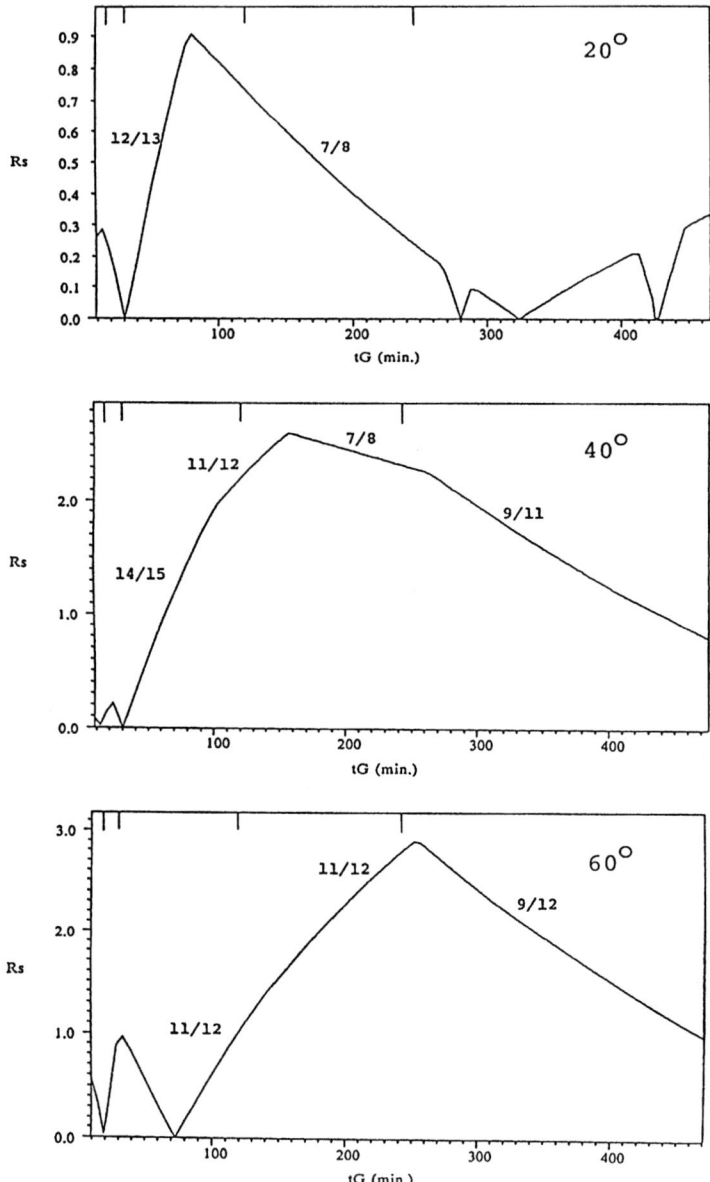

FIGURE 15
rhGH Trypsin digest. Resolution as a function of gradient time and temperature. Other conditions as in Figure 9c. Numbers refer to critical (least resolved) peak pairs. (From Chloupek, R.C., Hancock, W.S., Marchylo, B.A., Kirkland, J.J., Boyes, B.E., and Snyder, L.R., *J. Chromatogr.*, 686, 31–45, 1994. With permission.)

following causes: a poor choice of conditions for the initial experimental runs used as input to the computer, malfunctioning equipment, errors in the assumed values of column dead volume or equipment hold-up (dwell) volume, underperforming data systems, or a failure in any of the various assumptions that are incorporated into the model used for computer simulation. The DryLab software includes a number of validity checks that can recognize some of these errors and bring them to the attention of the user. However, it must be stated that computer simulation in most cases requires data of a higher quality than is needed for traditional trial-and-error HPLC method development. This means that the experiments used for computer simulation should be carried out with reasonable care, using equipment whose performance has previously been verified. With computers, the operating rule is always: garbage in, garbage out. And the "garbage out" usually smells worse than the "garbage in".

4.2.2 Convenience of Experimental vs. Simulated Separations

Computer simulations are obviously more convenient to carry out than "real" separations, once the required experimental data have been entered into the computer. In the ideal case, the data system for the LC system will collect the necessary information from the initial two experimental runs and transfer these results directly into the computer simulation program. Before computer simulation can begin, however, the individual peptides must be matched between the two runs ("peak tracking"). Present DryLab software allows for fully automatic data transfer and peak tracking, but complex samples such as the r-tPA digest discussed here will usually require some manual intervention. The inconvenience of this procedure must then be weighed against the possible advantages of computer simulation.

The peak-tracking software used in the DryLab program is based on matching peaks of similar relative retention and area. Samples that contain >20 to 30 components will often result in some adjacent peaks of similar area, which then prevents an unambiguous match of these peaks. The use of different detection wavelengths (e.g., 220 and 280 for peptides) can be used to resolve most such situations, and a future goal is to allow such information to be used automatically rather than manually.

Peak tracking is made difficult when the relative amounts of minor peaks exceed some threshold. Matches made on the basis of area then become less reliable due to the overlap of major and minor peaks and changes in these overlaps with changes in gradient time. Therefore, the enzymatic digestion should be carried out so as to minimize the extent of minor components in the sample. The importance of minor peaks so far as peak tracking is concerned also increases when the number of major peptides in the sample is >20.

5 CONCLUSIONS

The HPLC separation of an enzymatic protein digest into its primary peptides can represent a formidable challenge to the chromatographer. Even the simplest of such samples, e.g., where the number of primary peptide $\leqslant 20$, may require several experimental runs in order to achieve an adequate separation with an acceptable run time. When the number of peptides $\geqslant 20$, a strictly experimental approach may be totally impractical. The required number of experiments is too large, and it is often difficult to keep track of the effects of a change in conditions on each peak in the chromatogram. This is especially true when attempts are made to vary peak spacing or selectivity.

"Standard" HPLC conditions are often used for the separation and analysis of protein digests: acetonitrile/water gradients with TFA added as a low-pH buffer. Experience has shown that these conditions are generally most favorable to good separation. However, peak overlap is often observed in the resulting chromatograms, even when very flat gradients (and long run times) are employed. Peak overlap can be minimized by optimizing the selectivity of the system. It has been found that changes in gradient steepness and temperature are powerful and convenient means for altering peak spacing and separation. Computer simulation, as described here, greatly facilitates the effective use of this approach to optimized protein-digest separations.

REFERENCES

1. Mant, C.T. and Hodges, R.S., Eds., *High-Performance Liquid Chromatography of Peptides and Proteins*, CRC Press, Boca Raton, FL, 1991.
2. Hearn, M.T.W., Ed., *HPLC of Proteins, Peptides and Polynucleotides*, VCH Publishers, New York, 1991.
3. Chloupek, R.C., Hancock, W.S., Marchylo, B.A., Kirkland, J.J., Boyes, B.E., and Snyder, L.R., *J. Chromatogr.*, 686, 45, 1994.
4. Stadalius, M.A., Quarry, M.A., and Snyder, L.R., *J. Chromatogr.*, 327, 93, 1985.
5. Davis, J.M. and Giddings, J.C., *Anal. Chem.*, 53, 418, 1983.
6. Herman, D.P., Billiet, H.A.H., and deGalan, L., *Anal. Chem.*, 58, 2999, 1986.
7. Hancock, W.S., Chloupek, R. C., Kirkland, J.J., and Snyder, L.R., *J. Chromatogr.*, 686, 31, 1994.
8. Bennett, H.P.J., *High-Performance Liquid Chromatography of Peptides and Proteins*, Mant, C.T. and Hodges, R.S., Eds., CRC Press, Boca Raton, FL, 1991, 319.
9. Mant, C.T. and Hodges, R.S., *High-Performance Liquid Chromatography of Peptides and Proteins*, CRC Press, Boca Raton, FL, 1991, 327.
10. Guo, D., Mant, C.T., Taneja, A.K., Parker, J.M.R., and Hodges, R.S., *J. Chromatogr.*, 359, 499, 1986.
11. Kratzin, H.D., Kruse, T., Maywald, F., Thinnes, F.P., Gotz, H., Egert, G., Pauly, E., Friedrich, J., and Yang, C.-Y., *J. Chromatogr.*, 297, 1, 1984.
12. Glajch, J.L., Quarry, M.A., Vasta, J.F., and Snyder, L.R., *Anal. Chem.*, 58, 280, 1986.

13. Dolan, J.W., Lommen, D.C., and Snyder, L.R., *J. Chromatogr.*, 485, 91, 1989.
14. Chloupek, R.C., Hancock, W.S., and Snyder, L.R., *J. Chromatogr.*, 594, 65, 1992.
15. Mant, C.T., Burke, T.W.L., Zhou, N.E., Parker, J.M.R., and Hodges, R.S., *J. Chromatogr.*, 485, 365, 1989.
16. Mant, C.T. and Hodges, R.S., *High-Performance Liquid Chromatography of Peptides and Proteins,* CRC Press, Boca Raton, FL, 1991, 705.
17. Quarry, M.A., Grob, R.L., and Snyder, L.R., *Anal. Chem.*, 58, 907, 1986.
18. Snyder, L.R. and Stadalius, M.A., *High-Performance Liquid Chromatography, Advances and Perspectives,* Vol. 4, Horvath, Cs., Ed., Academic Press, New York, 1986, 195.
19. Ghrist, B.F.D., Snyder, L.R., and Cooperman, B.S., *HPLC of Biological Macromolecules,* Gooding, K.M. and Regnier, F.E., Eds., Marcel Dekker, New York, 1990, 403.
20. Ghrist, B.F.D. and Snyder, L.R., *J. Chromatogr.*, 459, 43, 1988.
21. Snyder, L.R. and Dolan, J.W., *LC.GC Mag.*, 7, 524, 1989.
22. Snyder, L.R. and Quarry, M.A., *J. Liq. Chromatogr.*, 10, 1789, 1987.
23. Ghrist, B.F.D., Cooperman, B.S., and Snyder, L.R., *J. Chromatogr.*, 459, 1, 1988.
24. Stuart, J.D., Lisi, D.D., and Snyder, L.R., *J. Chromatogr.*, 485, 657, 1989.
25. Lisi, D.D., Stuart, J.D., and Snyder, L.R., *J. Chromatogr.*, 555, 1, 1991.

CHAPTER 3

THE USE OF UV-SPECTRAL INFORMATION IN THE EVALUATION AND INTERPRETATION OF TRYPTIC MAPS

Hans-Jürgen P. Sievert

CONTENTS

1 Introduction 58
   1.1 Tryptic Mapping — The Challenges 58
   1.2 On-Line Spectroscopy to the Rescue 60
   1.3 Topics Covered 60

2 Spectral Matching 62
   2.1 Definitions and Terms 62
   2.2 The Match Limit Problem 64
   2.3 Two Approaches 64

3 Predicting the Match Limit: The Theoretical Approach 65
   3.1 Dissimilarity and Noise 65
   3.2 The Reduced Dissimilarity Score 67
   3.3 T14 as an Example 68
   3.4 Advantages and Disadvantages of the Theoretical
        Approach 71

4 Peak Purity Analysis 71
   4.1 Why Worry About Peak Purity? 71
   4.2 How to Assess Peak Purity 72

5	Describing the Match Limit: The Statistical Approach	73
	5.1 The Match Discriminator	73
	5.2 The T14 Example Revisited	74
	5.3 Advantages and Disadvantages of the Statistical Approach	74
6	Peak Identification	76
	6.1 Use of Spectral Libraries in HPLC	76
	6.2 Reverse Library Search	76
	6.3 Peak Identification Based on the Reduced Dissimilarity Score	77
	6.4 Peak Identification Based on the Discriminator Value	79
7	Peak Tracking and Correlation	80
	7.1 Applications for Peak Tracking	80
	7.2 Peak Correlation	80
	7.3 The Two-Column Example	81
8	Other Examples for Utilization of Spectral Information	84
	8.1 What is Wrong With This Peak?	84
	8.2 Where Is the Sugar?	85
	8.3 Beyond Peak Purity: Spectral Stripping	87
9	Use of Derivative Spectra and Spectral Smoothing	89
	9.1 Use of Derivative Spectra	89
	9.2 Smoothing of Spectra	90
	9.3 Examples	90
10	Conclusion	92
	10.1 UV vs. Mass Spectra	93
	10.2 Need for a Spectral Data Base of Peptide Fragments	93
	10.3 Need for Dedicated Data Analysis Software	94
	10.4 The Multidisciplinary Approach	94
References		95

1 INTRODUCTION

1.1 TRYPTIC MAPPING — THE CHALLENGES

In a book about peptide mapping, nothing needs to be said at this point about the process of tryptic mapping itself. However, if we look at Figure 1, which depicts the tryptic map of recombinant human growth hormone

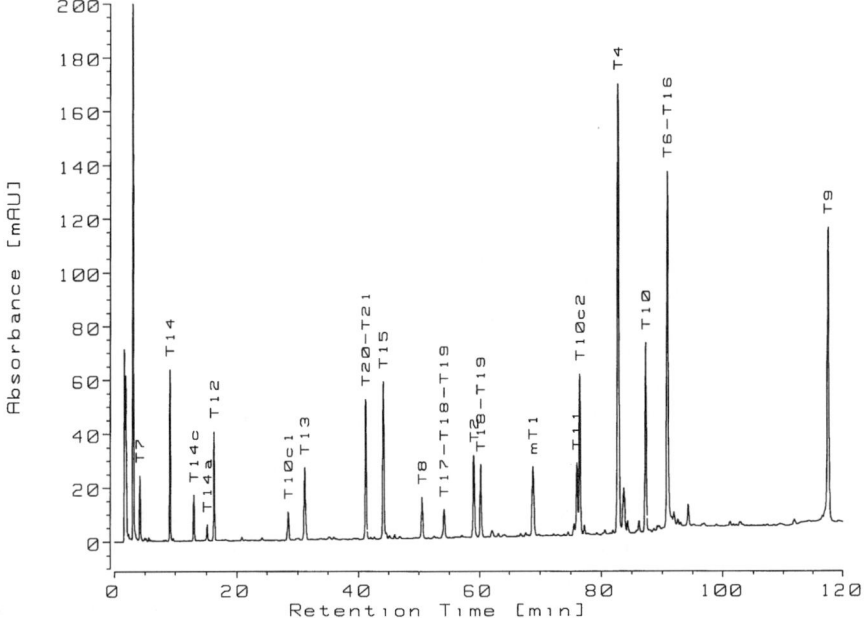

FIGURE 1
Tryptic map of recombinant human growth hormone. Conditions for tryptic digestion as described in Reference 1. Column: 4.6 × 150 mm Nucleosil C18, 5 µm, 100 Å, column temperature 40°C; mobile phase: 50 mM NaH$_2$PO$_4$, pH 2.85 (solvent A), acetonitrile (solvent B), 1 mL/min, linear gradient 0 to 60% B from 0 to 120 min; sample size 25 µL, containing 1.5 nM (40 µg) of Protropin® digest.

(r-hGH) with all fragments conveniently labeled, it is very easy to forget how much time and painstaking effort went into the assignment and confirmation of peak identities. Each peak had to be collected from the HPLC effluent and had to be analyzed subsequently by such diverse methods as fast atom bombardment mass spectrometry (FAB MS), N-terminal Edman sequencing, and amino acid hydrolysis, to name only a few.

Glycosylation, incomplete tryptic cleavage, nonstandard cleavage products, interference from disulfide bonds, and the possible presence of proteinaceous and other impurities from the manufacturing process make the unambiguous identification of each chromatographic peak in a tryptic map a real challenge. Furthermore, any change in the chromatographic method, such as the use of a different mobile phase, a new or different column, or a different separation principle, may necessitate repeating the complete characterization of each peak. Needless to say, any analytical technique that would facilitate this laborious process will be of great interest to the protein chemist.

1.2 On-Line Spectroscopy to the Rescue

The availability of on-line spectroscopic methods provides a whole new additional domain of information, which can aid greatly in the determination and confirmation of peak identity. In addition to retention time information, which in a complex tryptic map is only of limited use, mass spectra or ultraviolet (UV) spectra can be employed to gain further insight into the chemical makeup of a chromatographic peak.

The introduction of electrospray and related liquid chromatography–mass spectrometry (LC-MS) technology makes possible the use of molecular weight information for the identification of peptide fragments. Indeed, electrospray LC-MS has seen an enormous increase in use during the last few years, as illustrated elsewhere in this book. However, the technique requires expensive equipment, an experienced operator, and at this point cannot be considered suitable for routine, everyday use.

HPLC with diode array detection (HPLC–DAD), on the other hand, is a much simpler and less expensive technique that can also aid in the identification of tryptic peptides. It suffers primarily from the fact that there is a preconceived notion that peptide spectra do not contain sufficient detail for successful compound identification. While it is true that UV spectra do not provide absolute identification, they can be used to compare the spectral properties of an unknown peak to those of a set of well characterized standards and thus to arrive at the most likely identification.

1.3 Topics Covered

To provide an up-to-date perspective in this chapter on the discriminatory power of UV spectra, and to convincingly demonstrate their potential to distinguish extremely small spectral differences, we will begin by focusing on the process of spectral comparison or matching and discuss the primary problem encountered there, namely, that of assigning a meaningful threshold for the classification of two UV spectra as being identical or different.

This chapter begins by describing how the use of theoretical noise modeling can aid in the definition of a realistic match threshold. Next, the application of similarity profiles is discussed to ascertain peak purity as a necessary requirement before peak identification can take place. A second approach to spectral matching, based on statistical analysis, is then presented that can be utilized to define an unbiased discriminator value indicative of the difference between two spectra.

Attention is then given to the problem of peak identification and the different approaches outlined are applied to the task of tracking peaks under different conditions of chromatography. Lastly, several examples illustrating other problem areas in tryptic mapping where UV spectra can be useful, will be provided.

TABLE 1
Tryptic Peptides of r-hGH

Name	From	To	Amino Acid Sequence
T1	1	8	FPTIPLSR[a]
T1c	1	6	FPTIPL[a]
T2	9	16	LFDNAMLR
T3	17	19	AHR
T4	20	38	LHQLAFDTYQEFEEAYIPK
T5	39	41	EQK
T6	42	64	YSFLQNPQTSLCFSESIPTPSNR[b]
T7	65	70	EETQQK
T8	71	77	SNLELLR
T9	78	94	ISLLLIQSWLEPVQFLR
T10	95	115	SVFANSLVYGASDSNVYDLLK
T10c1	95	99	SCFAN
T10c2	100	115	SLVYGASDSNVYDLLK
T11	116	127	DLEEGIQTLMGR
T12	128	134	LEDGSPR
T13	135	140	TGQIFK
T14	141	145	QTYSK
T14a	141	145	qTYSK[c]
T14c	141	143	QTY
T15	146	158	FDTNSHNDDALLK
T16	159	167	NYGLLYCFR[b]
T17	168	168	K
T18	169	172	DMDK
T19	173	178	VETFLR
T18-T19	169	178	DMDKVETFLR[d]
T17-T18-T19	168	178	KDMDKVETFLR[d]
T20	179	183	IVQCR[e]
T21	184	190	SVEGSCGF[e]

[a] Contains an additional N-terminal methionine in the case of met-hGH (Protropin®).
[b] Linked by disulfide bridge to form T6-T16.
[c] q Denotes pyroglutamic acid.
[d] Incomplete tryptic cleavage.
[e] Linked by disulfide bridge to form T20-T21.

Most examples presented in this chapter are taken from the analysis of the tryptic map of recombinant human growth hormone (r-hGH). Since reference is made to a number of different tryptic fragments, we give the composition of the various standard and nonstandard tryptic peptides of r-hGH in Table 1. The peptide sequences shown in the table should facilitate the interpretation of the peptide spectra shown in subsequent figures.

2 SPECTRAL MATCHING

2.1 DEFINITIONS AND TERMS

Spectral matching is concerned with describing the similarity or dissimilarity between two spectra of interest in terms of a single number that is derived algorithmically from the digital data available for both spectra. Unlike visual comparison of spectra, which can be biased and subjective, numeric spectral matching will always generate the same results if the same algorithm is applied. Selection of the proper mathematical procedure and a thorough understanding of its capabilities and limitations, however, are still an important task.

The most commonly employed procedure for spectral matching is based on the pairwise correlation of the absorbance values obtained at each discrete wavelength for the two spectra compared. The correlation coefficient, r, is usually squared to eliminate negative correlation, and can further be multiplied by 1000 to arrive at a more manageable number. A "match factor", MF, defined in this way is a measure of similarity and ranges from 0 for spectra that are completely dissimilar to 1000 for identical spectra:

$$MF \equiv 1000 \cdot \frac{\Sigma(s_i - \bar{s})(u_i - \bar{u})}{\sqrt{\Sigma(s_i - \bar{s})^2 \cdot \Sigma(u_i - \bar{u})^2}} \qquad (1)$$

where $\bar{s}$ and $\bar{u}$ denote the mean absorbances and s_i and u_i the individual absorbances for spectra s and u, respectively.

While MF is the most commonly used measure of spectral similarity, for the discussion in this chapter we will employ another indicator of dissimilarity derived in a more intuitive fashion. Let us define the magnitude, {s}, of a spectrum, s, as

$$\{s\} \equiv \sqrt{\sum_{i=1}^{N} (s_i^2)} \qquad (2)$$

where s_i are the discrete absorbance values at the ith wavelength and the spectrum contains N data points in the wavelength region considered; {s} is a measure that describes the average absorbance across the wavelength region of interest. It is akin to the area under the spectral curve and is related to the variance

$$\sqrt{\Sigma(\hat{s}_i - \bar{s}_i)^2}$$

found in the denominator of Equation 1.

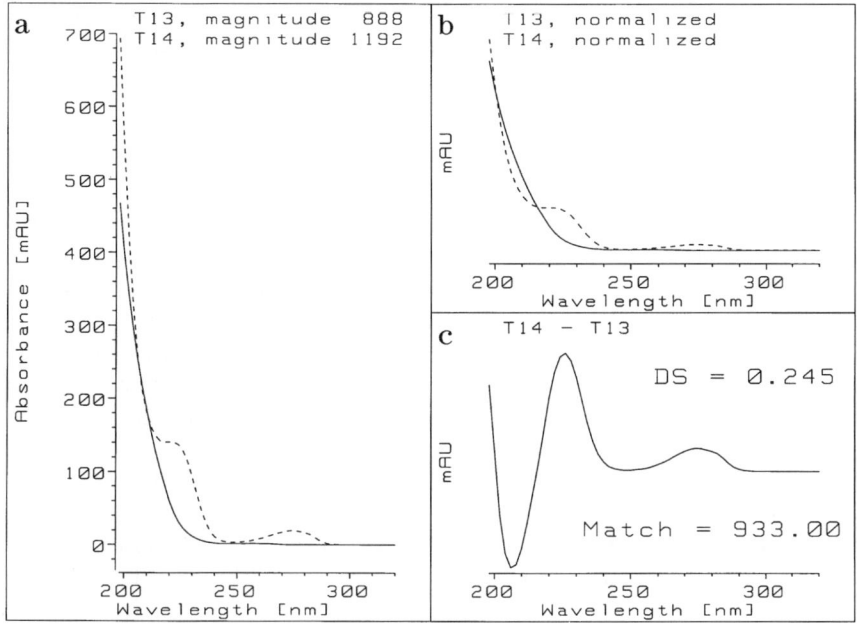

FIGURE 2
Spectral dissimilarity derived from the difference of two normalized spectra.

The spectral magnitude can be considered as a convenient factor that allows us to normalize any spectrum to unit size. The definition of a normalized spectrum, $\hat{s}$, is given in Equation 3. More information about the relative merits of different normalization procedures can be found elsewhere.[2]

$$\hat{s} = \frac{s}{\{s\}} \qquad (3)$$

A straightforward measure of spectral dissimilarity can now be derived as the magnitude of the difference spectrum obtained from two normalized spectra. Thus the "dissimilarity score", DS, for two spectra, s and u, can be defined in the following fashion:

$$DS \equiv \sqrt{\Sigma(\hat{s}_i - \hat{u}_i)^2} \qquad (4)$$

DS can range from a value of 0, characteristic of identical spectra, to a value of $\sqrt{2}$ for spectra that exhibit the largest possible difference.

Figure 2 illustrates the use of the dissimilarity score, DS. Two spectra, T13 and T14, from the tryptic map of r-hGH are compared in panel a, where they are drawn to scale. Their magnitudes are 888 and 1192, respectively. Dividing each spectrum by its magnitude yields the normalized spectra of panel b. If we subtract T13 from T14 after normalization, we obtain the difference spectrum shown in panel c. The dissimilarity score calculated from Equation 4 is 0.245, which corresponds to a match factor of 933.00 based on Equation 1.

2.2 THE MATCH LIMIT PROBLEM

Irrespective of which matching procedure we use, there is always one fundamental problem that needs to be solved: what is the threshold value or "match limit" that separates those match scores that classify two spectra as different from those that would let us consider them identical?

Apart from obvious factors such as proper wavelength calibration and careful matching of other instrument parameters, and the chemical environment imposed by the separation method, there are several other factors that prevent us from simply declaring any non-zero difference between two spectra as being significant.[2] Most of those can be controlled or compensated for — with one noticeable exception — instrument noise.

Noise is derived from various sources within the instrumentation and has the overall effect of causing random fluctuations in the absorbance values measured for a specific spectrum. Thus, two spectra acquired for the same compound at different times will appear slightly different. Even successive spectra taken from a pure chromatographic peak differ not only in concentration, but also exhibit those random fluctuations as illustrated in Figure 3.

The T12 peak, shown in Figure 3a in terms of its spectral S/N ratio, contains 25 spectra of varying concentration that are overlaid in panel c. If the spectra are normalized and the normalized average peak spectrum is subtracted, a number of difference spectra result as depicted in panel b. The difference spectra represent the presence of random noise. Due to the different scaling factors applied during normalization, the noise spectra also exhibit differences in amplitude, as can be seen in panel d where the magnitude of each noise spectrum is given as a function of time (solid line). This trough-like profile shows an inverse relationship to the peak profile in panel a. The dashed line, which gives the inverse of the S/N ratio from panel a for the T12 peak, confirms this relationship. Since, for a given set of instrument parameters, noise is constant, its influence becomes more pronounced as the compound concentration and, with it, absorbance decrease in the front and rear portions of the peak.

2.3 TWO APPROACHES

To overcome the limitation imposed by the presence of noise and to establish a meaningful match limit, we will present two approaches that have been employed by us.[2,3] The first one relies on actually determining how much noise is present and correcting the match score accordingly. Since this approach is based on some theoretical considerations concerning the influence

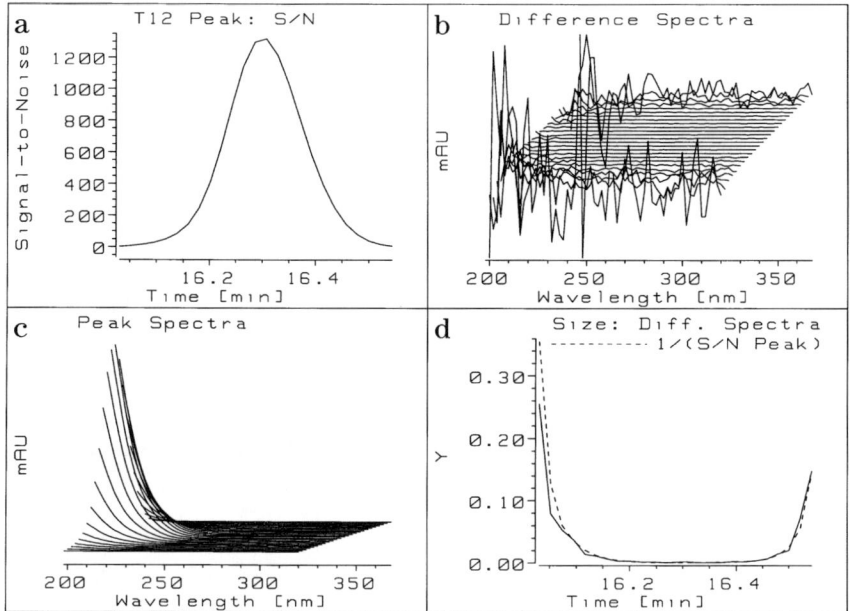

FIGURE 3
Influence of noise on spectra for the same compound at different concentrations.

of noise and tries to predict the match score that would solely be due to noise, we call it the predictive or theoretical approach.

If we do not want to concern ourselves with noise modeling and measurements, we can use the reproducibility of the matching process itself to set statistical limits to differentiate between identical and different spectra. Because we use statistics to describe the actual behavior of the matching process, we call this the descriptive or statistical approach.

Let us now take a closer look at these two solutions for dealing with the match limit problem. We will try to describe their relative merits and limitations and to give some practical examples illustrating their use.

3 PREDICTING THE MATCH LIMIT: THE THEORETICAL APPROACH

3.1 Dissimilarity and Noise

As already pointed out, noise makes two otherwise identical spectra seem different. If we were to compare a real, noisy spectrum to an "ideal", noise-free one, the observed difference would be the noise background of the

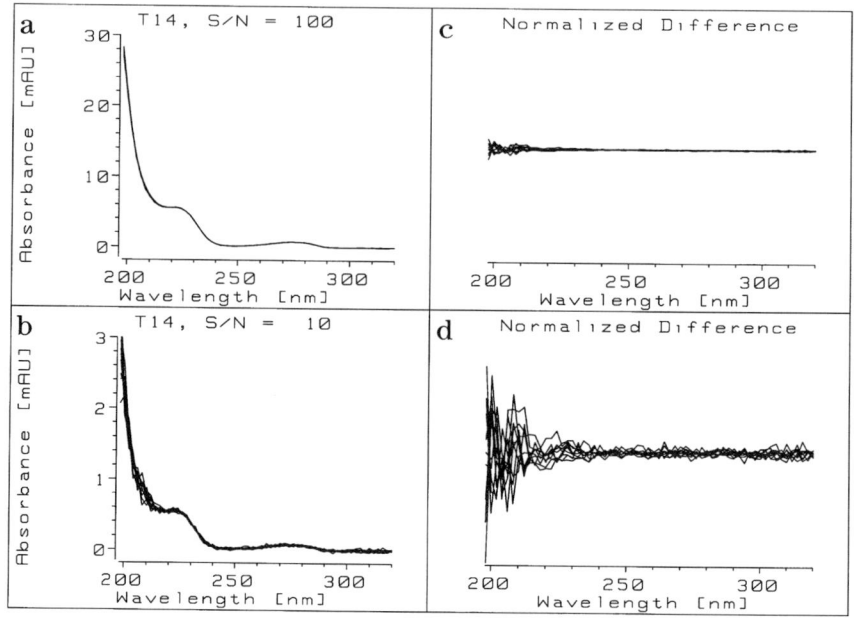

FIGURE 4
The apparent dissimilarity between multiple spectra of the same compound is related to noise.

instrument contained in the noisy spectrum. Since, in the definition of *DS*, a spectrum is divided by its magnitude for normalization, the noise content of the normalized spectrum will decrease as its magnitude increases. Because instrument noise itself is basically constant for a given set of analysis conditions, the apparent dissimilarity observed for spectra of different S/N ratio is proportional to the magnitude of this noise and inversely proportional to the magnitude of the noisy spectrum itself.

Figure 4 shows the comparison of 10 identical T14 spectra with different noise content drawn to scale at S/N ratios of 100 and 10 in panels a and b, respectively. Visual comparison clearly indicates that the relative noise content of the two groups of spectra is different. This is further illustrated by the corresponding normalized difference spectra in panels c and d, both plotted at the same attenuation.

Based on this relationship, we define a noise threshold, *TH*, which describes the degree of dissimilarity for the comparison of otherwise identical spectra as it would be expected solely from the presence of instrument noise. A more detailed mathematical derivation of the noise threshold and its generalization to the case of two noisy spectra, will be presented elsewhere.[3]

$$TH \equiv \frac{\{n\}}{\{u\}} = \frac{1}{S/N}, \quad \text{where} \quad S/N = \frac{\{u\}}{\{n\}} \qquad (5)$$

Here, u denotes a noisy spectrum and n its noise component. As is readily apparent, TH is inversely proportional to the S/N ratio of spectrum u. This was already confirmed by the overlay of noise magnitude and the inverse S/N of the T12 peak, as depicted in Figure 3d.

3.2 THE REDUCED DISSIMILARITY SCORE

If we divide DS from Equation 4 by the noise threshold, TH, we should arrive at a measure of dissimilarity that is now independent of noise. A value of one or less indicates that the dissimilarity could be explained just by noise; values larger than one point to a residual difference not due to noise and, thereby, to a definite difference in the compounds that the spectra were obtained from.

$$RDS \equiv \frac{DS}{TH} = DS \div \frac{\{n\}}{\{u\}} \qquad (6)$$

The only problem with this definition is the fact that we cannot directly measure the actual noise spectrum n contained in spectrum u. We can, however, take several baseline spectra elsewhere in the chromatogram and calculate an estimate of n, or more precisely, $\{n\}$, from these.

Figure 5 shows a plot of $\{n\}$ for the first 50 spectra, presumably representing the spectral baseline, from a tryptic map. Each spectrum was corrected by interpolating the solvent background from its two neighboring spectra. Obviously, $\{n\}$ fluctuates around a mean value with a certain standard deviation. The highest value that could typically be encountered will be within a confidence interval above this mean. Thus, the highest RDS that could realistically be attributed to noise would be this upper limit L, divided by the mean value for $\{n\}$:

$$L = \frac{mean(\{n\}) + T \cdot sdev(\{n\})}{mean(\{n\})} \qquad (7)$$

where T is a factor determined by the desired degree of confidence.

In the example in Figure 5 the mean noise is 0.430 with a standard deviation of 0.110. For the 99.73% confidence interval with N = 50, T equals 3.16. The upper noise limit is at $T \cdot \sigma$ above the mean, which calculates to be

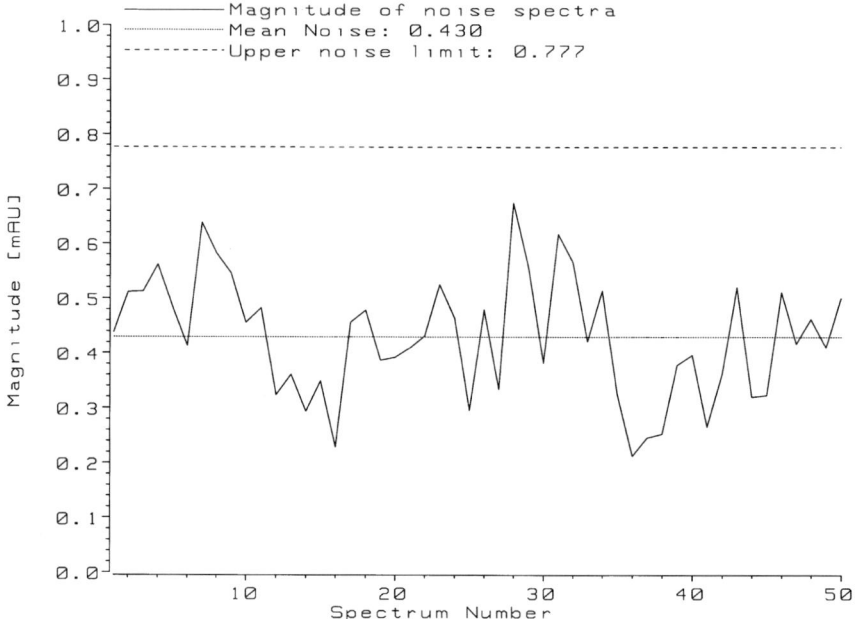

FIGURE 5
Determination of mean system noise, {n}, and upper noise limit.

0.777, or 1.81 times the mean. An *RDS* larger than 1.81 for the comparison of two spectra would, therefore, be indicative of a true spectral difference.

3.3 T14 AS AN EXAMPLE

Let us now look at a concrete example for the application of the reduced dissimilarity score. As seen in Table 1, one of the peptides in the tryptic map of r-hGH, T14, is also found in two nonstandard modifications. T14c is a chymotrypsin-like cleavage product of T14 and, in T14a, Glu has been converted to pyroglutamic acid. Since all three fragments contain tyrosine as the primary chromophore, their spectral distinction is somewhat of a challenge.

Figure 6 shows the three peptide spectra in question. Dissimilarity numbers and match factors for the three possible pairwise comparisons are given. The most similar spectra are encountered with the T14-T14a pair, which also is most similar in chemical makeup.

To determine whether it is even possible to identify each peptide unequivocally, we show in Figure 7 the three peaks for these fragments, each peak overlaid with RDS profiles generated from the comparison of all peak spectra with a standard spectrum for each of the three peptides. We refer to these kinds of profiles as similarity curves. By further averaging those RDS

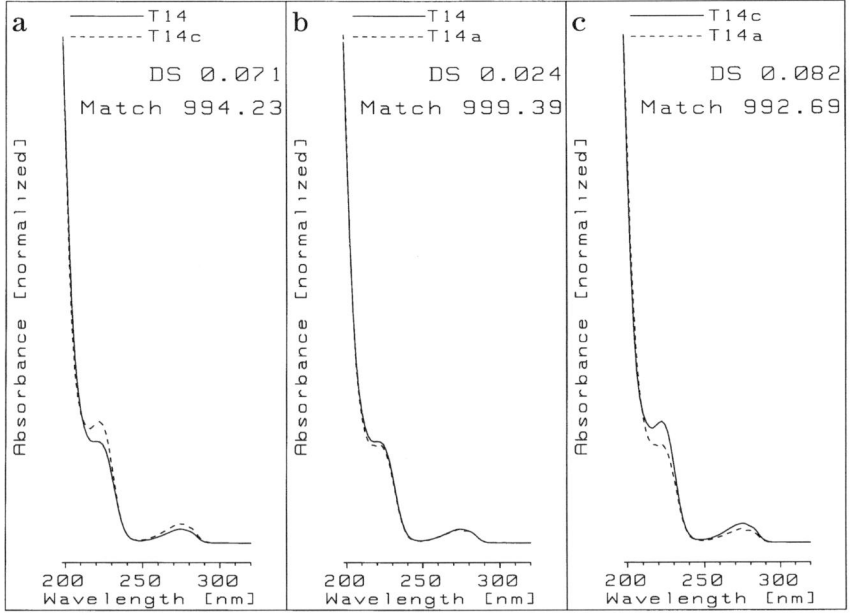

FIGURE 6
Spectral comparison of three T14 peptide fragments.

values that exceed the upper noise limit, we can assign a dissimilarity number, DN, to each attempted match.

$$DN = \frac{\Sigma g \cdot (RDS_i - L)}{N} \quad (8)$$

where

$$g = \begin{cases} 1 & \text{when } (RDS_i - L) \geq 0 \\ 0 & \text{when } (RDS_i - L) < 0 \end{cases} \quad (9)$$

From visual inspection of the similarity curves and from the values found for DN, we can conclude that each peptide can be positively identified. In all cases, the gap between the correct match and the closest incorrect match is sufficiently large to prevent misidentification.

Table 2 further supports this claim by providing data on the reproducibility of the dissimilarity number, DN. Data from five replicate injections (A to E) for the three peaks of Figure 7 are presented here. As can be seen, all correct matches have average values of 0.025 or less. The difference

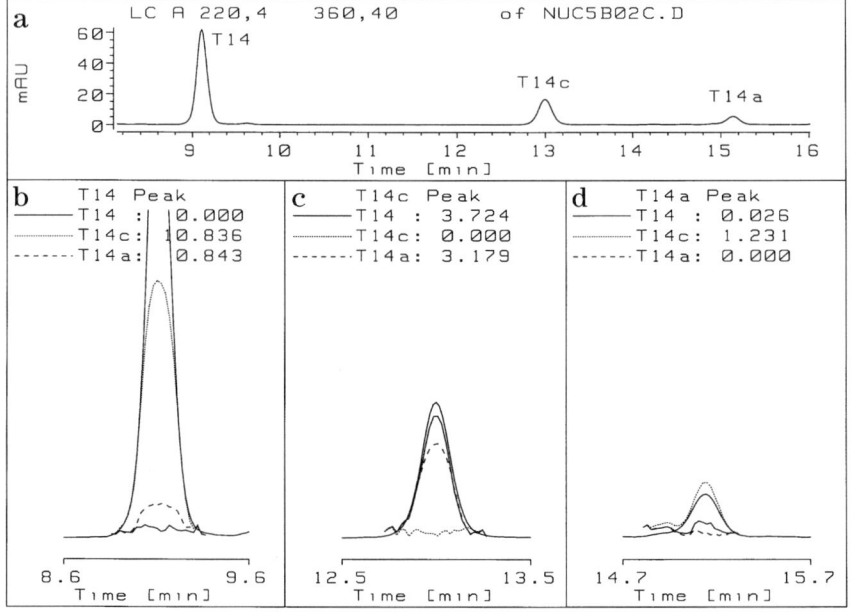

FIGURE 7
Identification of three T14 cleavage products. Panel a shows a section of the tryptic map for r-hGH analyzed as described in Figure 1, panels b to d highlight the different T14 fragments overlaid with similarity profiles used for identification.

TABLE 2

Reproducibility of Peak Identification Based on Value of *DN*

Peak	Std	A	B	C	D	E	Mean	Sdev
T14	T14	0.000	0.042	0.000	0.010	0.073	0.025	0.032
	T14c	12.605	12.939	10.836	14.256	12.199	12.567	1.237
	T14a	1.206	1.353	0.843	1.411	1.139	1.190	0.223
	Purity	0.000	0.048	0.000	0.000	0.064	0.022	0.031
T14c	T14	4.983	4.738	3.724	5.246	4.564	4.651	0.579
	T14c	0.000	0.028	0.000	0.003	0.033	0.013	0.016
	T14a	4.254	4.132	3.179	4.498	3.911	3.995	0.503
	Purity	0.000	0.010	0.000	0.000	0.017	0.005	0.008
T14a	T14	0.034	0.096	0.026	0.124	0.240	0.104	0.087
	T14c	1.595	1.529	1.231	1.067	1.897	1.464	0.324
	T14a	0.000	0.002	0.000	0.023	0.000	0.005	0.010
	Purity	0.000	0.000	0.000	0.074	0.000	0.015	0.033

between the correct and lowest incorrect DN value is statistically significant and is consistent enough to make peak identification believable. The rows for 'Purity' will be explained in the next section.

3.4 Advantages and Disadvantages of the Theoretical Approach

While the theoretical approach provides a very powerful set of tools for spectral matching, it is not without some drawbacks. Noise data need to be generated for each data set, which means that sufficient spectra have to be acquired. Generally, acquisition of all spectra for a chromatographic analysis is strongly recommended. If similarity curves are to be generated, all spectra are definitely required.

Because spectral S/N and magnitude are important factors in the calculations, care has to be taken to track the noise propagation through any averaging, summation, or subtraction process. Thus, standard spectra, which are typically obtained by averaging several individual spectra, have to be multiplied by the square root of the number of spectra averaged to reflect the improvement in S/N.

On the positive side, the predictive approach does quantitatively remove the effect that noise has on spectral match scores. It generates RDS values which are truly independent of noise and can be compared in a meaningful fashion for data sets acquired under quite different conditions. Even though it is still not possible to define an absolute match limit, simply because spectral similarities for chemically different compounds can range on a sliding scale from complete identity to pronounced dissimilarity, a quantitative, defensible, and reproducible match limit can be derived that will give the same results for everybody who employs it.

4 PEAK PURITY ANALYSIS

4.1 Why Worry About Peak Purity?

Before we go on to the descriptive approach, we would like to discuss another application of the reduced dissimilarity score, namely, the analysis of peak purity. Our ultimate goal in tryptic mapping is to identify the compound represented by each chromatographic peak. In order to do that in a meaningful fashion, we first have to establish that a given peak does indeed contain only one compound. This is exactly what peak purity is concerned with.

If we had a peak where at several time points more than one compound was eluting, the spectra measured at those points would be linear combinations of the underlying pure compound spectra. It is quite obvious that any identification scheme based on the comparison of these peak spectra with a standard spectrum would then lead to erroneous results. Thus, before we can try to identify a peak we have to establish its purity.

4.2 How to Assess Peak Purity

One possible approach to purity analysis is quite similar to the one described earlier for the purpose of peak identification. Rather than using the dissimilarity number obtained from the comparison of peak spectra against a standard spectrum for identification, we can also view it as an indicator of peak purity. The standard with the lowest dissimilarity number is definitely the compound with the most similar spectrum and the most likely candidate for peak identification.

Alternatively, we can calculate a dissimilarity number from the similarity profile obtained by comparing all peak spectra against the average peak spectrum. This has the advantage that no additional data in the form of external standard spectra are needed. The dissimilarity numbers obtained in this way are given in Table 2 in the rows marked 'Purity'. As we can see, these numbers are very similar to the *DN* values obtained for correct identification.

To assure peak purity, the dissimilarity number for a given peak should not exceed an empirically determined threshold. In the case of the five data files used in Table 2, a threshold value of 0.115 was calculated from the purity numbers for those peaks confirmed to be pure.

We could thus expand our peak identification procedure in the following way. First, the DN value for the average peak spectrum is calculated. If it indicates that the peak is probably pure, dissimilarity numbers are next obtained for all standards that could be expected to match the current peak. If the lowest value is at or below the noise threshold L, we can assume that the corresponding standard identifies the peak in question.

The example given in Figure 8 illustrates this approach. The figure depicts two peaks, together with dissimilarity curves for the average peak spectrum, the expected standard spectrum, and the closest mismatch. The corresponding dissimilarity numbers show that the first peak is pure and can be correctly identified as the T8 peptide.

The second peak, on the other hand, has a purity number of 0.485 and clearly exceeds the threshold value. The spectrum with the closest similarity is T17-T18-T19, which is the expected identity of this peak, but its DN does not differ very much from the purity number and is clearly higher than the DN seen for the correct match of the T8 peak. The next closest match, T20-T21, has a DN high enough that it would not be considered a possible match candidate for this peak. Based on the similarity profile, we can further state that the impurity is present in the front section of the peak where the dissimilarity numbers for T17-T18-T19 are highest.

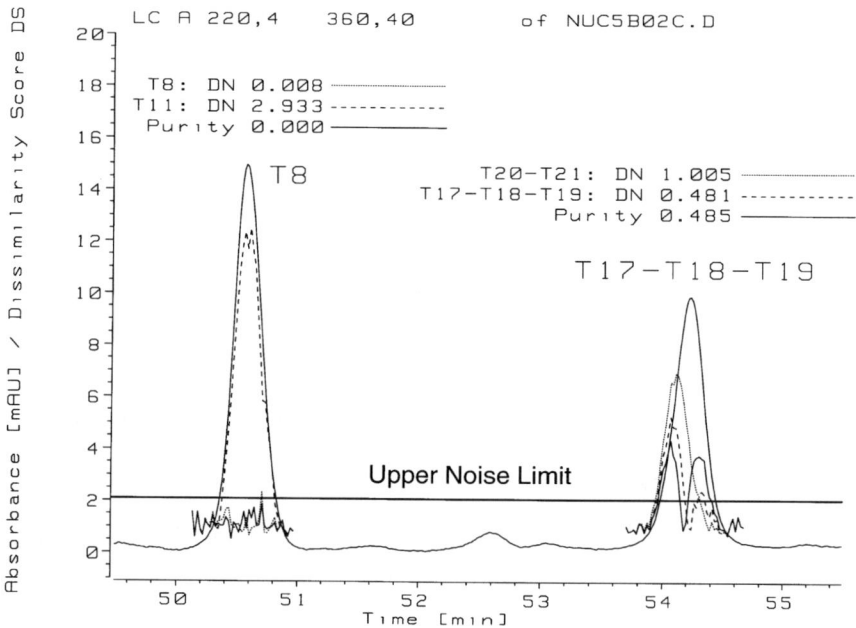

FIGURE 8
Determination of peak purity. A segment of the r-hGH tryptic map, analyzed as described in Figure 1, is overlaid with similarity profiles for the determination of peak purity and identity.

5 DESCRIBING THE MATCH LIMIT: THE STATISTICAL APPROACH

5.1 THE MATCH DISCRIMINATOR

Let us assume that we take a number of actual spectra for a compound U, either at several points across the apex of a chromatographic peak or derived for the same peak from a series of replicate injections, or as a combination, or both. The average, U_{av}, calculated from these spectra will be a better approximation of the ideal spectrum for U than each of the individual spectra U_i. By matching each U_i against U_{av}, we can determine how reproducible the matching process is in a situation where we know that all spectra involved represent the same chemical compound.

If we want to compare compound U with a standard S, we now match a representative spectrum S against all U_i and compare the resulting distribution of match factors with the ones obtained from the matches of U_{av} with each U_i. We can apply straightforward statistical criteria to the two distributions to assess the likelihood that they are identical or different.

This approach was first described by us in a more rudimentary fashion[2] and will be presented in more detail elsewhere.[3] We typically refer to the match of U_i against U_{av} as the automatch, AM; the comparison of S and U_i is called the crossmatch, CM. Since we are comparing S or U_{av} against the same data set, we can apply a T-test for paired variances to calculate a discriminator value, D:

$$D = \frac{|CM_{av} - AM_{av}|}{T \cdot \sigma} \quad (10)$$

D expresses the difference between the average AM and CM in terms of a multiple T of the pooled variance of the two data sets. T, in turn, defines a confidence interval based on Student's distribution for a desired degree of probability and a given number of degrees of freedom. A value of 1.0 for D is equivalent to stating that the likelihood that AM and CM constitute different distributions, and thus represent different compounds, is just at the desired probability level. Values larger than 1.0 would provide increased confidence in the difference between S and U.

5.2 The T14 Example Revisited

To illustrate the use of the discriminator value, let us consider the same example we used earlier: T14 and its nonstandard cleavage products. Figure 9 shows the AM and CM distributions obtained for the three peptides. AM is derived from the match of each peak spectrum against the average peak spectrum. CM scores were obtained from the comparison of each peak spectrum against a standard library spectrum.

The CM curves for the correctly matched standard in each case closely overlay the AM curves. For the T14a spectrum the closest CM curve is obtained for the T14 standard. In spite of the low S/N ratio of the T14a spectra, this curve is clearly separated from the AM curve.

The numeric D-values corresponding to Figure 9 are given in Table 3. As can be seen, values of D close to 1.0 are obtained for the match between each peak and its corresponding standard. The other two standards in each case result in higher D-values indicating that the spectral differences are sufficient for differentiation. Even the low-concentration T14a peak is identified correctly, although the D-values here are generally lower due to the decreased S/N for this peak.

5.3 Advantages and Disadvantages of the Statistical Approach

The statistical approach seems just as capable as the theoretical approach in its ability to correctly identify compounds with very similar spectra. One

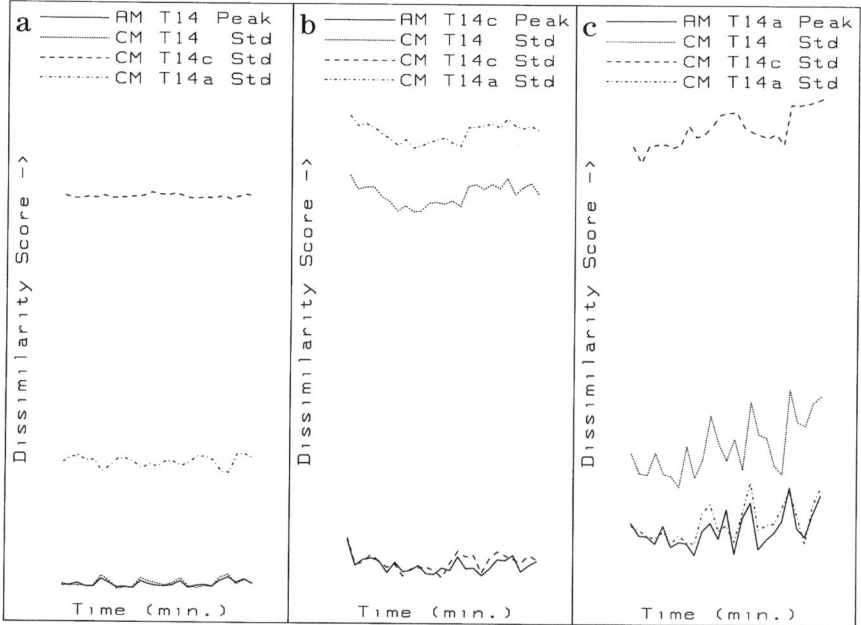

FIGURE 9
Automatch and crossmatch distribution for T14 fragments.

TABLE 3

D-Values for T14 Fragments

Peak	D-values for standards[a]		
	T14	T14c	T14a
T14	**0.56**	174.54	25.45
T14c	52.09	**0.86**	69.66
T14a	5.31	31.96	**1.10**

[a] All D-values were calculated for 25 individual spectra from each peak at a confidence level of 99.73%, T = 3.34.

advantage it offers is that no noise measurements are required and that no assumptions about theoretical behavior have to be made. It is self-calibrating, since each instance of the D-value is related to the reproducibility of the matching process for the current set of spectra. Because of that, it can be applied to any matching algorithm, be it based on the correlation coefficient, the dissimilarity curve, or any other approach.

The primary disadvantage is the need for multiple spectra, typically necessitating replicate analyses. The spectra employed need to be of approximately the same S/N ratio to avoid artificially high σ values stemming from the noise dependency of the match factors. Peak profiles are typically not possible with the statistical approach, making it less suited for purity analysis.

6 PEAK IDENTIFICATION

6.1 Use of Spectral Libraries in HPLC

In contrast to peak purity analysis, which could be undertaken without external standard data by using spectra derived from the peak in question, identification of a chromatographic peak requires the availability of standard spectra with known identity. These spectra are typically stored in a spectral library. Unlike mass spectra which describe a compound property, molecular weight, that is independent of the separation parameters employed, UV spectra can differ greatly as a function of mobile phase composition and pH.

To overcome this apparent limitation, the separation system used to acquire spectra for standards and unknowns has to be well defined and should be standardized. Rather than utilizing vast compound libraries as in mass spectrometry, UV spectra should be kept in smaller, separate libraries, each representing a particular separation system. Several examples from the recent literature underscore the need to clearly define the column and mobile phase used for a specific spectral library.[4,5]

If data acquired on two different instruments are to be compared, care must be taken that both instruments have been calibrated properly for wavelength accuracy. Typically, only data from instruments of the same manufacturer, with matched optical parameters such as slit width and diode bunching, should be employed for compound identification. Two fundamentally different approaches to the use of spectral libraries in HPLC can be described.

In the forward library search, an attempt is made to identify each compound in an unknown sample from a large library of standard compounds. For reverse library searching, a limited library of standard compounds, all of which are expected to be present in the sample, is searched and correlated against the unknown spectra in the current sample. Since the latter better describes the situation encountered with a tryptic map, we will limit ourselves to a brief discussion of reverse library searching.

6.2 Reverse Library Search

This search mode is an expansion of the normal calibration procedures used currently in HPLC. In addition to having a calibration table with information on expected retention time and response factors for each unknown, we now also have available to us the corresponding compound spectrum. Consequently, the library contains spectra only for those compounds that are known or expected to be present in the sample.

For each unknown sample we then try to assign a matching compound spectrum to every standard spectrum in the library. Peaks in the sample that do not match up against any standard, or peaks that are missing altogether, may or may not be significant and can be investigated further. As with conventional calibration procedures, the separation system used is standardized; standards and unknowns are analyzed under the same conditions, usually using the same instrument and column. An application of reverse library searching for the purpose of determining the similarity between the tryptic map of a reference standard and a sample from a current batch of recombinant DNA-derived human growth hormone has been described by us previously.[1]

6.3 Peak Identification Based on the Reduced Dissimilarity Score

Irrespective of the mode of library searching employed, peak identification is ultimately based on finding the best match between unknown and standard spectra. Of the two matching procedures described earlier, the reduced dissimilarity score lends itself more easily to peak identification of a single sample, since it only needs the spectral data available from that sample. Thus, dissimilarity numbers similar to the ones presented in Table 2 could be used to find the most likely correspondence for each standard spectrum. Conflicting assignments could be arbitrated in a number of ways — most obvious would be the use of retention time and peak response as described by us earlier.[1]

Figure 10 shows an example of the kind of information available from a reverse library search for purposes of peak identification. To each standard, one and only one unknown peak is assigned based primarily on the spectral similarity shown in the 'dMatch' column. If the peak assignment is ambiguous, spectral data can further be combined with the deviations between actual and expected retention time and response, given in the 'dRet' and 'dResp' columns, respectively. Retention time deviations are obtained by dividing the absolute difference between standard and unknown by the user-selected retention time delta, which was 0.1 min in this case.

Response deviations combine area and height response and are expressed as an absolute percentage of the corresponding area and height values for the standard and further divided by the response delta value which was 5% in the example presented. The retention time window of 5.0 min indicates that the search for matching unknown spectra was limited to a time window of this size centered upon the retention time of the standard.

```
Standard File L:NUC5STD.L                           5 Mar 93    8:41 am

Unknown in file NUC5B02A.D, 74 entries
Wavelength range from 198 to 320 nm
Match: Ln =  0.453, Retention: delta =  0.100, Response: delta =  5.000
Retention time window: 5.000 min
=============================================================================
  # Std Name      RetTime    Unk  RetTime    dMatch    dRet    dResp    Purity
-----------------------------------------------------------------------------
  1 T7              4.166     3    4.163     1.788    0.037    0.179    1.743
  2 T14             9.117     9    9.097     0.015    0.207    0.070    0.001
  3 T14c           12.999    11   12.996     0.000    0.036    0.159    0.000
  4 T14a           15.141    13   15.140     0.000    0.013    0.612    0.000
  5 T12            16.289    14   16.269     0.014    0.191    0.074    0.000
  6 T10c1          28.449    18   28.425     0.000    0.236    0.172    0.000
  7 T13            31.188    19   31.142     0.025    0.463    0.093    0.029
  8 T20-T21        41.207    24   41.169     0.122    0.374    0.118    0.172
  9 T15            44.133    27   44.087     0.000    0.456    0.106    0.000
 10 T8             50.555    32   50.495     0.003    0.604    0.365    0.000
 11 T17-T18-T19    54.216    34   54.167     1.245    0.482    1.103    1.501
 12 T2             59.121    36   59.058     0.023    0.634    0.039    0.003
 13 T18-T19        60.242    37   60.201     0.015    0.418    0.167    0.000
 14 T1             68.933    43   68.841     0.002    0.925    0.185    0.001
 15 T11            76.040    51   76.019     0.068    0.217    0.158    0.000
 16 T10c2          76.546    52   76.489     0.072    0.575    0.157    0.080
 17 T4             82.764    57   82.704     1.935    0.601    0.249    1.422
 18 T10            87.310    61   87.248     0.465    0.620    0.366    0.278
 19 T6-T16         90.871    63   90.799     2.449    0.723    1.863    0.822
 20 T9            117.582    73  117.493     3.287    0.899    4.482    5.016
=============================================================================
```

FIGURE 10
Sample report for reverse library search. The major peaks of the r-hGH tryptic map, analyzed as described in Figure 1, are identified based on a library file containing standard data for the 20 peptides shown.

The last column in the report gives the purity of the unknown peaks assigned to each standard. Peaks T7 and T17-T18-T19, while identified correctly, do contain additional impurities. The situation in the case of the last four peaks is more complex. It appears likely that the high purity numbers do not indicate the presence of an impurity but are related to problems with detector linearity. These four peptides are present at high concentrations, which could result in a distortion of the spectra taken in the apex region. Nonlinearity, rather than the presence of an impurity, is usually indicated when the peak height is large and the similarity curve exhibits a symmetric bulge in the center portion of the peak.

This is illustrated in Figure 11 where the similarity profile for the average spectrum (dotted line) and the apex spectrum (dashed line) of the T6-T16 peptide are given for a wavelength range from 198 to 320 nm in panel a. The two profiles are symmetric with respect to the peak apex. The average profile exhibits a maximum at the apex and then falls off to the sides, which implies that the peak apex differs from the rest of the peak. The apex profile shows

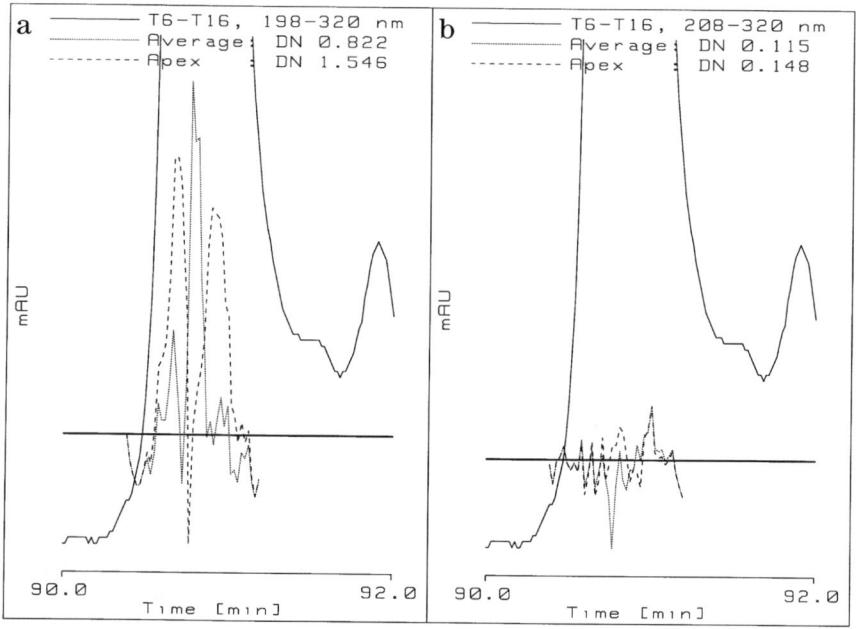

FIGURE 11
Similarity curves for a peak with nonlinear spectra.

a minimum at the apex with two side-lobe maxima which would be expected of a peak where the spectra in the apex region are distorted due to nonlinear behavior.

Further confirmation is obtained from the data in panel b, which differ only in their wavelength range which is 208 to 320 nm. The lowest 10-nm region of the spectrum, where peptides have the highest extinction coefficients and potentially exhibit nonlinearity, has been eliminated. Now both profiles are much more indicative of a pure peak and are almost completely superimposed.

Nonlinearity is a function of the absorbance level as well as the lamp intensity at each wavelength. Therefore, it is not possible to specify a maximum absorbance level above which linearity is compromised. In a conservative approach, absorbance data exceeding 500 mAU for the region below 200 nm, and absorbances exceeding 1000 mAU for wavelengths above 200 nm can potentially exhibit nonlinear behavior.

6.4 Peak Identification Based on the Discriminator Value

The same principles as outlined for the use of the RDS in peak identification could be applied to the discriminator value. Two major drawbacks with this approach, however, are the need for replicate data files and the fact that not all peak spectra are utilized in the evaluation procedure. Our earlier work was heavily based on using the discriminator value, but the more complete treatment of noise interference as implemented in the reduced similarity score has resulted in the latter becoming the method of choice.

7 PEAK TRACKING AND CORRELATION

7.1 Applications for Peak Tracking

Any time a chromatographic separation results in retention times for one or several peaks that differ from those of a reference standard, the question of peak identification resurfaces. Thus, column aging, and the switch to a new column of the same type from a different or even the same manufacturer, can affect retention times, elution order, and number of chromatographic peaks to the point where identification might be required again. This is especially true in cases where a column of an altogether different type is employed. If UV spectra are available, than the spectral information may be sufficient to track the peaks and establish their new position unequivocally, without the need for a more involved identification procedure.

7.2 Peak Correlation

The example we will use to demonstrate the application of peak tracking also serves to illustrate another aspect of spectral identification that can be quite useful when dealing with unknown tryptic maps. Before identification even becomes an issue, it would often be desirable to be able to determine which chromatographic peaks in a tryptic map are relevant and, therefore, warrant subsequent identification.

One approach that we will present is based on utilizing the spectral data to automatically correlate peak spectra from replicate analyses of a given map. Those peaks that are present in all replicate analyses and exhibit identical spectral make-up are considered to be significant and should be identified. The advantage of this procedure is that it can be fully automated and that it is superior to any visual, retention time-based scheme of correlation. Furthermore, no knowledge of peak identity is needed for correlation, as long as overall spectral similarity is used as the criterion. The peaks identified in this fashion could still be artifacts, albeit reproducible ones, that are not derived from any tryptic fragments, but then that is a possibility that is always present.

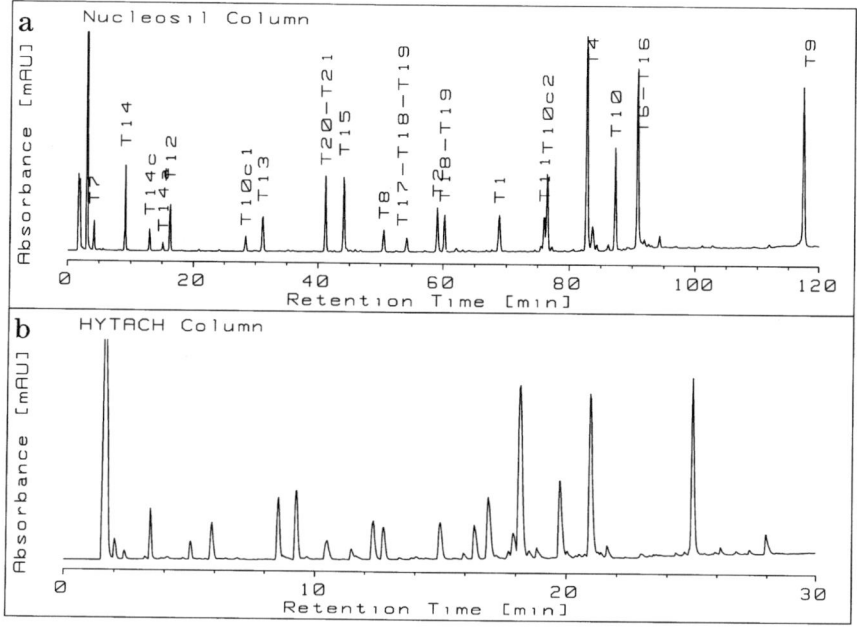

FIGURE 12
Tryptic map of r-hGH analyzed on two different reversed phase columns. Tryptic digestion as described in Reference 1. Panel a: chromatographic conditions as given in Figure 1. Panel b: column: 4.6 × 105 mm HYTACH C18 NPS, 2 μm, column temperature 40°C; mobile phase: 25 mM NaH_2PO_4, pH 2.85 (solvent A), 60% acetonitrile-40% 25 mM NaH_2PO_4, pH 2.85 (solvent B), 0.45 mL/min, linear gradient 0 to 45% B from 0 to 20 min, 45 to 100% from 20 to 30 min; sample size 10 μL, containing 600 pM (15 μg) of Protropin® digest.

7.3 THE TWO-COLUMN EXAMPLE

The example given in Figure 12 shows the tryptic map of r-hGH, analyzed on a conventional Nucleosil column (panel a) and on a nonporous HYTACH column (panel b). While there are numerous, fairly obvious correspondences, the overall chromatographic profile is different enough, especially with respect to analysis time, to raise the question of proper identification for the tryptic fragments.

The first stage employed in the tracking process was the automated correlation of five replicate analyses with the Nucleosil column. Apex spectra for the first replicate were compared to the peak spectra from each of the other four runs. Dissimilarity scores and other measures of similarity, as presented in Figure 10, were then calculated and used to correlate the peaks. Since retention times should be quite reproducible in this case, only peaks in a narrow time window were investigated. Apex spectra for those peaks

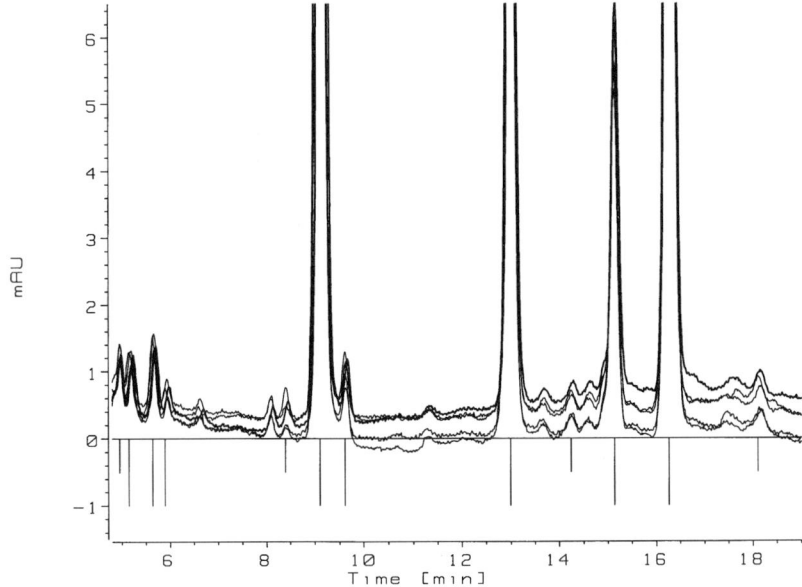

FIGURE 13
Correlation of peak data from multiple injections.

occurring in all five replicate injections were then averaged and stored into a library which was used for the subsequent identification step.

Next, peaks of low intensity were removed from the standard library and the remaining peaks were either labeled with the name of the corresponding tryptic peptide or marked as unidentified. Figure 13 illustrates this process. It displays a short section of the chromatographic traces for all five replicates and evaluates peak data from the second replicate (heavy line). Those peaks that were found to occur reproducibly in all replicates are marked with long negative tick marks. Peaks marked with the short ticks were not present in all injections.

Once a library with standard spectra for the Nucleosil data had been obtained, it was used to identify peaks in each of the five HYTACH maps. Only those peaks that were labeled as reproducible from correlation of all HYTACH replicates were matched. Since it was not necessarily expected that all peaks from the Nucleosil map would be present, and since the retention time differences were substantial, we used a forward library search procedure rather than the reverse search.

In each HYTACH map, for every reproducible peak above a certain height threshold the first few best matches were generated against the Nucleosil standard library. As can be seen in Figure 14, where peak assignment based on the overall best match is shown for the first replicate run, all major peptide fragments were identified correctly. T7 and T14 are missing, presumably these peptides are not sufficiently retained on the HYTACH column and elute in the solvent front.

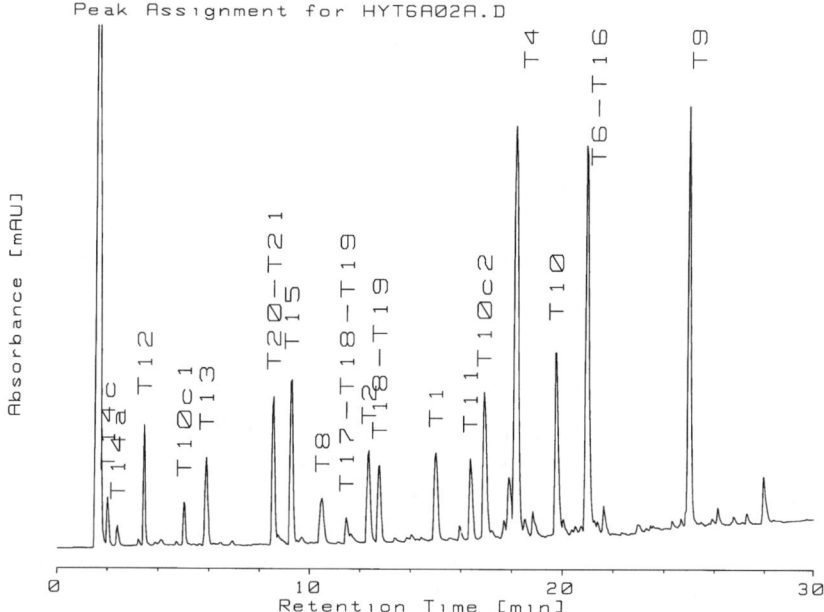

FIGURE 14
Peak tracking of tryptic fragments after switch from a Nucleosil to a HYTACH column.

TABLE 4

Comparison of Purity Data for Two Peaks Analyzed on Two Different Columns

Peak	Purity Number					
	A	B	C	D	E	Mean
Nucleosil Column:						
T8	0.000	0.051	0.000	0.009	0.008	0.014
T17-T18-T19	1.501	1.012	0.479	0.851	1.000	0.968
HYTACH Column:						
T8	1.774	2.698	3.390	1.975	4.309	2.829
T17-T18-T19	0.000	0.000	0.000	0.000	0.000	0.000

Table 4 shows purity data for the neighboring peaks T8 and T17-T18-T19. The latter was found to be impure in the Nucleosil runs. As can be seen, purity data are very reproducible from replicate to replicate, and clearly indicate that with the HYTACH column the T17-T18-T19 peak is now pure while the T8 peak is clearly impure. Whether this indicates that the contaminant moved in elution position from under T17-T18-T19 to T8 will be investigated in the next section when we talk about spectral stripping.

8 OTHER EXAMPLES FOR UTILIZATION OF SPECTRAL INFORMATION

So far we have tried to present a systematic analysis of the different ways in which UV spectral data can be used for purposes of peak purity analysis and peak identification with special emphasis on tryptic mapping. In this section we will present some examples that illustrate in a more general way how spectral data have been helpful in the elucidation of tryptic maps and in solving specific problems.

8.1 What is Wrong With This Peak?

The first example is from our initial work with r-hGH. The commercially available Protropin® has a methionine in the N-terminal position of the first peptide, T1, and will be referred to as met-hGH for better clarity. We were working with a second-generation form of recombinant human growth hormone, r-hGH, where this methionine had been removed by the signal peptidase during protein biosynthesis. When analyzing the tryptic map of r-hGH, it was noted that the T10 fragment yielded less consistent match numbers with a trifluoroacetic acid (TFA) gradient system than with a phosphate gradient.

Closer investigation of the TFA gradient data showed that the traditional match factor for each peak spectrum against the downslope spectrum, when plotted against retention time, decreased towards the front of the peak, as indicated in Figure 15a.

The three spectra at the peak positions marked in panel a were then normalized and overlaid as shown in panel b. Close inspection of the aromatic region plotted in panel c indicated that the upslope spectrum might have a lower aromatic content than the apex and downslope spectra. If the upslope portion of the peak did indeed contain an aliphatic impurity then, by normalizing in the aromatic region and subtracting the downslope from the upslope spectrum, we should be able to eliminate the main T10 component and obtain a difference spectrum characteristic of the impurity.

Figure 16a shows that normalization in the aromatic region does indeed lead to an observable difference in the lower UV region. The difference spectrum is given in panel b (solid line) together with the closest match from the standard library, T1 (dashed line). Mass spectroscopy data confirmed that T1c, a nonstandard cleavage product of T1 (see Table 1), does elute under the T10 peak.

Panel c presents the results of a multicomponent analysis of the T10 peak using T10 and T1c standard spectra from a phosphate map, where they are resolved chromatographically, as the two peak components. The dotted and dashed lines describe the relative concentration of T10 and T1c, respectively, further confirming the presence of T1c in the front section of the T10 peak.

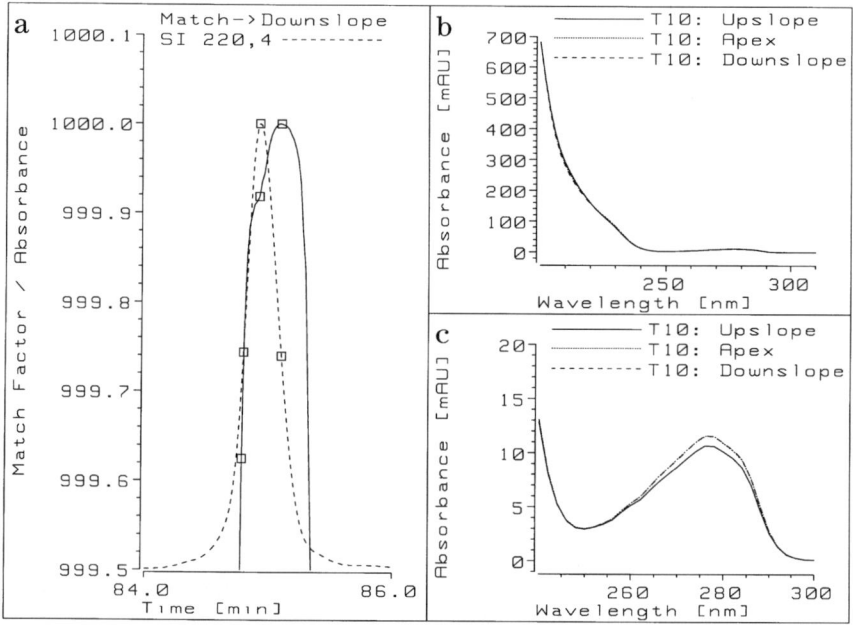

FIGURE 15
Spectral analysis of the T10 peak of r-hGH analyzed with a TFA gradient.

8.2 Where Is the Sugar?

Our next example is taken from the tryptic map of recombinant tissue plasminogen activator (Activase®, r-tPA). The largest tryptic fragment for this protein, T8, contains specific amino acid sites which potentially could undergo deamidation or glycosylation. Either modification should result in a spectrally very similar product with slightly different chromatographic behavior. Would spectral data be of any use in solving this problem?

T8 has a molecular weight of approximately 2900 and contains 27 amino acids, including 1 tyrosine and 4 phenylalanine residues (see Table 5, footnote b). This gives it the rather unique spectrum shown in Figure 17a. Due to the high phenylalanine to tyrosine ratio, the region between 250 and 285 nm is nearly flat. Since deamidation or glycosylation would have little or no effect on the spectral shape of any related fragments, the spectrum of Figure 17a can be used to look for any peaks in the tryptic map that exhibit spectral characteristics similar to T8.

Tryptic digestion is typically done at a pH of 8.6, which is sufficiently high to encourage deamidation. Digestion was therefore also undertaken at a lower pH of 7.5 to reduce the likelihood of deamidation. In the two tryptic maps obtained in this fashion, only two peaks with a T8-like spectrum were found. They eluted next to each other and are labeled as L1 and L2 for the

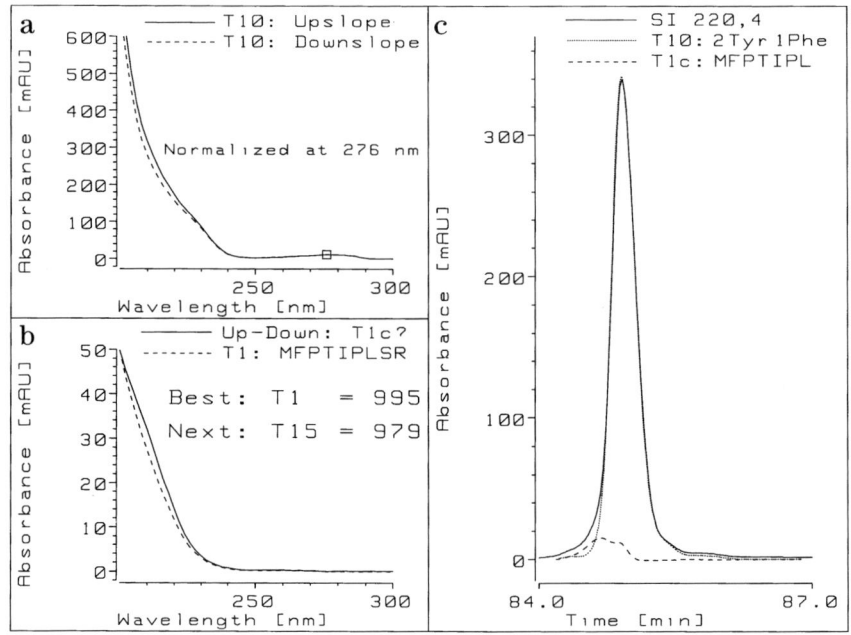

FIGURE 16
Contaminant of T10 peak is identified as T1c.

TABLE 5

Match and Purity Analysis of the T8 Peaks in r-tPA

Standard	Sample Peak			
	H1[a]	H2[b]	L1[a]	L2[b]
H1	0.175	5.525	0.359	21.088
L2	11.261	1.032	2.049	1.126
Purity	0.060	0.186	0.199	0.872

[a] Presumed identity T8-T9; CFNGGTCQQALYSFDFVCQCPEGFAGK-CCEIDTR.
[b] Presumed identity T8; CFNGGTCQQALYSFDFVCQCPEGFAGK.

low-pH map and as H1 and H2 for the high-pH map. The second peak, L2, is present at higher concentration in the low-pH map whereas H1 is the main peak at the higher pH.

We then investigated purity and match profiles for the two maps. The more intense peak at each pH (H1 and L2) was used as a target spectrum. The results are shown in Table 5. At either pH, the first peak showed a better match against H1 and the second peak resembled L2 more closely. The

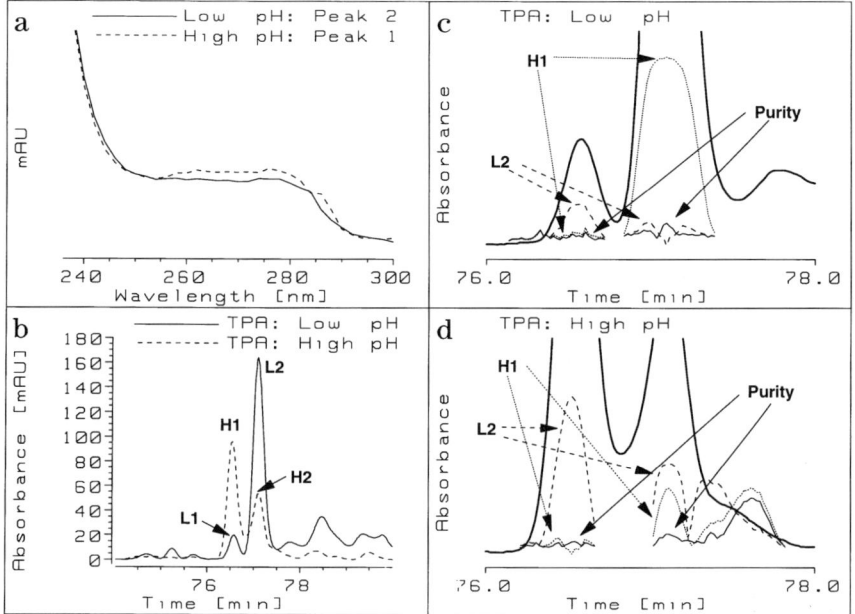

FIGURE 17
Analysis of the T8 peak from r-tPA. Chromatographic conditions: column 3.9 × 150 mm Nova-Pak C18, 4 µm, 60 Å, column temperature 40°C; mobile phase: 50 mM NaH$_2$PO$_4$, pH 2.85 (solvent A), acetonitrile (solvent B), 1 mL/min, linear gradient 0 to 27% B from 0 to 90 min, 27 to 57% B from 90 to 120 min; sample size 25 µL, containing 1.5 nM (100 µg) of Activase® digest.

dissimilarity numbers for the other standard in all cases were sufficiently high to make the identification unambiguous.

As it turned out, deamidation was not an issue for T8 under either digestion condition. The carbohydrate content of the peptide is also constant; T8 is always completely fucosylated. However, T8 is not always completely cleaved from the next peptide, T9. The resulting incomplete cleavage product, T8-T9, elutes before the T8 peak. At the lower pH, cleavage is more complete and less T8-T9 is found. At the higher pH, which is the standard for tryptic digestion, incomplete cleavage is more pronounced and T8-T9 is present in excess of T8. The difference in amino acid composition between T8 and T8-T9 is due to seven aliphatic amino acids, which explains the spectral similarity of the two peptides (see Table 5, footnote a).

8.3 Beyond Peak Purity: Spectral Stripping

When a peptide peak in a tryptic map is found to be impure, subsequent spectral analysis becomes very difficult and, at best, tentative. Since the impurity typically is not known, peak deconvolution based, for instance, on

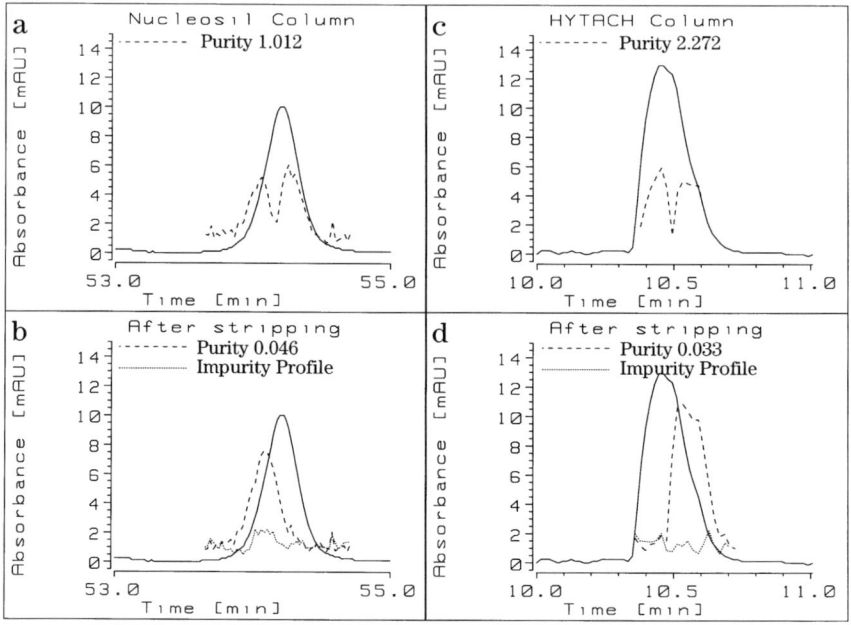

FIGURE 18
Analysis of impure peaks after removal of main component by peak stripping.

multicomponent analysis (MCA) is not possible. If the identity of the primary peak component is known, however, its spectral contribution can be stripped from all peak spectra and the resulting impurity spectrum can be further analyzed.

To strip one component from a series of spectra, an approach similar to the one described by Lacey[6] for the elimination of background spectra was employed. This process has to be applied to peak spectra as well as to standard spectra, and results in a series of modified spectra that are mathematical constructs and have no direct physical meaning. These spectra do, however, retain their respective similarities.

By applying this process to the T17-T18-T19 peak established as impure in Figure 8, it became possible to demonstrate the presence of the impurity in the front section of the peak and to establish that its spectral characteristics most closely resembled those of T4. Figure 18 illustrates this process in panels a and b, where the purity profile is shown before and after spectral stripping, respectively. Panel b also gives a relative concentration profile for the location of the impurity.

Applying the same procedure to the T8 peak of the tryptic map analyzed with the HYTACH column yielded similar results (Figure 18, panels c and d). This time the impurity elutes in the rear portion of the peak (dashed line in panel d) and again resembles most closely the T4 peptide. This could

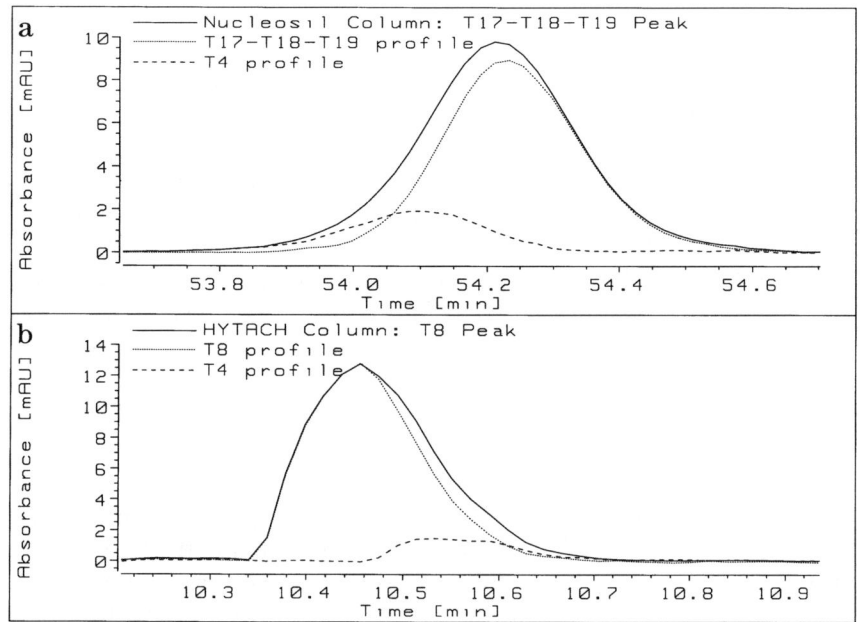

FIGURE 19
Multicomponent analysis of two impure peaks.

support our earlier hypothesis that in switching from the Nucleosil to the HYTACH column the impurity eluting under the T170-T18-T19 peak migrates to the T8 peak.

Since we now have a candidate spectrum for the impurity, we can use MCA to decompose the peak profile into its two components as shown in Figure 19. Using the T4 spectrum as approximation for the unknown impurity, the concentration profiles shown further support our hypothesis and confirm the results of the spectral stripping procedure.

9 USE OF DERIVATIVE SPECTRA AND SPECTRAL SMOOTHING

9.1 Use of Derivative Spectra

Traditionally, UV-spectroscopists have relied on the use of derivative spectra for a number of reasons. They allow for better visual determination of absorbance bands, wavelength maxima, inflection points, and shoulders. Depending on the derivative order applied, derivative spectra can also exhibit reduced background interference and scatter. The inevitable increase in noise that is associated with numeric derivatization could partially be overcome by sampling over longer time intervals.

For HPLC, the situation is quite different. Sampling times for spectra are limited due to the dynamic flow environment. Background interference is largely reduced by virtue of separation in the time domain. Digital data processing eliminates the need for visual interpretation of spectral data. In principle, the spectral matching process could benefit from the use of derivatives, since a derivative spectrum not only is related in a unique way to the zero-order spectrum, but also describes the relationship between neighboring data points. Since for most matching procedures the order in which the absorbance values are presented is irrelevant, this property of derivative spectra could be useful.

9.2 Smoothing of Spectra

As indicated above, digital derivatization increases the noise in the spectral data. For spectral matching, where we are interested in the S/N ratio of a spectrum, the signal size is reduced more than the noise and S/N is therefore degraded. This can be seen in Figure 20 where spectral noise and peak size are shown for zero-, first-, and second-order spectra as exemplified by the T17-T18-T19 peak from r-hGH. Noise for the derivative spectra is somewhat reduced, as seen in panel a, but overall S/N decreases by a factor of four for each derivative. This also manifests itself in an overall reduction of the ability to establish the presence of the known impurity with higher-order derivatives (similarity curves, dashed line, panels b to d).

Since digital data processing is not limited to derivative formation, but also lets us apply various types of different smoothing procedures to our data, the question arises whether digital smoothing could improve the usefulness of zero-order spectra and overcome the increased noise in derivative spectra for an overall enhancement in our ability to differentiate very similar spectra. As we will see in the next section, smoothing can have a beneficial effect on absorbance spectra, but does not suffice to overcome the noise increase associated with derivative spectra.

9.3 Examples

The first example expands on the data presented in Figure 20. For three different peaks from the tryptic map of r-hGH that are all known to be impure, the purity numbers are plotted for each spectral order as function of the width of the smoothing interval. The polynomial smoothing procedure was based on the algorithms first described by Savitzki and Golay.[7]

As can be seen in panels a through c of Figure 21, the purity number for the zero-order spectrum increases in all cases to a maximum for a smooth of

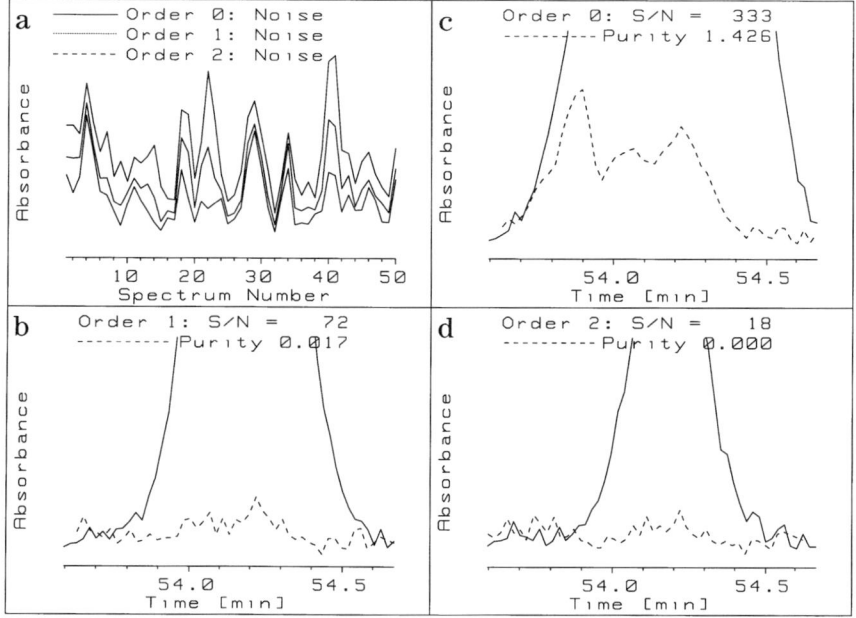

FIGURE 20
Effect of spectral derivative on S/N ratio and purity assessment of the T17-T18-T19 peak from r-hGH.

about 15 points. While a similar effect is noted for some of the derivative data, most noticeable in panel b, the discriminating power of first- and second-order derivatives is greatly reduced as evidenced by the reduction in the purity number. The effect is most pronounced for the T7 peak. Since all three peaks have very similar spectra — T7 and T8 are aliphatic peptides, T17-T18-T19 contains one phenylalanine — this finding must be related to the nature of the impurity.

The second example tries to account for the fact that the performance of the matching process depends on the nature of the two spectra compared. We present dissimilarity scores for the pairwise comparison of different fragments derived from the T14 peptide of r-hGH. This peptide was our model system for the discussion of matching procedures.

In Figure 22 dissimilarity numbers for the three most similar pairs are plotted for the zeroth- through second-order spectra as a function of the width of the smoothing interval. To make the data more comparable, the dissimilarity curves were divided by the numbers obtained for the self match for each order.

Again, there is some improvement in the zero-order data. In all cases the dissimilarity numbers, and thereby the perceived spectral differences, increase

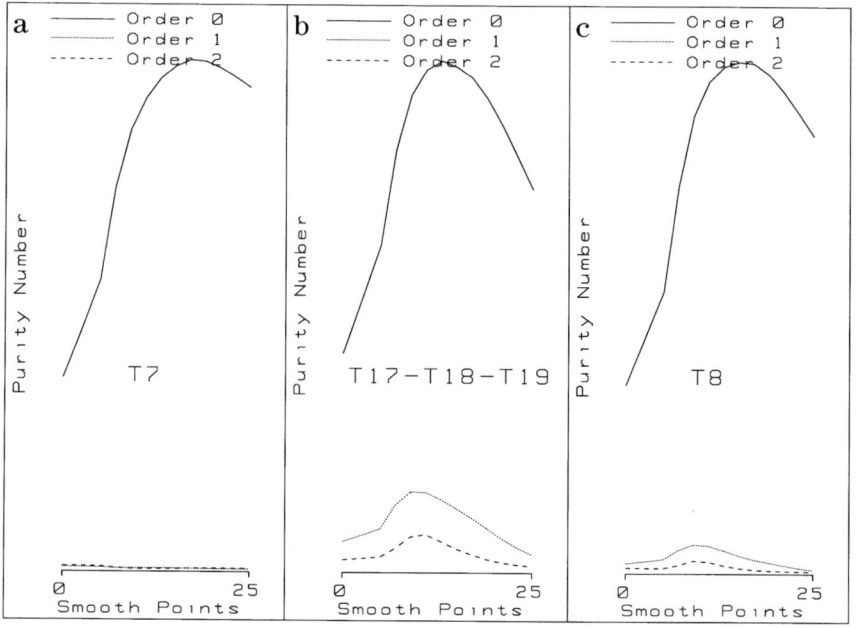

FIGURE 21
Effect of derivative formation and smoothing on purity analysis of several peaks.

as a function of the smoothing interval. The derivative curves all show reduced dissimilarity numbers, indicating that the matching process has not been helped by the spectral processing. Smoothing does definitely improve the dissimilarity numbers, but they do not approach those obtained for the zero-order spectra.

To conclude this section, we feel that the use of derivative spectra does not provide any benefits for spectral matching. This is supported by numerous additional examples not presented here. Smoothing of zero-order spectral data, on the other hand, seems to improve spectral recognition. We plan to pursue this matter further to determine whether the observed behavior can be generalized and whether smoothing is always indicated.

10 CONCLUSION

In this chapter we have tried to present not just examples for the use of UV-spectral data for the analysis of tryptic maps, but to provide a more general theoretical foundation for the applicability of the various procedures described. With the inherent skepticism of the traditional protein chemist

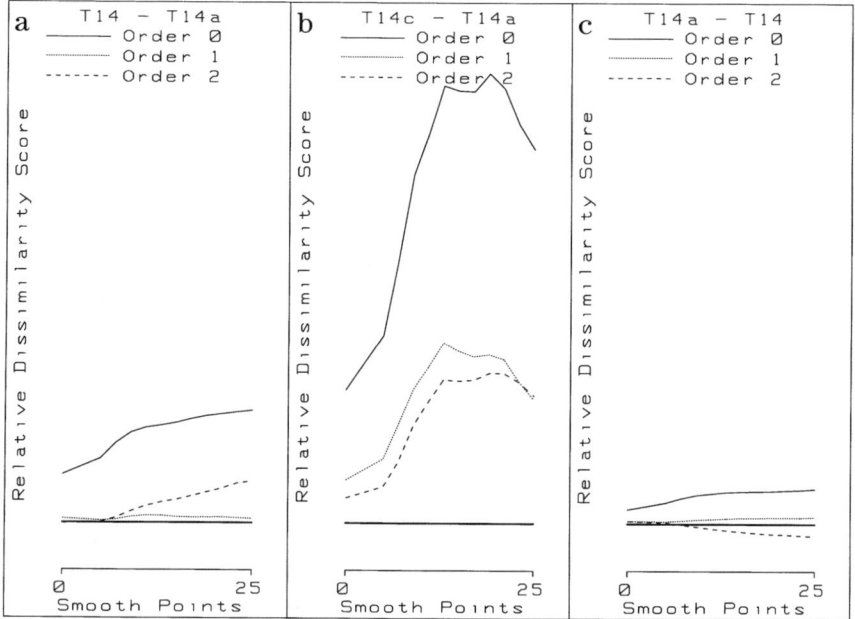

FIGURE 22
Effect of derivative formation and smoothing on the recognition of T14-derived peptides.

towards the usefulness of peptide spectra for purposes of identification, it is important to demonstrate in a defensible way the advantages and limitations associated with their use.

10.1 UV vs. Mass Spectra

As indicated in the introduction, UV spectra are no substitute for the information that can be obtained from mass spectral data. With the rapid growth of electrospray MS and related techniques, we will see increased use of mass spectroscopy in the area of peptide analysis. However, for the foreseeable future, utilization of on-line UV spectra is going to play a prominent role in bioanalytical applications. Ideally, the information provided by both techniques could be combined to give an even better understanding of the analytical problems that need to be solved.

10.2 Need for a Spectral Data Base of Peptide Fragments

There have been several attempts to use the spectra of individual amino acids to derive the composition or identity of a peptide using MCA. Due to the great similarity of the aliphatic amino acid spectra, this approach is limited to the determination of the three aromatic amino acid residues.

We propose here that a better approach might be to start a collection of spectra for peptides of known composition. A spectral data base containing a sufficient number of peptides could be used to establish a number of subgroups using chemometric techniques such as cluster analysis or principal component analysis. This would enable us not so much to identify an unknown peptide, but to assign it to a subgroup with known properties. Since the composition of expected tryptic or other digestion products is known *a priori*, this group assignment could be sufficient in many cases to determine the most likely identity.

10.3 NEED FOR DEDICATED DATA ANALYSIS SOFTWARE

While most diode array manufacturers provide software to manipulate spectra, analyze peak purity, and do library searches, there are currently no software utilities geared specifically to the protein chemist. One of the primary challenges in the automated processing of tryptic maps and other chromatographic data of similar complexity is the assignment of suitable baseline points where reference spectra are taken for background correction.

What is needed is a set of utilities that allow for interactive exploration of peptide spectral data as well as for the unattended batch processing of routine chromatograms. Only with ready access to the vast amount of information available from a diode array detector will the spectral analysis of tryptic maps become a routine task.

10.4 THE MULTIDISCIPLINARY APPROACH

Beyond data analysis software for spectral data, there is a great need for data reduction tools that encompass a number of different disciplines. UV-spectral data could be combined with information from retention prediction, as outlined by the group of Hodges,[8] and then coupled with mass spectral data either from the same analysis or a separate run. Other possible sources of information include data from fluorescence, radioactivity, or on-line low-angle laser light-scattering detectors. The correlation of data of this nature will provide more information than the sum of all the individual bits and parts.

We hope that we have presented enough material to motivate other researchers to apply similar principles to the analysis of spectral data and, at some time in the future, embrace a multidisciplinary approach to the solution of bioanalytical problems.

REFERENCES

1. Sievert, H.-J.P., Wu, S.-L., Chloupek, R., and Hancock, W.S., *J. Chromatogr.*, Automated evaluation of tryptic digest from recombinant human growth hormone using ultraviolet spectra and numeric peak information, 499, 221, 1990.
2. Sievert, H.-J.P. and Drouen, A.C.J.H., Spectral matching and peak purity, in *Diode Array Detection in HPLC*, Huber, L., George, S.A., Eds., Marcel Dekker, New York, 1993, 51.
3. Sievert, H.-J.P. and Drouen, A.C.J.H., New approaches to automated peak purity analysis and peak identification in HPLC utilizing diode array detection, manuscript in preparation.
4. Turcant, A. and Premel-Cabic, A., Toxicological screening of drugs by microbore high-performance liquid chromatography with photodiode-array detection and ultraviolet spectral library searches, *Clin. Chem.*, 37, 1210, 1992.
5. Bogusz, M. and Wu, M., Standardized HPLC/DAD system, based on retention indices and spectral library, applicable for systematic toxicological screening, *Anal. Toxicol.*, 15, 188, 1991.
6. Lacey, R.F., Elimination of interferences in spectral data, *Appl. Spectrosc.*, 43(7), 1135, 1989.
7. Savitzki, A. and Golay, M.J.E., Smoothing and differentiation of data by simplified least squares procedures, *Anal. Chem.*, 36(8), 1627, 1964.
8. Mant, C.T., Burke, T.W.L., Zhou, N.E., Parker, J.M.R., and Hodges, R.S., Reversed-phase chromatographic method development for peptide separations using the computer simulation program ProDigest-LC, *J. Chromatogr.*, 485, 365, 1990.

CHAPTER 4

THE USE OF CAPILLARY ELECTROPHORESIS FOR PEPTIDE MAPPING OF PROTEINS

Eugene C. Rickard and John K. Towns

CONTENTS

1 Introduction 97
2 Use of Peptide Mapping by CE for Quality Control 99
3 Applications 105
4 Conclusion 114
References 116

1 INTRODUCTION

The quality control system for proteins that are used as pharmaceutical products is defined by a combination of the physicochemical characterization of the product itself, the specific production process, and the biological potency. The physicochemical characterization must incorporate analytical tests that establish the identity, purity, and potency of the material to the satisfaction of the regulatory agencies.[1-4] It also is crucial to demonstrate that the material has a consistent composition from lot to lot and that the material remains essentially unchanged during storage. These requirements necessitate development of characterization tests that possess high specificity and accuracy while being robust and reliable. Structural characterization to confirm identity is a formidable challenge in the development of quality control strategies for proteins.

The structure of small molecules usually can be verified with a few spectroscopic techniques such as NMR, IR, mass spectrometry, and/or UV in conjunction with elemental analysis. However, substantiating the proper structure for a protein is much more complex because (1) available spectroscopic techniques do not provide definitive structural data, and (2) protein structure includes not only molecular composition (primary structure) but other highly complex features — secondary, tertiary, and in some cases quaternary structural elements — that are crucial to the protein's activity. In addition, some classes of proteins — such as glycoproteins and phosphorylated proteins — contain microheterogeneity introduced by post-translational modifications. Finally, it is important to unambiguously discriminate the desired drug substance from structural variants that are very similar chemically to the main component. Clearly, no single analytical test will address all of these identification issues for proteins. Therefore, a wide variety of techniques that examine different molecular attributes such as size, charge, and hydrophobicity will be required to verify identity. Peptide mapping is perhaps the most powerful and universally used technique to corroborate the presence of the correct primary sequence and, when nonreducing conditions are used, it also can be used to confirm the correct disulfide bond formation.

Peptide mapping (or peptide fingerprinting) is a relatively direct technique. Proteins are subjected to enzymatic digestion or cleavage by chemical reagents to break them down into discrete fragments. These peptide fragments are then differentiated via a suitable separation technique to give a distinctive peak pattern. Matching of the peak profile pattern customarily includes comparison of the relative retention times, but may also include a comparison of peak areas or other attributes. If the pattern of the fragments from the protein being tested matches the pattern of fragments prepared by the digestion of a reference protein, then the two proteins are most likely the same. Historically, RP-HPLC (reversed-phase HPLC) has been used for separation of the peptide fragments. However, capillary electrophoresis (CE) is uniquely suited to provide complementary information based on an orthogonal separation principle. That is, RP-HPLC separates peptide fragments due to differences in hydrophobicity whereas CE separates fragments based on their charge and size. It has been convincingly shown that there is no correlation between the elution order for peaks separated by RP-HPLC and CE.[5-8] Thus, CE is equally valuable for peptide mapping and, when used in conjunction with another technique such as RP-HPLC or mass spectrometry, provides essentially unambiguous verification of structure.

2 USE OF PEPTIDE MAPPING BY CE FOR QUALITY CONTROL

Specificity (sometimes referred to as selectivity), efficiency, reproducibility, robustness (sometimes referred to as ruggedness), speed, and minimal operator attention are desirable characteristics for quality control methods used to confirm identification. Thus, techniques such as RP-HPLC that have high separative power, give reproducible chromatograms, and are easily automated are preferred for peptide mapping. On the other hand, chromatographic methods have limitations. These limitations include long analysis times and the possibility that not all the peptide fragments will be separated by any one technique no matter how useful it seems.[9] In addition, a chromatographic separation may be "wedded" to the specific chromatographic column used in the method development phase since columns vary due to uncontrolled (or deliberate) changes in the coating chemistry or stationary support. Thus, long-term irreparable drifts in chromatographic performance may be observed. These drifts can drastically alter the separation achieved for the peptide map and may result in the need to rework the analytical method.

In contrast, capillary electrophoresis:

1. Gives a unique specificity that complements RP-HPLC methods,
2. Provides fast, efficient separations of the peptide fragments,
3. Uncouples the separation column from the resulting peptide map so that the separation is unaffected by a change in column, and
4. Can be easily automated.

However, CE also suffers from some limitations. The initial difficulty in rapidly developing optimized separations has been largely overcome through explorations of the factors that affect separation.[10] Others have used a variety of separation conditions based upon empirical choices designed to balance factors such as conductivity, analyte-wall interactions, pH control, separation time, and resolution.[11-14] Knowledge gained from these experiences makes it relatively easy and rapid to develop a separation procedure. Second, CE is more sensitive to changes in the sample matrix and separation conditions than RP-HPLC.[11,14,15] Third, although the column itself is stable, sample components can interact with the capillary wall to produce changes in peak migration times, peak shapes, and peak areas. Although this problem is reduced for peptide mapping in comparison to that observed for separation of large, hydrophobic proteins, buffer additives or coating of the capillary may be necessary.[11,13,14] Finally, relatively little experience has been obtained for peptide maps by CE in a quality control environment.

The separative power or resolution of CE in free solution is a function of the efficiency and selectivity of the system. CE is well known for its highly efficient separations even though the peak capacity is less than that of RP-HPLC.[14] Specificity in CE derives from differences in electrophoretic mobilities, specifically upon the ratio of charge to size. Although the sizes of the fragments in a peptide map are fixed (for a specific protein and a specific digestion procedure), their charges can be manipulated through control of the pH. Most fragments contain terminal carboxyl and amino residues and some contain an ionizable side chain group. Thus, the net charge on a fragment can vary from at least +1 to −1 depending upon the pH and the pK_a values. Since the pK_a values vary between different amino acid residues, and even vary for a specific residue depending upon its distinctive chemical environment, the degree of ionization will vary from residue to residue at any specific pH.[14,16] Thus, the pH can be adjusted to maximize the differences in charges to achieve the desired or optimum specificity.[10,14] The choice of the buffer system depends upon the desired pH, current levels that can be tolerated, effect on analyte-wall interaction, etc. General guidelines for the selection of buffers are available;[7,11,14] some specific considerations are discussed below.

Robustness refers to how dependent a particular method is on changes in operating conditions that may be encountered from laboratory to laboratory, from sample to sample, and with time. Operationally, this refers to how dependent the migration times, peak shapes, and peak areas are on experimental conditions when the separation is performed under the same nominal conditions. The variability contributed by the digestion procedure is one of the most important factors; however, digest variability will be ignored in this discussion because it would affect any technique used for the separation. The remaining variability is contributed by the separation process. Table 1 lists some of the common factors that affect robustness, including within-laboratory repeatability and between-laboratory reproducibility.

A change in electroosmotic flow velocity is the predominant contribution to the variability of migration time.[12] The electroosmotic flow velocity is directly dependent upon the zeta potential at the capillary surface-buffer interface. It is affected by the ionic strength of the buffer, the components of the buffer, and the composition of the capillary surface. Higher ionic strength buffers and buffer components or capillary coatings that reduce the surface charge will reduce the zeta potential and the electroosmotic flow velocity. Changes in the capillary surface produced by adsorption of sample constituents usually can be reversed by an appropriate washing procedure. Temperature is another factor that influences migration time. Increases in temperature reduce the viscosity of the solution and that leads to higher electroosmotic and electrophoretic flow velocities.

The major contributor to peak area variation is poorly controlled injection volume. The use of one or two internal standards allows the analyst to minimize the effects due to electroosmotic flow variation and normalizes for

TABLE 1
Factors that Affect Robustness, Reproducibility, and Repeatability of Peptide Maps

Factor	Variable(s)	Consideration(s)	Potential Effect(s)
Completeness of digest	Digest conditions	Tight control of digest times, temperaures, and concentrations	Variable peak pattern in appearance, and size
Zeta potential	Ionic strength of buffer Buffer composition Capillary surface	Electroosmotic flow velocity directly related to zeta potential	Variable electroosmotic flow and velocity
Temperature	Current Cooling efficiency	Increased temperature reduces viscosity and increases velocities	Changes in buffer viscosity and migration times
Injection volume	How performed Use of internal standards	Control of volume of sample injected	Variable peak areas
Sample matrix	Ionic strength Composition	Higher conductivity than separation buffer Different composition than separation buffer	Decreased resolution and possible distortion of peak shapes
Amount and volume of sample injected	Sample concentration Injection volume	Large sample loads Large injection volumes	Distorted peak shapes
Buffer capacity	Buffer strength Buffer composition Buffer pH	Separation buffer strength not sufficiently higher than analyte concentration	Lack of pH control leading to distorted peaks
Analyte-capillary wall interactions	Buffer ionic strength Buffer additives Capillary coatings	Minimize interaction of sample components with capillary wall	Distortion of peak shapes Loss of material

changes in injection volume. This procedure can improve the precision of the migration time and peak area measurements by a factor of ten or more.[12,17,18] For quantitative measurement of peak areas, the areas have to be corrected for their migration times to eliminate the bias introduced by differential migration velocities.[12] Thus, control of buffer composition, capillary surface composition, adsorption of sample components, column temperature, and injection volume as well as the use of internal standards will give increased reproducibility for migration times and peak areas. Additional factors that influence quantitation are discussed in other references.[12,13,19]

The peak shape can be influenced by several factors. Detrimental effects on the peak shape can result from the sample matrix, the amount of volume of the injected sample, the buffer capacity and ionic strength of the separation buffer, and interaction of the analyte with the capillary wall.

CE is sensitive to the ionic strength of the sample matrix in a manner that is analogous to the sensitivity of RP-HPLC on the sample solvent strength. When the conductivity of the sample matrix is higher than that of the separation buffer, then the electric field strength within the sample zone is lower than it is in the remainder of the capillary. This produces a decrease in efficiency and resolution, both of which are proportional to the field strength. Besides poorer separations, peak shapes are usually distorted. Both phenomena are seen in Figure 1.[20] A similar effect may be observed when the sample is prepared in a buffer that has a different composition than the separation buffer. In contrast, when the conductivity of the sample is lower than that of the separation buffer, then the increased electric field strength in the sample zone produces a focusing effect that concentrates the sample. This focusing effect, sometimes referred to as "sample stacking", allows higher injection volumes to be used when samples are prepared in water or in a dilute buffer than would be possible with samples prepared in the separation buffer. The net results are that the concentration sensitivity is increased with very little sacrifice in peak efficiency. Similar increases in concentration sensitivity can be achieved with pH-mediated stacking[11,14] or with whole capillary injection.[14]

In addition to effects from the sample matrix, CE peak shape is dependent upon the volume and mass of the sample loaded onto the column. Significant band broadening begins to occur when the sample volume is greater than about 1% of the capillary volume.[14] This corresponds to 2 to 3 nL for a 100 cm × 50 μm i.d. capillary. The amount of sample applied to the column can produce an effect in two different ways. First, high sample loads distort the electric field within the sample zone that results in a broadened peak and a loss of resolution.[14] Second, it is possible to produce such high concentrations of analyte within the peak that the sample precipitates. Finally, the way that the sample is loaded — hydrodynamic or electrokinetic — can influence the electropherogram obtained.[12,14] For example, electrokinetic sample loading discriminates between charged species dependent upon their charge and the composition and conductivity of the sample matrix. Although there is no discrimination in components, the total quantity of sample loaded in a hydrodynamic mode is influenced by the viscosity of the sample.

The buffer strength and ionic strength of the buffer must be much higher than that of the sample (at least 200- to 300-fold higher) to maintain pH control and constant conductivity throughout the sample zone.[10,14] Otherwise, peak broadening and/or distortion will occur. Figure 2 shows the effect of increasing buffer concentration (sodium phosphate, pH 2.44) for a sample dissolved in 0.03% trifluoroacetic acid.[21] Several effects are apparent as the buffer strength, and therefore the ionic strength, increases. First, at the lowest buffer concentration (0.025 M), the ionic strength of the sample is approximately the same as the buffer. This gives peak distortions due to a lack of

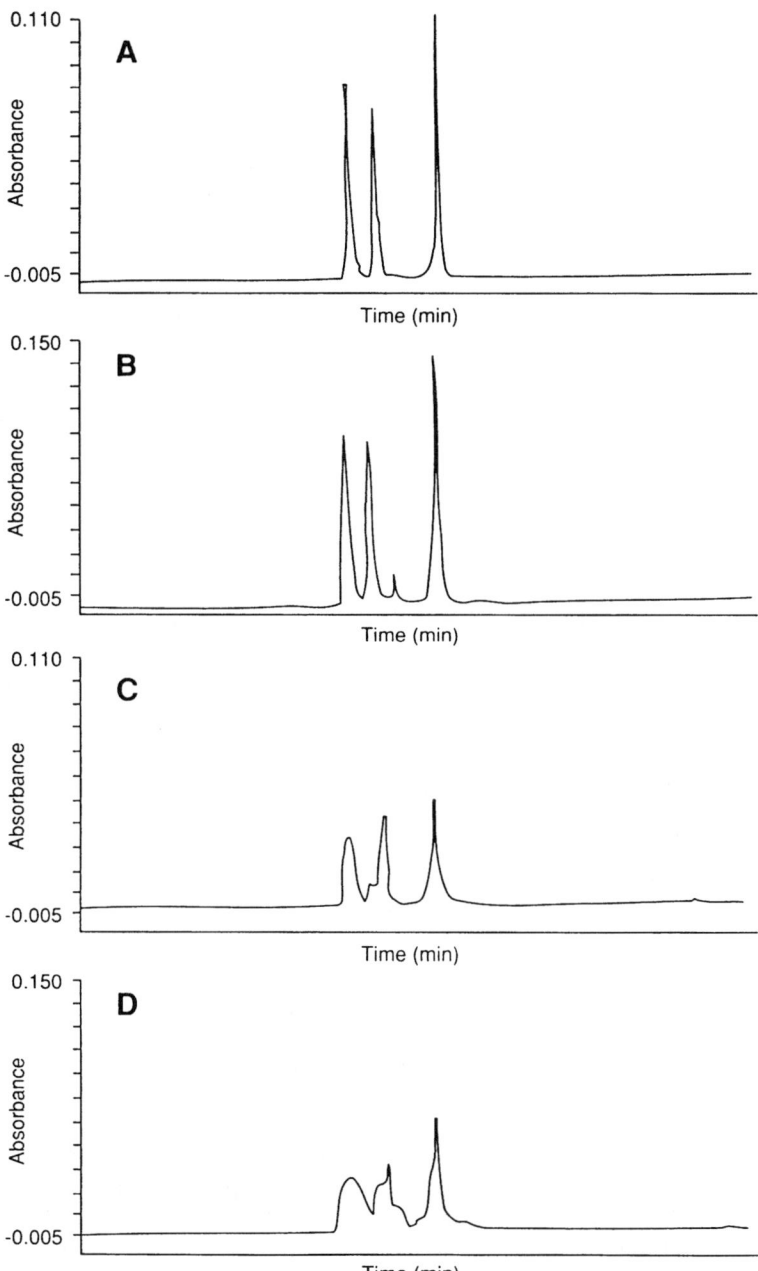

FIGURE 1
Effect of salts in the sample matrix. (A) 30 mM sodium chloride, 10 s injection; (B) 30 mM sodium chloride, 20 s injection; (C) 100 mM sodium chloride, 10 s injection; and (D) 100 mM sodium chloride, 20 s injection. Data from Reference 20.

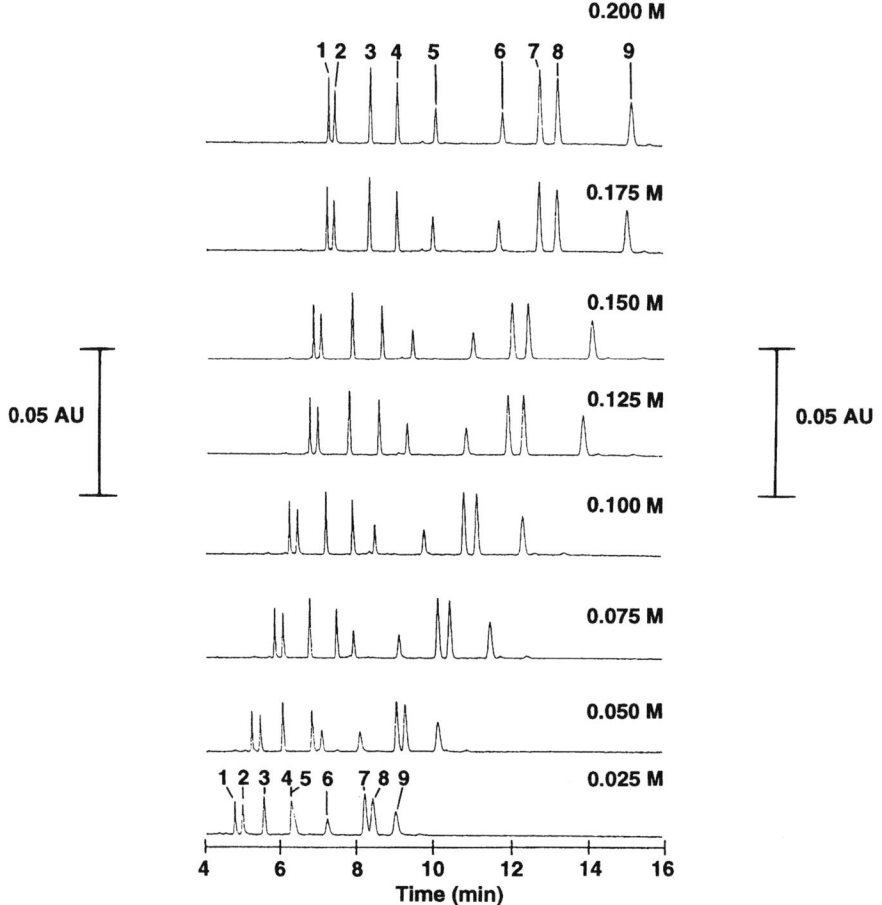

FIGURE 2
Effect of indicated buffer ionic strength on peak shape and migration time of peptides. Buffer: pH 2.44 sodium phosphate; 30 kV; 20°C; 200 nm detection; 57 cm total (50 cm window) × 50 μm i.d. capillary. Peak identification: (1) dynorphin; (2) bradykinin; (3) angiotensin II; (4) TRH; (5) LHRH; (6) bombesin; (7) leu-enkephalin; (8) met-enkephalin; and (9) oxytocin. Data from Reference 21.

adequate buffering and possibly nonuniform electric fields within the sample zone, as described above. The peaks become sharper and move to longer migration times with increasing buffer concentration. This is most likely due to better buffering, focusing of the initial injection and to a reduction of the electroosmotic flow.

The interaction of proteins with bare silica capillaries has been discussed extensively. Some techniques used to minimize this problem can be found in the literature.[11-14] Some specific approaches are described below in Section 3.

Overall, CE has demonstrated the characteristics — specificity, efficiency, speed, and automation — that are necessary for a viable tool within a quality control laboratory. In addition, it gives information that is not available from other techniques and complements that available from conventional mapping techniques such as RP-HPLC. However, robustness and reproducibility have not been addressed extensively for peptide mapping. These issues must be clarified before peptide mapping by CE can move into the quality control environment. Generally, when a direct comparison is made with RP-HPLC, CE gives essentially equivalent power for the confirmation of identity in less time, although its specific strengths are different from those of RP-HPLC.

3 APPLICATIONS

Having described the principles and characteristics involved in peptide mapping by CE, it is helpful to follow up with selected applications. These examples illustrate the power of CE in the separation of digest fragments for peptide mapping utilizing a wide variety of detection modes and auxiliary methodology. These peptide maps demonstrate the unique differences in separation specificity that make CE a viable alternative and/or complementary technique to existing chromatographic methods.

CE has been widely used for the analysis of proteins by peptide mapping. A large number of the studies give a peptide map without attempting to identify the peaks, but others have identified the peaks and compared the maps to those obtained by RP-HPLC. One of the earliest examples of the high resolving power of CE with peak identification was the mapping of recombinant human growth hormone (hGH).[3,5,6,8] The application of CE for peptide mapping is further illustrated in the analysis of the peptide fragments produced by tryptic digestion of recombinant human interleukin-6 (IL-6).[22] Figure 3 shows the amino acid sequence of the protein and the 23 possible cleavage sites using trypsin. Figure 4 shows the separation of these fragments by both RP-HPLC and CE. Almost all the 23 different fragments were resolved by CE in approximately 20 min. All the peaks in the electropherogram were identified by reinjection of peptides that had been purified by RP-HPLC or by ion exchange chromatography and sequenced. The advantages of peptide mapping with CE over RP-HPLC in the analysis of IL-6 include better durability of the capillaries compared to RP-HPLC columns and shorter separation times. The reproducibility of the retention times in CE was found to be comparable to that of RP-HPLC. A big advantage to the CE methodology was that little development effort was expended. In both the hGH and IL-6 cases, the CE and RP-HPLC separations have similar resolutions although some fragments give overlapping peaks by either technique.

Peptide mapping by CE can also be used to detect and investigate degradation products or protein variants. This is illustrated in Figure 5 in the characterization of genetic hemoglobin variants.[23] CE tryptic mapping was performed for each globin, so that complete variant characterization can be

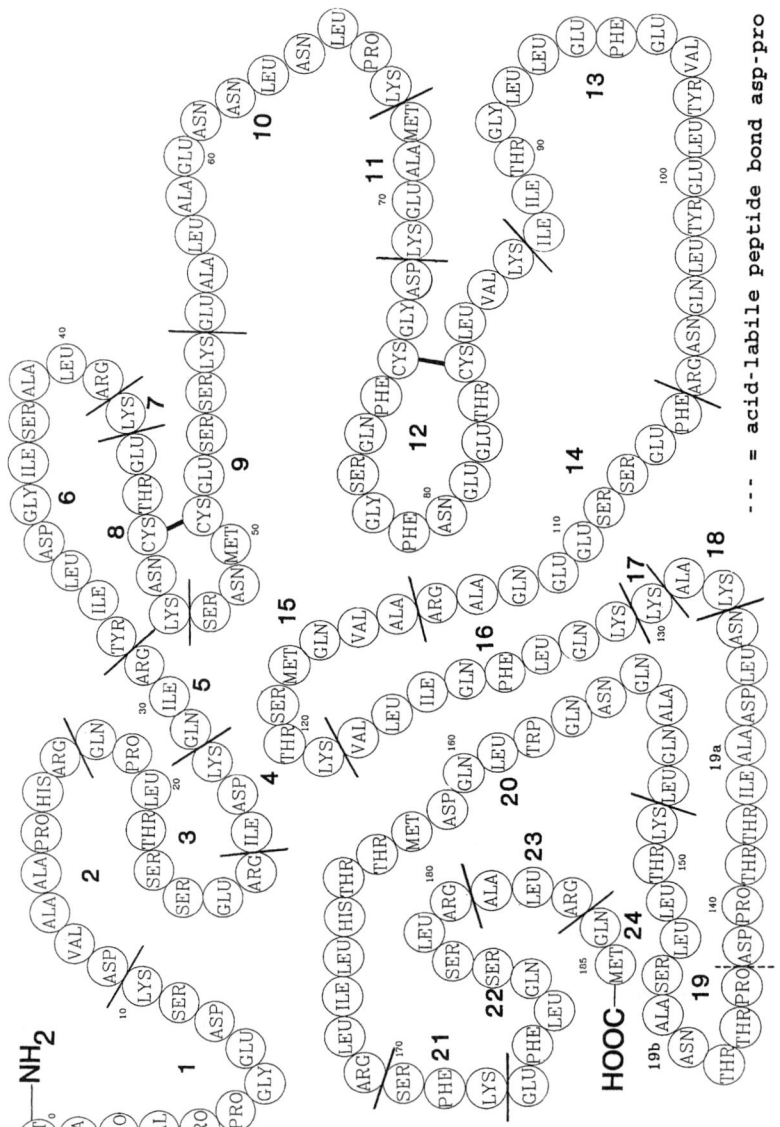

FIGURE 3
Trypsin cleavage sites of rh IL-6. (From Schnyder, D., Bach, D., and Helk, B., Tryptic Mapping of RH IL-6 By CE, paper presented at HPCE Meet., San Diego, CA, Feb. 1991. With permission.)

USE OF CAPILLARY ELECTROPHORESIS

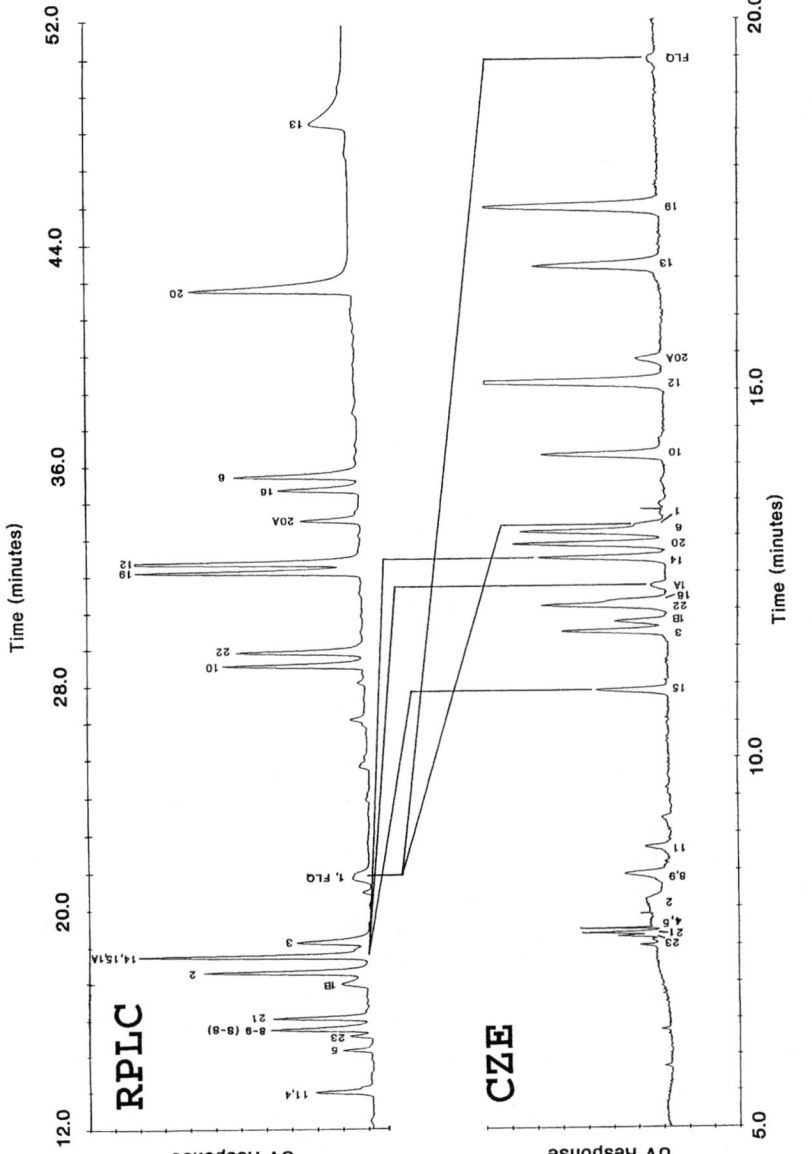

FIGURE 4

Comparison of RP-HPLC and CE tryptic maps of IL-6. Peptide peaks are identified by a number indicating the position in the protein sequence. Data from Reference 22.

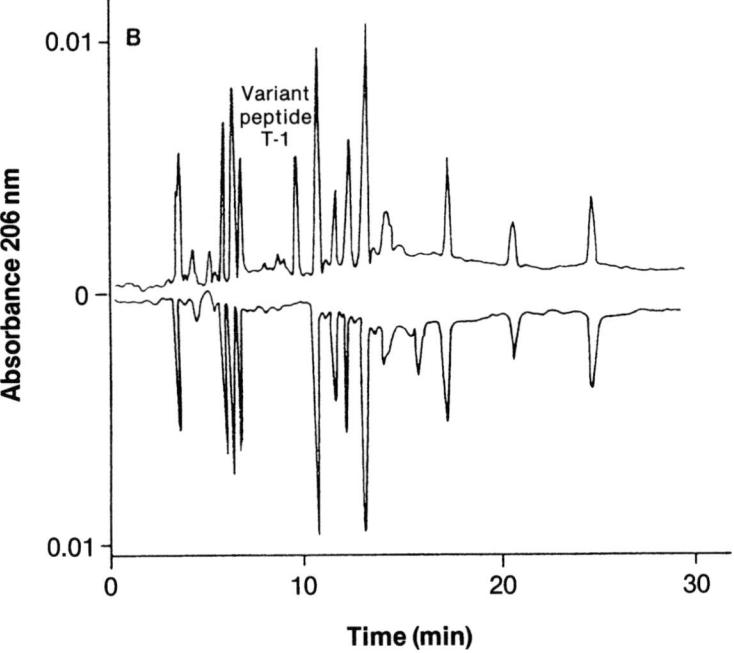

FIGURE 5
Tryptic maps by CE of a variant San Jose β-globin (top) compared with the normal β-globin. Data from Reference 23.

achieved by direct comparison of the variant tryptic map with the corresponding normal one (Figure 5). The CE map shows a similar pattern for the β-variant globin, except that one of the peptides (peptide 121-132) was absent, thus indicating where in the protein sequence the modification took place. FAB/MS (mass spectrometry using fast atom bombardment) analysis confirmed that the glutamic acid at position 121 of the β-chain is replaced by lysine. Coupling electrophoretic data with analysis of enzymatic digests by FAB/MS allowed for a simple and rapid analytical method for the structural identification of abnormal human hemoglobins.

The wide variety of detection modes available for CE is largely attributed to the work regarding detection for column liquid chromatography. However, the detection of narrow zones of low analyte concentrations in peptide mapping within capillaries of 10 to 75 μm internal diameter places a heavy burden upon the sensitivity of the CE detection system used. Of the various detection modes available, the most popular detection method for peptide mapping is UV/Vis, which is available in all the commercial instruments. The additional chemical selectivity offered by multiwavelength UV/Vis absorbance detection also has been exploited. Schlabach and Sence simultaneously monitored molecular absorption at two wavelengths (200 and 280 nm) to analyze a trypsin digest of β-lactoglobulin A digest.[24] The recent emphasis on characterization of trace proteins and peptides demands detection techniques of much higher sensitivity. Tagging peptides with fluorescent moieties is one way to improve detection. The use of fluorescamine in CE separations of tryptic peptides was first reported by Jorgenson and Lukacs.[25] Derivatization with this reagent was found to be both effective and convenient. Fluorescence detection remains one of the most sensitive methods for monitoring molecules separated in a capillary column.

Although most digests are performed using trypsin, Cobb and Novotny[26] used three different protein digestion reagents — trypsin, chymotrypsin, and cyanogen bromide — for the analysis of human serum albumin (HSA). These reagents clip at different amino acid residues, producing a unique mixture of peptides depending on which reagent is used. The total digestion of reduced and alkylated HSA by trypsin should yield 75 fragments. The large fragment mixture results in a very complex peptide map (Figure 6A). The use of cyanogen bromide greatly reduces fragmentation. Cyanogen bromide cleavage at the six methionine residues results in three fragments with intact disulfide bonds (Figure 6B) or seven fragments when the disulfide bonds are reduced and alkylated (Figure 6C).

The use of cyanogen bromide, which cleaves at the carboxyl end of methionine residues, results in larger peptide fragments due to the scarcity of methionine residues in most proteins, but the larger fragments have a greater tendency to adsorb to the inner wall of the capillary. This adsorption hampers both the efficiency of the separation and the reproducibility of the migration times and peak heights from run to run. As also happens in RP-HPLC separations, our experience has been that the larger peptides in a map often exhibit poorer peak shapes and lower recoveries as a result of adsorption to the column surface. The high surface-to-volume ratio of the fused silica capillaries used in CE creates a strong potential for sample-wall interactions and impairs performance in peptide mapping. To combat this limitation, often it is enough to simply add a species such as morpholine[10] or lysine[27] to compete for sites on the capillary wall. Additives, however, may not always do the trick and it may therefore be necessary to reduce wall interaction by deactivating the capillary wall. Coating the interior surface of the capillary with an uncharged hydrophilic material reduces or eliminates electroosmotic flow and minimizes adsorption of larger peptide fragments. The disadvantage is that the capillary coating must be extremely stable under typical separation conditions, as a loss of coating leads to degradation of efficiency and reproducibility.

Castagnola et al. found that initial peptide separations of myoglobin by CE on unmodified capillaries gave nonreproducible results.[28] The measured

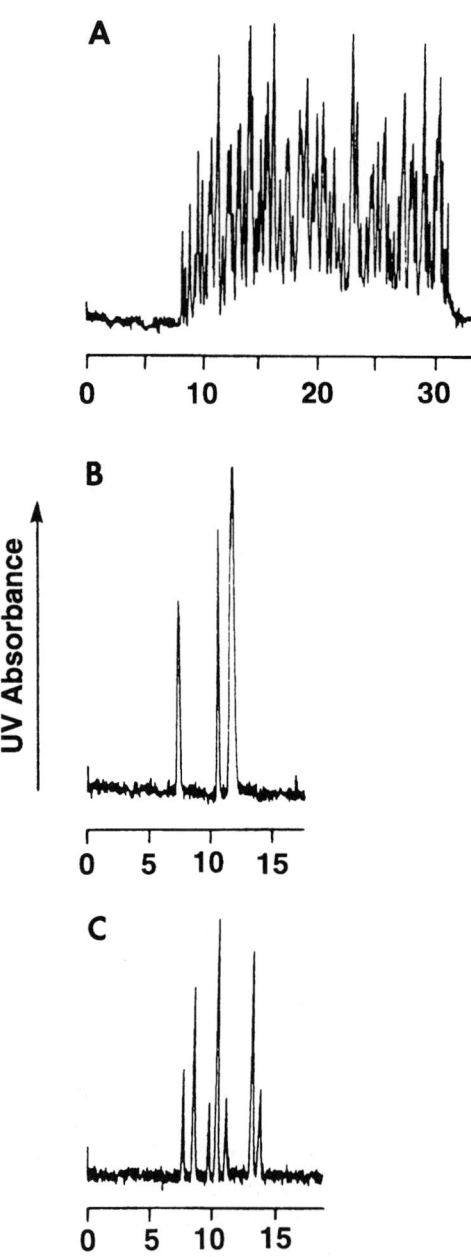

FIGURE 6
Digests of human serum albumin: (A) tryptic digest of reduced and alkylated HSA. Cyanogen bromide digest of HSA with (B) intact disulfide bridges and (C) reduced and alkylated disulfide bridges. Data from Reference 26.

migration time variability was greater than 8%, and resolution parameters (theoretical plate number and resolution) were very poor. Additives to the separation buffer to compete with charged sites on the capillary wall provided better resolution, but the observed reproducibility was still unsatisfactory. Modification of the capillary by a monolayer of acrylamide provided high resolution and reproducible conditions for the separation of approximately 8 pmol of tryptic peptides from horse myoglobin. In the CE tryptic mapping and submapping of human AGP (α_1-acid glycoprotein), all separations were performed on a capillary with a hydrophilic coating on the inner walls. The hydrophilic coating minimized solute-wall interactions and also permitted the electrophoresis of basic proteins at acidic pH with high separation efficiencies. Furthermore, with the coated capillaries, the electroosmotic flow decreased by a factor 3.5, allowing for more resolved peaks and better reproducibility.[29] Submapping of glycosylated and nonglycosylated tryptic fragments of the glycoprotein by CE was facilitated by selective isolation of the glycopeptides through solid-phase extraction. In addition, the electrophoretic map and submaps of the whole tryptic digest and its Con A (concanavalin A) fractions allowed the elucidation of the microheterogeneity of the glycoprotein. Thibault et al. analyzed tryptic peptides of glucagon using noncovalently coated capillaries and found them to be useful for minimizing sample adsorption onto the walls of the capillary.[30]

Cobb and Novotny[31] demonstrated the detection of a single amino acid modification in a protein by analysis of the difference between tryptic maps of phosphorylated and dephosphorylated β-casein. To accomplish this, the system utilized trypsin immobilized on agarose gel and placed in a small reactor column to reproducibly digest as little as 50 ng of protein. Besides the shorter analysis times for CE, the sensitivity improved nearly two orders of magnitude for CE in comparison to microcolumn HPLC. This improved sensitivity was attributed to the absence of gradient elution and organic modifiers in the buffer system. The analysis of phosphorylated and dephosphorylated samples showed that the peaks exhibit shifts in retention time between the two forms. Dephosphorylation effectively reduces the net negative charge of the peptide containing phosphoserine residues, thus reducing the migration times of those peaks in the dephosphorylated peptide map.

The peptide mapping of β-casein also provided an example of indirect fluorescence detection. The analysis of a tryptic digest of β-casein found 13 of the expected 16 peptides.[32] Laser-induced fluorescence detection gives excellent sensitivity, but relatively few molecules fluoresce naturally. Fluorescent derivatization can be used to tag an analyte, but this is not always possible, and the derivatization reaction alters the analyte's structure permanently. Indirect fluorescence detection transfers much of the sensitivity advantage of fluorescence detection to nonfluorescent analytes without the need for sample derivatization. Detection is based upon charge displacement of the charged fluorescent buffer additive and is not based upon any absorption or emission property of the analyte. The separation of the β-casein digest was generally reproducible in terms of retention time and peak height. Subfemtomole quantities of tryptic digest mixtures were separated within 3 min, with mass limits of detection a factor of 180 lower than those of UV absorbance detectors.

A further example of the use of peptide mapping is the analysis of proteins that are post-translationally modified by glycosylations at either/or both the N or O sites. The conventional analytical approach has been to characterize the protein and the carbohydrate portion of the glycoprotein separately and then to combine the results. An improved approach involves the development of analytical methodology that addresses both the translationally expressed protein and the posttranslationally modified protein simultaneously by affinity CE peptide mapping. The model protein system was recombinant human erythropoietin (rHuEPO) as shown in Figure 7.[33] The protein requires little sample handling, other than endoproteinase digestion, and no derivatization reactions are required. The separation employs an ion-pairing agent to increase peptide resolution, to decrease analyte wall interactions, and to evaluate glycopeptide microheterogeneity.

The total tryptic map is segregated into two regions, nonglycosylated and glycosylated peptides. Reproducibility of the peptide map is excellent; the map results in baseline separation of 16 tryptic peptides and one doublet peak — 18 tryptic peptides are identified within the first 30 min from a theoretical possible of 21 peptides, assuming complete digestion. The other three peptides are associated with three N-glycosylation positions: N24 plus N38 are associated with the same tryptic peptide; N83 and the O-linked site (serine 126) reside on separate tryptic fragments. Considerable microheterogeneity is associated with the carbohydrate structure(s), as indicated by the number of peaks observed in the glycosylated region of the electropherogram. As least 12 glycopeptide forms were partially or totally separated. The map furthermore allows for the evaluation of the microheterogeneity associated with the three rHuEPO glycoproteins. CE evaluation of glycopeptide microheterogeneity appears to be simpler, faster, and just as sensitive as other more frequently employed methods for glycopeptide characterizations. The goals are to further develop CE analysis of peptide mapping and to extend this strategy to glycopeptide mapping to definitively identify all carbohydrate structures from the corresponding CE migration time.

Separation of complex peptide mixtures was one of the early demonstrations of the analytical power of capillary zone electrophoresis. However, there are methods that may be employed to further enhance specificity of the tryptic map. While migration rates of various peptides can be optimized

USE OF CAPILLARY ELECTROPHORESIS

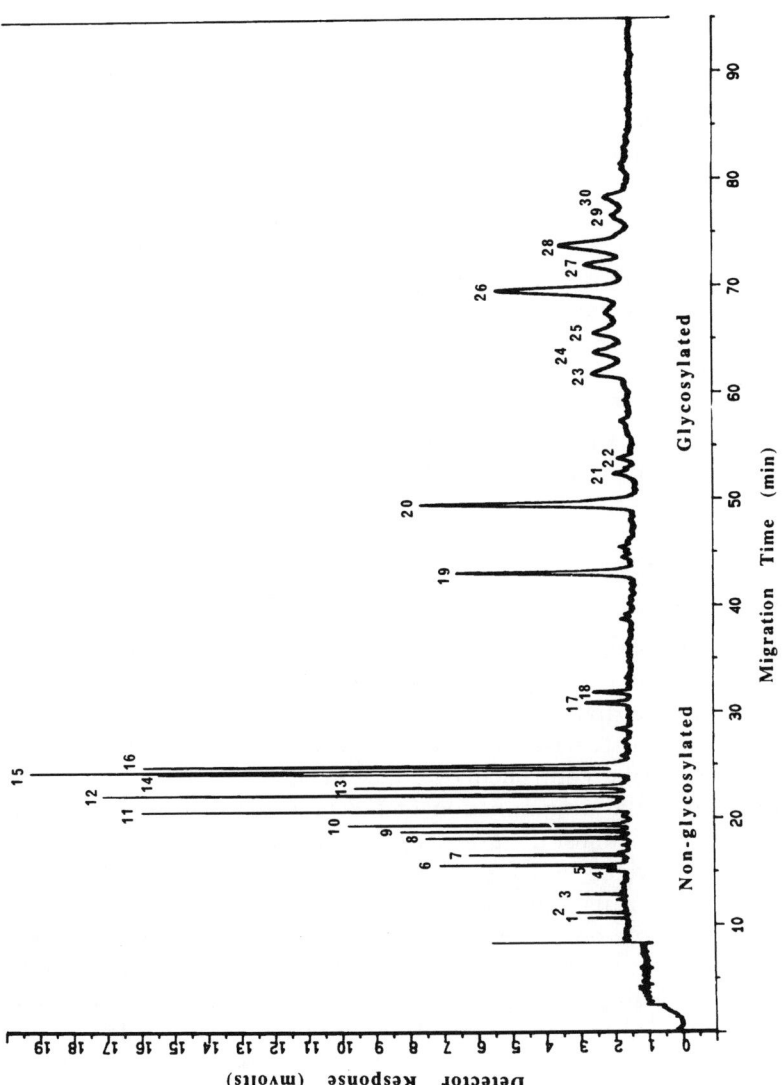

FIGURE 7
CE profile of trypsin-digested rHuEPO. The nonglycosylated peptide and the glycopeptide sections of the map are designated and the corresponding CE peaks are numbered according to migration time. Data from Reference 33.

through an appropriate pH adjustment, the use of micellar electrokinetic capillary chromatography (MECC) can be beneficial in separating substances with similar net charge values. Through the addition of micelle-forming surfactants or inclusion-forming compounds (such as cyclodextrins) to the buffer medium, a dynamic partition mechanism of solute separation is established. In optimizing resolution of fluorescamine-labeled peptides, cyclodextrin additives were found to be beneficial in obtaining highly efficient separations of the peptides produced from the tryptic digestion of cytochrome c; sensitivity was enhanced by fluorescence labeling.[34]

A system that further enhanced the selectivity of the tryptic mapping of proteins is a comprehensive two-dimensional (2-D) separation system.[35,36] This system utilized reversed-phase chromatography as the first-dimension separation and capillary zone electrophoresis as the second-dimension separation. The 2-D system has a much greater resolving power and peak capacity than either of the two systems used independently of each other. Figure 8 shows the three-dimensional data representation of the 2-D tryptic map for the analysis of fluorescent-labeled peptide products from a tryptic digest of cytochrome c.[36] These data provide a means of viewing peak profiles in either separation dimension, and contour mapping of the 3-D data provides a fingerprint of the protein digest.

4 CONCLUSION

Capillary electrophoresis is uniquely suited to yield separations of protein digests. The separation can be readily optimized by variation of the pH and composition of the separation buffer. That is, the strength of CE is based upon its ability to separate based on charge (and size), and the charges of the digest fragments can be controlled by adjusting the pH. That allows the analyst to maximize the discrimination of CE in a way that gives it equal power to the ability to adjust the polarity of the mobile phase in RP-HPLC. The information provided by CE is orthogonal to that given by RP-HPLC and, thus, is complementary in nature.

Capillary electrophoresis provides many of the characteristics desirable for use in a quality control environment, such as good specificity, high efficiency, short analysis times, and the ability to be automated. CE has been applied for analytical characterizations. Additional studies are needed, however, to demonstrate that it possesses sufficient robustness and reproducibility when used for peptide mapping.

USE OF CAPILLARY ELECTROPHORESIS

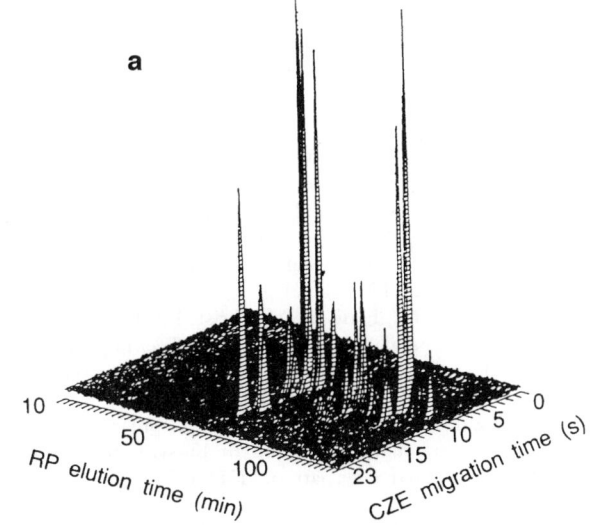

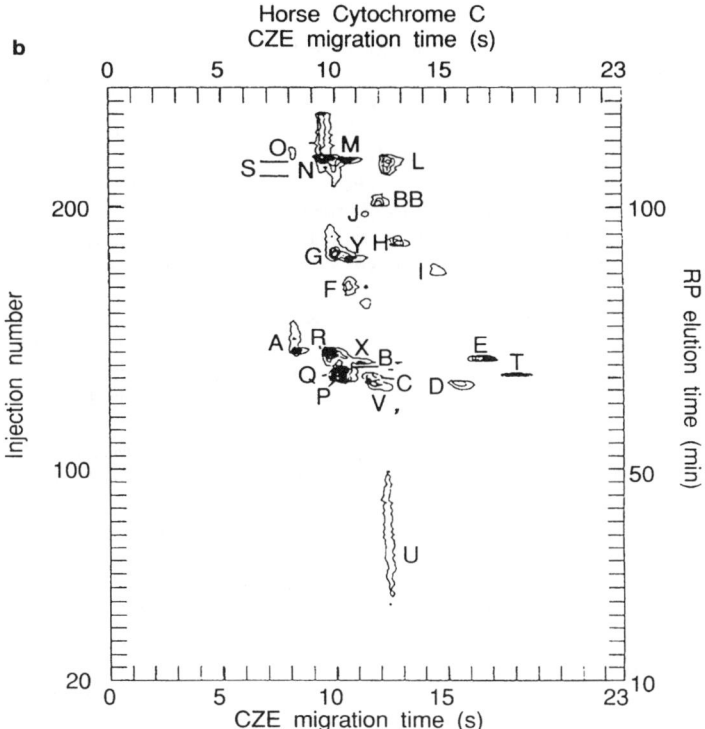

FIGURE 8
Chromatoelectropherogram of fluorescamine-labeled tryptic digest of cytochrome c. (a) 3-D plot and (b) contour plot of same data set. Data from Reference 36.

REFERENCES

1. Garnick, R.L., Solli, N.J., and Papa, P.A., *Anal. Chem.*, 60, 2546, 1988.
2. Farid, N.A., Atkins, L.M., Becker, G.W., Dinner, A., Heiney, R.E., Miner, D.J., and Riggin, R.M., *J. Pharm. Biomed. Anal.*, 7, 185, 1989.
3. Riggin, R.M. and Farid, N.A., Analytical chemistry of therapeutic proteins, in *Analytical Chemistry: Capillary Electrophoresis and Chromatography*, ACS Symp. Ser. No. 434, Horvath, C. and Nikelly, J.C., Eds., American Chemical Society, Washington, D.C., 1990, 113.
4. Kovach, P.M., Riggin, R.M., Farid, N.A., Ghrist, B.F.D., Janis, L.J., Nielsen, R.G., Rickard, E.C., and Sittampalam, G.S., Analytical Techniques for the Quality Control of Biomolecules, presented at the FDA Drug Chemistry, Manufacturing and Controls Seminar, Rockville, MD, Feb. 21, 1991.
5. Nielsen, R.G., Riggin, R.M., and Rickard, E.C., Capillary Electrophoresis of Fragments Produced by Enzymatic Digestion of Biosynthetic Human Growth Hormone, presented at the First International HPCE Meeting, Boston, MA, April 10, 1989.
6. Frenz, J., Wu, S.-L., and Hancock, W.S., *J. Chromatogr.*, 480, 379, 1989.
7. Rickard, E.C. and Towns, J.K., High resolution separation of macromolecules, in *Methods in Enzymology*, Karger, B.L. and Hancock, W.S., Eds., Academic Press, Orlando, in press, chap. 38.
8. Nielsen, R.G., Riggin, R.M., and Rickard, E.C., *J. Chromatogr.*, 480, 393, 1989.
9. Becker, G.W., Tackitt, P.M., Bromer, W.W., LeFeber, D.S., and Riggin, R.M., *Biotech. Appl. Biochem.*, 10, 326, 1988.
10. Nielsen, R.G. and Rickard, E.C., *J. Chromatogr.*, 516, 99, 1990.
11. Schwartz, H.E., Palmieri, R.H., Nolan, J.A., and Brown, R., Separation of peptides, in *Introduction to Capillary Electrophoresis of Proteins and Peptides*, Beckman Instruments, Inc., Fullerton, CA, 1992, chap. 3.
12. Moring, S., Quantitative aspects of capillary electrophoresis analysis, in *Capillary Electrophoresis Theory and Practice*, Grossman, P.D. and Colburn, J.C., Eds., Academic Press, New York, 1992, chap. 3.
13. Colburn, J.C., Capillary electrophoresis separations of peptides: practical aspects and applications, in *Capillary Electrophoresis Theory and Practice*, Grossman, P.D. and Colburn, J.C., Eds., Academic Press, New York, 1992, chap. 9.
14. Weinberger, R., *Practical Capillary Electrophoresis*, Academic Press, New York, 1993.
15. Grossman, P.D., Background concepts, in *Capillary Electrophoresis Theory and Practice*, Grossman, P.D. and Colburn, J.C., Eds., Academic Press, New York, 1992, chap. 1.
16. Rickard, E.C., Strohl, M.M., and Nielsen, R.G., *Anal. Biochem.*, 197, 197, 1991.
17. Dose, E.V. and Guiochon, G.A., *Anal. Chem.*, 63, 1154, 1991.
18. Lee, T.T. and Yeung, E.S., *Anal. Chem.*, 63, 2842, 1991.
19. Dose, E.V. and Guiochon, G.A., *Anal. Chem.*, 64, 123, 1992.
20. Satow, T., Machida, A., Funakushi, K., and Palmieri, R.L., *J. High Resolut. Chromatogr.*, 14, 276, 1991.
21. McLaughlin, G., Biehler, R., Anderson, K, and Schwartz, H.E., Capillary Dimensions with P/ACE™ 2000 Series Instruments: 50-μm Vs. 75-μm-i.d. Capillaries, Beckman Technical Information Bulletin TIBC-106, Beckman Instruments, Inc., Palo Alto, CA, 1991.
22. Schnyder, D., Bach, G., and Helk, B., Tryptic Mapping of RH IL-6 by CE, presented Feb. 1991 at the HPCE-91 meeting, San Diego, CA, abstr. PT-24, manuscript in preparation.

23. Ferranti, P., Malorni, A., Pucci, P., Fanali, S., Nardi, A., and Ossicini, L., *Anal. Biochem.*, 194, 1, 1991.
24. Schlabach, T. and Sence, R., *Spectra*, 149, 47, 1990.
25. Jorgenson, J.W. and Lukacs, K.D., *Anal. Chem.*, 53, 1298, 1981.
26. Cobb, K.A. and Novotny, M.V., *Anal. Chem.*, 64, 879, 1992.
27. Bushey, M.M. and Jorgenson, J.W., *J. Chromatogr.*, 480, 301, 1989.
28. Castagnola, M., Cassiano, L., Rabino, R., Rossetti, D.V., and Bassi, F.A., *J. Chromatogr.*, 572, 51, 1991.
29. Nashabeh, W. and El-Rassi, Z., *J. Chromatogr.*, 536, 31, 1991.
30. Thibault, P., Pleasance, S., and Laycock, M.V., *J. Chromatogr.*, 542, 483, 1991.
31. Cobb, K.A. and Novotny, M., *Anal. Chem.*, 61, 2226, 1989.
32. Hogan, B.L. and Yeung, E.S., *J. Chromatogr. Sci.*, 28, 15, 1990.
33. Rush, R.S., Derby, P.L., Strickland, T.W., and Rohde, M.F., *Anal. Chem.*, 65, 1834, 1993.
34. Liu, J., Cobb, K.A., and Novotny, M., *J. Chromatogr.*, 519, 189, 1990.
35. Bushey, M.M. and Jorgenson, J.W., *Anal. Chem.*, 62, 978, 1990.
36. Bushey, M.M. and Jorgenson, J.W., *J. Microcolumn Sep.*, 2, 293, 1990.

CHAPTER 5

PACKED CAPILLARY HPLC — ELECTROSPRAY IONIZATION MASS SPECTROMETRY

John T. Stults, Beth L. Gillece-Castro, William J. Henzel, James H. Bourell, Kathy L. O'Connell, and Lydia M. Nuwaysir

CONTENTS

1 Background 120
   1.1 Mass Spectrometry 120
   1.2 Capillary HPLC 122
   1.3 Sample Preparation 123

2 LC-MS Applications 124
   2.1 Peptide Mapping 124
   2.2 Identification of Modified Peptides 126
       2.2.1 LC-MS With Postacquisition Processing 127
       2.2.2 Chromatogram Comparison After Modification Removal 128
       2.2.3 Specific Scan Modes 128
       2.2.4 Selective Isolation of Modified Peptides 129
   2.3 Peptide Sequencing 131
   2.4 Protein Identification by Peptide-Mass Database Searching 134
   2.5 Protein Mixtures 136

3 Summary: Today and Tomorrow 137

References 138

1 BACKGROUND

Success with packed-capillary reversed-phase high performance liquid chromatography (RP-HPLC) has increased dramatically in recent years. The ability to characterize picomole amounts of peptides and proteins by mass spectrometry and Edman microsequencing has been coupled with the ability of chromatography to match these sensitivities. The power of the capillary HPLC for separation of small amounts of material with very high resolution was recognized early,[1] but the technique was rarely used outside of relatively few research laboratories. The need for isolation of low levels was particularly evident in the field of protein analysis, where the detection limits for routine automated Edman degradation dropped from approximately 1 nmol in the early 1980s[2] to nearly 1 pmol a decade later.[3,4] Detection limits for mass spectrometry likewise dropped to the low picomole/high femtomole range in the late 1980s, a result of the developments of new ionization techniques such as continuous-flow fast atom bombardment (CF-FAB),[5] electrospray ionization (ESI),[6] and matrix-assisted laser desorption/ionization (MALDI).[7,8] In addition, the low flow rate provided by capillary HPLC was required initially for efficient on-line coupling of RP-HPLC to a mass spectrometer. At the same time, straightforward methods for preparing packed-capillary columns were developed and efficient columns from commercial sources were introduced. These columns provided access to the separation technique for all laboratories.

1.1 Mass Spectrometry

Direct coupling of RP-HPLC to mass spectrometry was made initially through direct liquid introduction[9] or a moving belt interface.[10] Both techniques utilized electron impact or chemical ionization and were incompatible with peptides. Thermospray bridged this gap and was heavily used during the late 1980s.[11-14] Continuous-flow FAB (CF-FAB) permitted on-line separation of peptides with FAB mass spectrometry detection.[5,15,16] The accessible mass range (<3000 Da) and picomole sensitivity gave this technique great promise. It has been used successfully in a number of laboratories,[5,17-28] but its use became routine in relatively few locations due in part to difficulty in long-term maintenance of a stable signal over multiple chromatograms. Flow rate restrictions (1 to 5 µl/min), imposed by the limitations in the pumping capacity of most mass spectrometer vacuum systems, also limited widespread use of CF-FAB. These flow rate restrictions prompted many laboratories to use capillary RP-HPLC. Although CF-FAB continues to be utilized in many labs, electrospray ionization (ESI)[6,29-32] was rapidly adopted by many mass spectrometrists and protein chemists as the simplest and most versatile technique for interfacing with RP-HPLC. Despite the overwhelming dominance of ESI, comparisons of the two techniques show that the sensitivity of CF-FAB remains equal to or greater than ESI for small peptides.[33-36] The remainder of this chapter will focus exclusively on ESI.

Initially, the flow rate requirement of ESI was restricted to 0.5 to 2 µl/min. The use of a nebulizing gas,[37] heated interface capillary,[38] or ultrasonic nebulizer[39] later permitted flow rates up to 1 ml/min. Nonetheless, its ease of use at low microliter flow rates and the need for the utmost in sensitivity had firmly established the use of capillary columns. The main requirement for ESI and CF-FAB is the use of volatile buffer systems. A water-acetonitrile-trifluoroacetic acid system is most often used.

Ions are formed in ESI for a wide variety of compounds including small organic molecules, carbohydrates, oligonucleotides, peptides, and proteins. The spectra of peptides and other small molecules show attachment of one or two protons $(M+H)^+$ or $(M+2H)^{2+}$, while large molecules display a characteristic series of multiply charged ions. Typically, a proton is attached for approximately each 1000 Da of molecular mass. As a result, the mass-to-charge ratio of each peak (the property actually measured by the mass spectrometer) is usually in the range of 500 to 2000, even for large proteins, so common mass analyzers such as quadrupole mass spectrometers are used for mass analysis. The molecular mass is easily calculated from any two multiply charged peaks. More than two peaks permit an averaged mass calculation with an indication of statistical significance. Detection limits are generally 10^{-6} to 10^{-7} M (0.1 to 1 pmol/µl).

The LC-MS experiments described below with electrospray ionization were performed with quadrupole mass spectrometers, due primarily to their ease of use and widespread availability. Other types of mass spectrometers are used for capillary LC-MS as well. The sector instrument permits high-resolution measurements and a generally higher mass range.[40-44] Its use with electrospray ionization was initially limited by difficulties in interfacing the atmospheric pressure source with the high-voltage optics of the instrument. These technical hurdles are now surmounted and instruments are commercially available. The Fourier transform ion cyclotron resonance (FT-ICR) mass spectrometer has also been interfaced with ESI.[45-47] An advantage of this instrument is its ultrahigh resolution, ion trapping capabilities for very high sensitivity, and analysis by collision-induced dissociation (CID or MS/MS). A third analyzer that has been used for ESI is the quadruple ion trap.[48] This instrument also has capabilities for ultrahigh resolution, high sensitivity, and MS/MS. A fourth analyzer that measures mass by time-of-flight has also been demonstrated.[49] It offers advantages of high sensitivity, unlimited mass range, and high spectrum acquisition rates. The latter three analyzers are only beginning to be used with ESI and their utility is expected to grow significantly in the near future.

TABLE 1

Characteristics of Various Column Diameters

Diameter (mm)	Flow rate (μl/min)	Sample range (pmol)	Rel. peak conc.	Ease of use
4.6	500–1000	10^2–10^5	1	++++
2.1	100–200	50–10^4	5	++++
1.0	40–100	5–10^3	20	+++
0.32	1–10	0.5–500	200	++

1.2 Capillary HPLC

A small-diameter HPLC column offers several advantages. As the diameter decreases, the analyte peak concentration in the column increases in proportion to the inverse square of the diameter.[50,51] The optimum flow rate decreases as does the solvent consumption with decreased diameter. The low solvent consumption produces significantly less environmental waste and makes the use of more expensive solvents feasible, although this latter point has not been widely exploited. At small capillary diameters the resolution also improves.[52] Table 1 shows the characteristics of HPLC columns in normal use.

Capillary HPLC is not without its difficulties. Column-to-column reproducibility for commercially packed columns have not matched those of 2.1 and 4.6 mm i.d. analytical columns. The loading capacity of the column decreases proportionally as the amount of packing material decreases. Small capillaries are more prone to clogging. Extra-column band broadening at unions is particularly pronounced at low flow rates, and leaks are more difficult to observe visually. Despite these drawbacks, capillary columns are routinely used with few problems arising, and their advantages are worth the effort required to become familiar with their idiosyncrasies.

Solvent gradients at low flow rates are difficult to generate due to the requirement for submicroliter per minute flow in each pump (e.g., total flow of 5 μl/min requires 0.05 μl/min for pump B at 1% B). The most straightforward approach for generating low flow rates is to split the flow from a gradient formed at higher flow rates. This approach involves higher solvent consumption but allows use of most HPLC pumping systems. Changes in viscosity as the organic concentration changes will cause the flow rate to change during the gradient, although it should be reproducible from run to run. The change in split ratio can be partially compensated by increasing the flow rate during the gradient. Alternatively, a second column can be incorporated to keep the split ratio constant.[53] Another approach is to pre-form the gradient and store it in a long capillary before use.[54] We normally use dual syringe pumps operated at 50 to 100 μl/min, followed by solvent splitting to generate the gradient at 2 to 5 μl/min. We also find the gradients formed with conventional reciprocating piston pumps operated at these flow rates or higher, followed by splitting, are satisfactory. Successful operation of capillary columns at higher flow rates (45 µl/min) has been demonstrated recently by use of perfusion particles.[55] We have used these columns for rapid desalting of proteins. The higher flow rate and perfusion mechanism allow rapid gradients that yield considerably shortened separations (e.g., 12 min vs. 60 min) without sacrificing chromatographic resolution.

Capillary columns may be obtained commercially from a number of sources (e.g., LC Packings, San Francisco, CA; Keystone Scientific, Bellefonte, PA; Micro-Tech Scientific, Sunnyvale, CA; Michrom BioResources, Auburn, CA), although many laboratories routinely pack their own columns with a great deal of success. The advantages of packing one's own columns include significantly lower cost and greater flexibility in the choice of column dimensions and packing materials. A number of different methods have been devised for column preparation.[51,54,56-59] They differ mainly in the process for forming a frit to hold the packing material in the column.

Sample injection is typically done with a conventional HPLC injector. Injection loop sizes are typically small (<5 µl). Fortunately, the retention mechanism for peptides fits an absorption/desorption model, rather than simple partitioning.[60,61] Thus, most peptides will be retained even when large volumes of sample are injected, although some small peptides may elute. As a result, larger injection volumes can be used with few deleterious effects. The main limitation becomes the time required to load the sample: a 100-µl sample requires 20 to 100 min to load at a 5- to 1-µl/min flow rate, respectively, before the gradient begins. The use of a higher flow rate during loading can reduce this time to the extent that a higher back-pressure can be tolerated. The tubing volume of the system can also yield a significant delay in the gradient. For this reason, the pre-column split is normally located close to the injector and the column is connected directly to the injector. The retention mechanism for peptides also means that column lengths greater than 150 mm have little effect on the resolution,[62] allowing short columns to be used. As a result, higher sample recovery is possible due to the minimized irreversible sample adsorption on the small quantities of packing material that are used in short packed capillary columns.

The column is normally connected to an ultraviolet (UV) absorbance detector that precedes the mass spectrometer. If sufficient material is available, the eluant can be split before the mass spectrometer and fractions collected, as is routinely done with larger columns.

1.3 Sample Preparation

Sample preparation for capillary LC-MS is an area that is just beginning to be explored. Specifically, chemical and enzymatic cleavages of proteins at low (<10 pmol) levels are necessary to pursue peptide mapping at high sensitivity. The most simplistic approach, conventional digestion in dilute solution, does not work well because the reaction rate is diffusion limited.

Furthermore, dilute solutions of peptides and proteins frequently adsorb to walls of containers, pipette tips, etc., adding to losses.

Several approaches have been used to solve this problem. One is to scale down the digestion solution to ≤10 μl and thereby keep surface areas low and protein concentrations high. A further key to low-level digestions is to use higher than normal enzyme-to-substrate ratios (e.g., E:S 1:10).[63] The use of a modified trypsin that is resistant to autolysis is important for higher enzyme concentrations. Enzymes with modified lysines are presently marketed by Promega and Boehringer-Mannheim.

A second approach for successful digestions of low-level samples is the use of immobilized proteases.[13,64] The immobilized protease offers advantages of resistance to autolysis and adaptability to small volumes. Alternatively, Hunt and co-workers use trypsin that is adsorbed to the wall of a small capillary tube to digest 1 to 10 pmol of protein.[56] A concentrated solution of trypsin is drawn into the capillary. The trypsin solution is removed and a solution of the protein to be digested is loaded. The quantity of trypsin that has been adsorptively immobilized in the capillary is sufficient for digestion.

A third approach involves *in situ* digestion of proteins in polyacrylamide gels or on polyvinylidene difluoride (PVDF) membranes following gel electroblotting. The in-gel procedure involves partial dehydration of the gel, followed by rehydration with the protease-containing buffer solution.[65] Peptides are released into the digestion buffer. Alternatively, the *in situ* digestion of the electroblotted protein can be performed on the PVDF membrane. This procedure is performed after Coomassie blue staining to identify the protein location. The membrane spot is excised, and *in situ* reduction/alkylation and digestion is performed.[66-70] The peptides are released into the digestion buffer that contains 10% acetonitrile and analyzed by LC-MS. Peptides not released into the digest buffer may be extracted from the membrane with DMSO for subsequent analysis.

2 LC-MS APPLICATIONS

2.1 Peptide Mapping

Peptide separations that involve LC-MS are most commonly used to confirm the proper biosynthesis of proteins generated by recombinant DNA techniques.[71,72] In addition, peptide maps may also be used to confirm the amino acid sequence of a native protein which has a known cDNA sequence. In either case, a list of expected peptides are generated from the putative protein sequence and matched with the observed peptides. Efficient computer algorithms perform this matching function well.[73,74]

A typical chromatogram is shown in Figure 1. The total ion current (TIC) trace shows the summation of all ion intensity at each scan. It provides

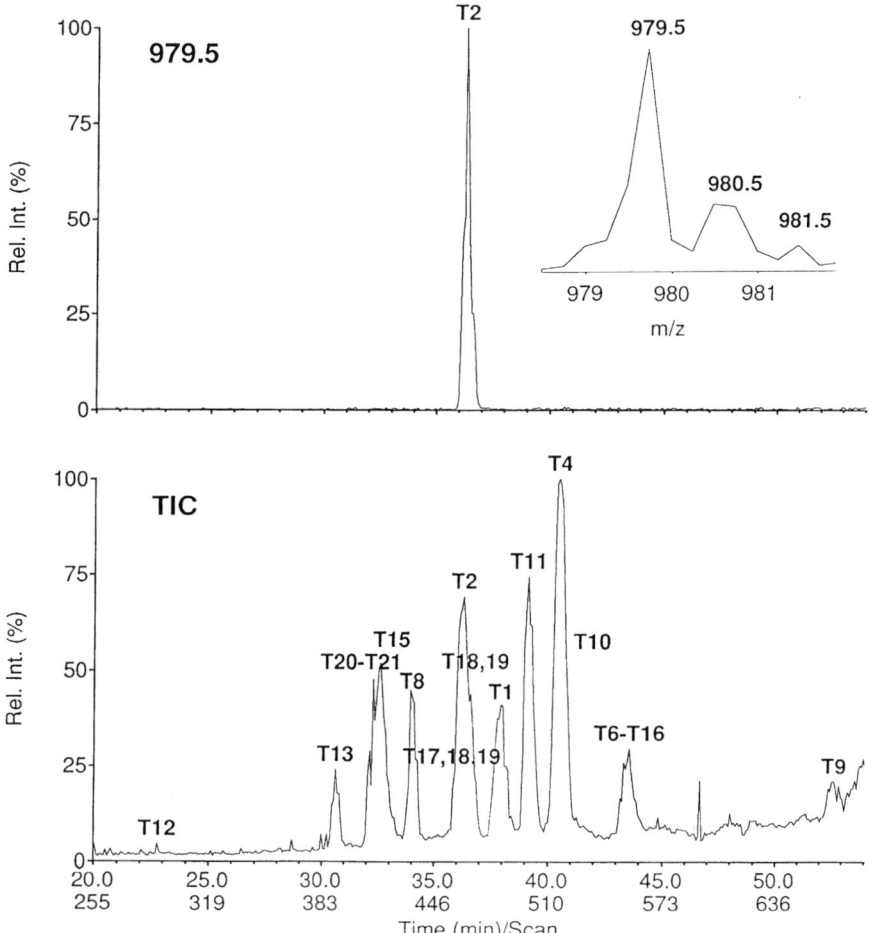

FIGURE 1
Packed capillary LC-MS of the tryptic digest of recombinant human growth hormone (hGH, MW = 22,256). The lower trace is the total ion current (TIC) chromatogram produced by 10 pmol of digested protein. The column was 180 μm × 15 cm, C18 (LC Packings). Solvent A was 2% acetonitrile/0.1% TFA. Solvent B was 90% acetonitrile/0.085% TFA. The gradient was 2 to 60% B in 30 min, with a hold at 60% B for 15 min. Data were processed with the program ENHANCE (Sciex) to improve signal-to-noise and the apparent resolution. The upper trace is the reconstructed ion current trace for peptide T2, $(M+H)^+$ = 979.5. The inset shows the isotope resolution that permits assignment of this peak to a singly charged ion.

an indication of the presence of compounds eluting from the column and it is in good agreement with the UV profile that was obtained with an analytical column (4.6 mm).[75] Relative peak heights vary because the UV absorbance and ionization efficiency are unrelated physical characteristics of the peptides. Differences in ionization efficiency between peptides, plus slight variations in instrument response over time, mandate the use of internal standards when using mass spectrometry for quantification. A mass chromatogram (extracted ion current profile) for the T2 peptide of hGH is also shown in Figure 1, with excellent S/N. At low sample levels the TIC frequently displays a low S/N due to the inclusion of background ions from the solvents and from column bleed. An alternative display, such as a base-peak chromatogram (displays the abundance of the most abundant peak in each spectrum as a function of time),[76] provides improved apparent S/N.

The quadrupole mass spectrometer is normally operated at sufficient mass resolution to separate isotope peaks up to about m/z 1000. The isotope separation is important for determination of the charge state of peptide ions, which typically have one to three protons but which may show only one peak, and thus precludes the use of molecular mass determinations from multiple ions. Where higher sensitivity is required, isotope separation may be sacrificed with the concomitant loss in information.

The resolution used in Figure 1 was sufficient ($m/\Delta m \sim 1000$) to permit the identification of singly charged ions by the isotope separation. The latter point is important because doubly charged ions are frequently observed in typical tryptic digests due to the two basic sites on the peptides (N-terminus plus Lys or Arg at the C-terminus). Small peptides often display just one peak, which may be singly or doubly charged, making determination of both the ion mass *and* charge difficult unless the isotope spacing permits identification of the singly charged species. Larger peptides often produces ions with several different charge states that allow mass and charge determination from the multiple peaks.

Lower S/N becomes apparent with lower sample amounts, as shown in Figure 2. The S/N of the mass spectra can be improved in part by lowering the resolution and step size of the mass spectrometer, albeit with a loss in selectivity and information content. Figure 2 shows the TIC and mass chromatogram for T2 obtained for a 1 pmol of tryptic peptides of hGH at lower resolution. At the resolution used to acquire the data in Figure 1, no peaks were evident in the TIC of the 1 pmol sample (data not shown).

Detection limits of approximately 1 pmol per peptide in a mixture can be routinely obtained. This amount is for an entire tryptic digest that is carried out at a higher level (>100 pmol) from which an aliquot is taken. When less protein is available, inefficient digestion and sample losses become problematic (see above). The highest sensitivity is achieved only by optimization of all aspects of sample preparation, sample introduction, and instrument performance.[56,77]

2.2 IDENTIFICATION OF MODIFIED PEPTIDES

In many instances, one of the goals of an experiment is to determine the presence (or absence), locations, and identities of posttranslational modifications. Four approaches, outlined below, may be utilized to identify modified

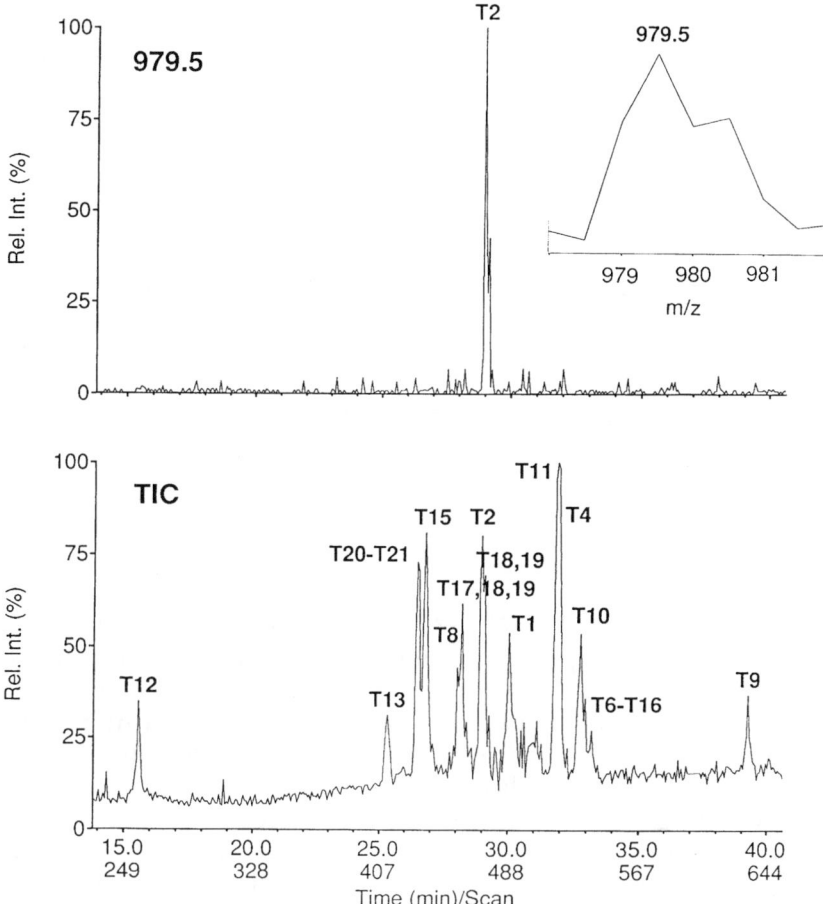

FIGURE 2
Packed capillary LC-MS of the tryptic digest of hGH for 1 pmol of digested protein. Separation conditions were the same as given for Figure 1 except a 10% hold at 2% B preceded the gradient. The TIC (lower trace) and reconstructed ion current trace for T2 (upper trace) are shown. The resolving power was lowered to produce a greater S/N, but with loss of the charge-indicative isotope separation.

peptides. The choice of approach depends upon the amount of protein available for analysis, the size (complexity) of the protein, and the availability of methods for analysis of specific modifications.

2.2.1 LC-MS With Postacquisition Processing

The most straightforward and universal approach is data acquisition by conventional LC-MS of masses for all components in a mixture. As noted above, a computer program can match the expected peptides with the observed masses. A second step involves searching for expected peptide masses that display a mass shift that is characteristic of a specific modification, e.g., +80 Da for phosphorylation.[78] Glycopeptides frequently display a diagonal pattern in a "contour" plot of the data of time vs. m/z.[71] The diagonal pattern arises from the microheterogeneity of the glycosylation; more heavily glycosylated peptides (higher mass) are slightly more hydrophilic and elute slightly earlier than peptides with smaller glycoforms.

Peptides that may contain posttranslational modifications can be located by a list of masses that do not match the list of expected peptides. The list of unmatched peptides can become complicated since unexpected peptides may also arise from partial or unexpected (low-frequency) bond cleavage by the protease, or from contaminant proteins, including the protease itself. For large proteins (>40 kDa), this approach can become unwieldy because the number of possible matches for a peptide within the mass accuracy typically attained with a quadrupole mass analyzer becomes large, especially once the constraint of enzyme specificity is removed. Confirmation of the peptide identity by tandem mass spectrometry or other methods becomes crucial in those instances.

2.2.2 Chromatogram Comparison After Modification Removal

One method to simplify the identification of modified peptides is by comparison of the peptide maps for a protein before and after the removal of the modification. Asn-linked carbohydrate may be removed by treatment of the protein (or the peptide digest) with PNGase F.[79] Phosphate may be removed with alkaline phosphatase.[80] A variation of this approach is the biosynthesis of a protein under conditions that prevent modification. For example, glycoproteins expressed in prokaryotic cells are not glycosylated. Alternatively, the protein could be mutated to eliminate the consensus sequences for Asn-linked glycosylation in eukaryotes.

2.2.3 Specific Scan Modes

Techniques for specific detection of modified peptides have been developed for glycopeptides and phosphopeptides. A particularly innovative approach makes use of the carbohydrate fragmentation that is generated by electrospray source excitation.[81-83] By raising the electric field in the desolvation region (typically the orifice or nozzle-skimmer voltage), the ions are accelerated through the surrounding gas molecules and undergo extensive collisions that cause fragmentation. The carbohydrates yield oxonium fragment ions that are indicative of their parent saccharide subunits (N-acetyl hexosamine, m/z 204; N-acetyl neuraminic acid, m/z 292; and the disaccharide hexose + N-acetyl hexosamine, m/z 366). During each scan a high orifice voltage is used to induce the fragmentation when scanning the mass region of the low-mass fragment ions. The orifice voltage is then lowered to its normal value for the remainder of the mass scan to identify peptide and glycopeptide molecular weights. Shown in Figure 3 is the glycosylation site identification of the natriuretic peptide receptor C (NPR-C) extracellular domain. Extracted ion chromatograms for m/z 204, 292, and 366 give the precise locations of the glycopeptides in the chromatogram. Capillary LC/MS with orifice potential stepping shows that of the 45 tryptic peptides in the digest only 4 peaks give fragments at m/z 204. Three peaks were identified as N-linked glycopeptides, T4, T35, and T23. The fourth peak was a nonglycosylated tryptic peptide which happens to form a carboxy-terminal dipeptide fragment at m/z 204. The extracted ion profile of m/z 366 had the three glycopeptide peaks, and the extracted ion profile for m/z 292 had two peaks for glycopeptides T4 and T35 (data not shown). Peptide T23 was glycosylated with high mannose structures which contain no N-acetylneuraminic acid (m/z 292). O-linked carbohydrates can be detected equally well by the source excitation method,[82,83] but this protein appears not to contain O-linked carbohydrates.

The source excitation method can also be used to determine sites of phosphorylation by examination of m/z 79 (PO$^-$) in the negative ion mode.[84,85] A separate method for the identification of phosphorylated peptides utilizes neutral loss scans for the loss of phosphoric acid from the peptide by collisional activation.[86] Typically, a loss of m/z 49 is monitored for doubly charged parent ions of tryptic peptides. However, this method is less universal because the neutral loss depends on the charge of the parent ion.

2.2.4 Selective Isolation of Modified Peptides

Affinity chromatography techniques that specifically isolate modified peptides permit a dramatic reduction in the complexity of mixtures and can increase the relative concentration of components of low abundance. In the case of phosphopeptides, an immobilized iron affinity column selectively retains phosphopeptides.[87-89] The retained phosphopeptide are eluted with ammonium ions directly into the mass spectrometer (Figure 4), or may be separated and subsequently analyzed by LC-MS. The positive ion mass spectra shown in Figure 4 were obtained by immobilized metal-ion affinity chromatography electrospray ionization of the tryptic digest of β-casein. The two peaks correspond to the two phosphorylated peptides, T6 with one phosphoserine, T1-2 with four phosphoserines. The upper spectrum was obtained from 10 pmol of digested protein. The lower spectrum was obtained from 30 pmol of β-casein that was loaded on an SDS-PAGE gel, electroblotted to PVDF, and digested *in situ* on the membrane with trypsin. Both sample preparation methods produced spectra which show only the two phosphorylated peptides from β-casein.

Alternatively, antibodies provide exquisite selectivity for molecule isolation by affinity techniques, although only a few are available that bind a

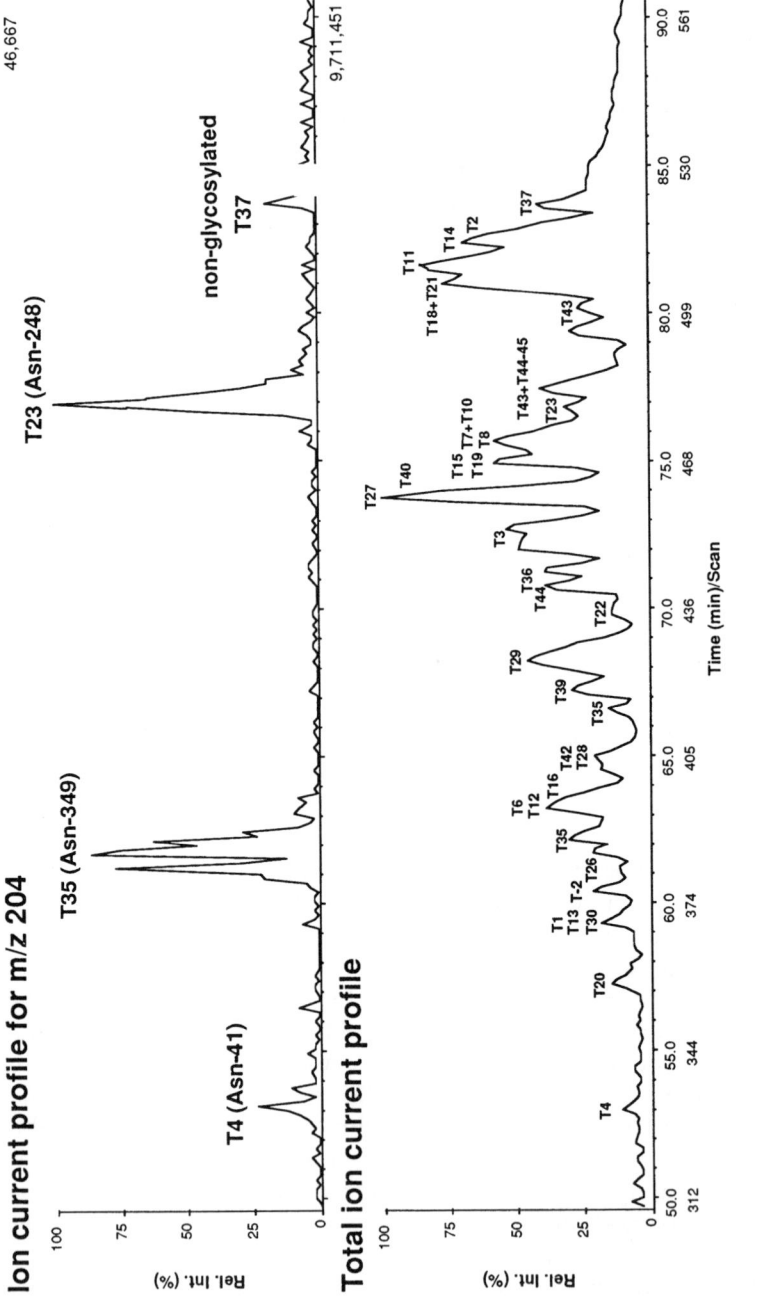

FIGURE 3

specific amino acid modification. One example is the antiphosphotyrosine antibody which can be used to isolate phosphotyrosine-containing peptides selectively.[90]

2.3 Peptide Sequencing

Confirmation of the identification of peptides in a peptide map often requires sequence determination. In particular, unexpected molecular weights in peptide maps of known proteins or maps of proteins with unknown sequence necessitate determination of their peptide sequences. Tandem mass spectrometry, either off-line[91] or on-line with HPLC,[55,77,92-94] is useful for peptide sequencing. The masses of the peptides determined by LC-MS are used for subsequent sequence ion generation by collision-induced dissociation (CID or MS/MS).

Sequence data can be obtained by tandem mass spectrometry on a triple quadrupole instrument.[91] This technique is usually performed with samples that are isolated fractions which are introduced as individual samples to the mass spectrometer. Sequencing isolated fractions is most often done due to the need to know the precursor ion mass for the molecule to be sequenced. Another approach is to determine the masses of all components in a preliminary chromatogram, then input those masses and their respective retention times for the data system to use in the subsequent MS/MS analysis.[76,77,94,95] A third approach that is possible with some data systems is to determine the precursor mass in real time and collect the fragmentation spectrum from the same chromatographic peak. Typically, five to ten times more material is required for the MS/MS measurement than for a single-dimensional mass spectrum. The larger requirement for sample is due to the distribution of the ion current, initially concentrated in the single precursor peak, among all the fragment ions. This dilution effect is somewhat offset by greatly reduced chemical background in the product ion spectrum.

FIGURE 3

Chromatograms for the identification of glycosylation sites of the natriuretic peptide receptor-C (NPR-C) extracellular domain. The lower trace is the total ion current chromatogram for the tryptic digest of NPR-C, 90 pmol. The separation was performed with a 320 μm × 10 cm C18 packed capillary column (LC Packings). Solvent A was 2% acetonitrile/0.05% TFA. Solvent B was 90% acetonitrile/0.0425% TFA. The column was washed with 0% B for 30 min following a 20-μl injection, then eluted with a 0 to 60% B gradient in 60 min. The upper trace is the reconstructed ion current profile for m/z 204 that corresponds to the fragment ion indicative of the monosaccharide HexNAc, derived from N- and O-linked carbohydrate. The data readily identify the glycopeptides in the mixture and clearly show that three of the four potential N-linked glycosylation sites are occupied. The peak splitting for T35 results from different glycoforms attached to this peptide. Peptide T37 coincidentally gives a fragment at m/z 204, but is not glycosylated. (From Stults, J.T., O'Connell, K.L., Garcia, C., Wong, S., Engel, A., Garbers, D.L., and Lowe, D.G., *Biochemistry*, 33, 11372, 1994. With permission.)

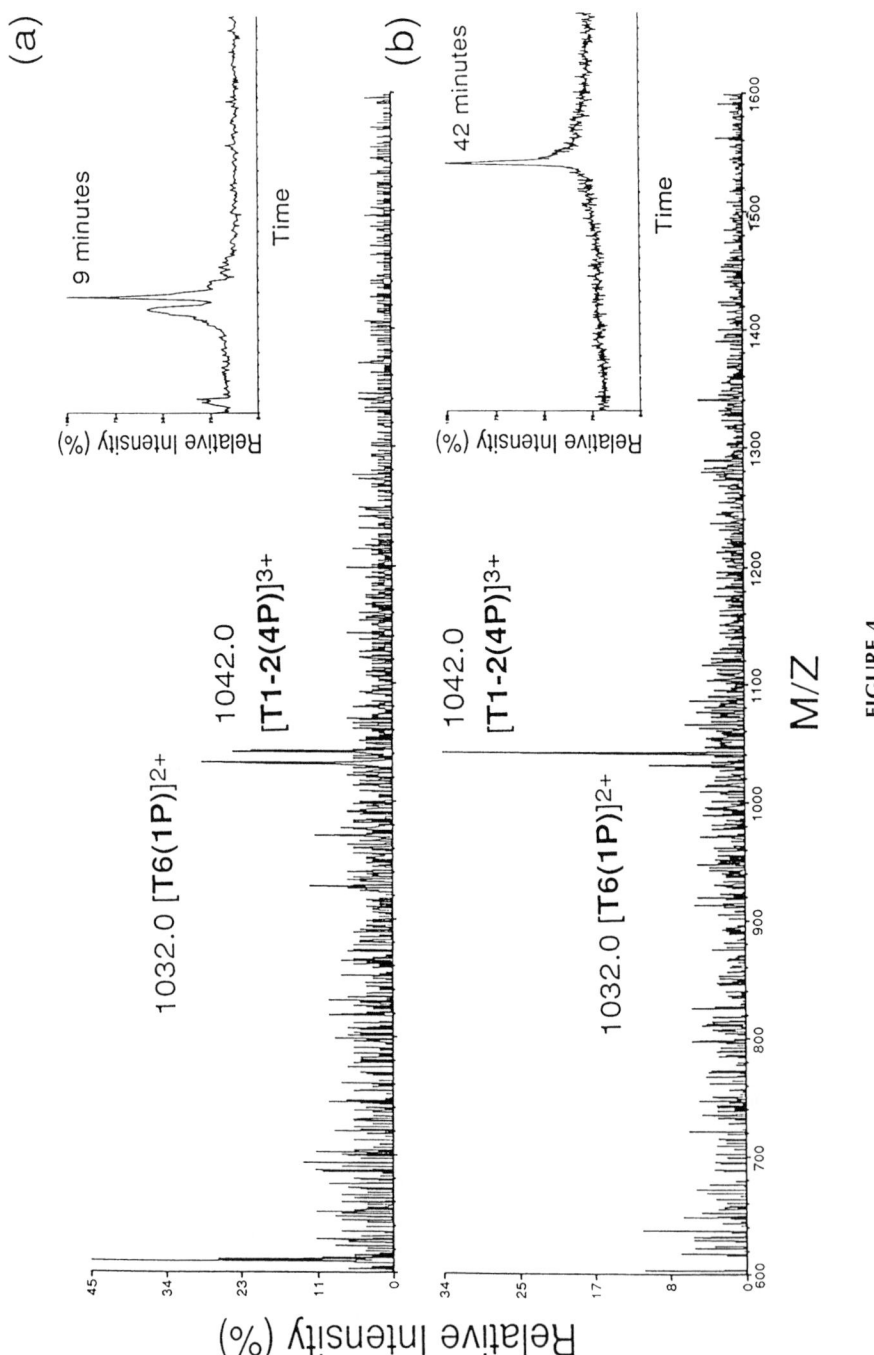

FIGURE 4

The fragmentation data that is normally obtained with a triple quadrupole may also be obtained with a single quadrupole via the source-excitation method identical to the method for observing glycopeptide fragments.[96] The spectra are often very similar to those obtained with the MS/MS experiment. One has to be particularly careful to perform this experiment only with pure samples, otherwise the precursor-fragment ion relationship may be ambiguous.

One application of high-sensitivity sequence analysis is the measurement of peptides that are bound by MHC Class-I molecules. A single MHC allele may present hundreds of different peptides, and the quantities available are in the low-to-subpicomole range. The peptides presented are typically nine amino acids in length and similar in composition, making their complete separation impossible. Mass spectrometry has proven to be an excellent tool for the analysis of these peptides.[77] Figure 5 shows the chromatogram of a mixture of peptides derived from a specific allele (HLA-B0702) of MHC Class-I molecules.[97] At least five peptides with m/z 520.5 are readily observed in this mixture, as is evident from the reconstructed ion current trace. The most abundant of these was selected as a precursor ion for a subsequent LC-MS/MS experiment (Figure 6).

Although in some instances the complete sequence is obtained with an MS/MS experiment, in many cases only a partial sequence is obtained. However, the verification of peptide identities in a tryptic map and the identification of sites of modification may often be satisfied with partial sequences. Hybridization probes for cDNA cloning require only a sequence of six amino acids, and protein sequence databases are readily searched with a partial sequence. Therefore, partial sequences are sufficient to solve many problems.

Obtaining peptide sequence by Edman degradation can also complement mass analysis. Conversely, peptide masses are also invaluable as an aid to Edman degradation.[98-101] For example, the peptide mass gives an estimate of the number of cycles that must be programmed for a peptide. It also gives an independent verification of the accuracy of a sequence and, where there is a question, it often allows identification of a residue at a cycle that otherwise

FIGURE 4
Positive ion mass spectra obtained by immobilized metal-ion affinity chromatography (IMAC)-electrospray ionization of the tryptic digest of β-casein. The two peaks correspond to the two phosphorylated peptides, T6 with one phosphoserine, and T1-2 with four phosphoserines. The upper trace is the spectrum obtained from 10 pmol of digested protein, isolated by Fe^{3+}-IMAC, eluted directly into the mass spectrometer with 2% acetonitrile/0.1% ammonium acetate, pH 10.5. The lower trace was obtained from 30 pmol of β-casein that was loaded on an SDS-PAGE gel, electroblotted to PVDF, digested *in situ* on the membrane with trypsin, and the resulting peptides loaded onto the IMAC column. Elution conditions were the same as for the upper trace following a longer column-wash procedure. The insets are the TIC traces for each experiment. (From Nuwaysir, L.M. and Stults, J.T., *J. Am. Soc. Mass Spectrom.*, 4, 662, 1993. With permission.)

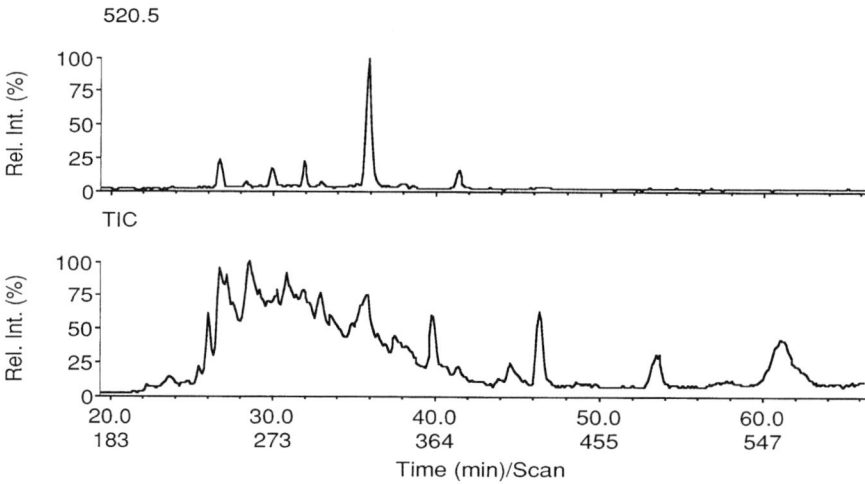

FIGURE 5
Capillary HPLC/MS of the mixture of peptides eluted from the MHC allele HLA-B0702. MHC complexes were immuno-purified from EBV-transformed human B cells. In this case the HLA-B0702 complex was purified and the peptides bound to it were eluted with acetic acid. After ultrafiltration, the concentrated acetic acid solution was injected onto a 180 μm i.d. × 150 mm column packed with 5 μm C_{18} stationary phase. After a 10 min wash at 15% acetonitrile, the aqueous acetonitrile gradient was increased from 15% to 40% acetonitrile in 50 min. The HPLC solvents also contained 0.05% trifluoroacetic acid. The lower trace is a total ion chromatogram (TIC) for a portion of the gradient. The top trace shows an extracted ion current profile of the ions at m/z 520.5. One major peptide, M_r 1039, gives a doubly charged ion ($[M + 2H]^{+2}$) at this mass-to-charge ratio.

might be ambiguous (e.g., Ser, Cys). The mass also confirms or rules out the existence of a posttranslational modification. When multiple peptides are present in a single sample, the mass of each component can be used to sort the sequence of each peptide from the mixture of residues at each cycle.

2.4 Protein Identification by Peptide-Mass Database Searching

A recent innovative use of masses from a protein digest is the identification of a protein from a sequence database.[102] The observed peptide masses are matched with lists of expected peptide masses that are generated by calculation of the digest fragment molecular weights for each protein in the database. Efficient matching is possible with as few as three or four peptides from a tryptic digest.

Figure 7 shows the LC-MS total ion current profile that corresponds to the tryptic digest of a single protein spot isolated by two-dimensional gel electrophoresis of the whole-cell lysate from *E. coli*.[102,103] The gel was

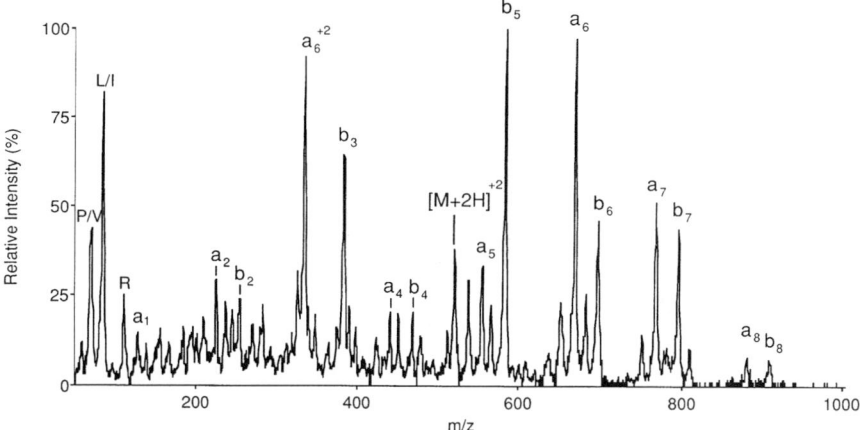

FIGURE 6
Capillary LC/MS/MS of one of the peptides eluted from HLA-B0702. Collision-activated dissociation spectra for several peptides were acquired during a second HPLC run identical to the one shown in Figure 5, and the spectrum shown is for the precursor ion at m/z 520.5. The ions were fragmented upon collision with argon atoms at 70 eV energy (laboratory frame of reference). The masses of the products reflect a composition including proline, valine, arginine, and leucine or isoleucine (leucine and isoleucine are indistinguishable in this experiment). Charge was retained on the N-terminal basic amino acids, giving rise to *a* and *b* fragment ions reflecting a sequence that matches a portion of the protein CD20, RPKSNIVLL, found in the protein sequence database.

electroblotted to a PVDF membrane and stained with Coomassie blue. The spot of interest was cut from the membrane and an *in situ* reduction/alkylation/digestion was performed as described above.[69] The peptides that eluted into the digestion buffer containing 10% acetonitrile were analyzed by LC-MS. The masses of the major peaks observed were input to the program FRAGFIT[102] to search the protein sequence database for a match of the peptide masses. A single *E. coli* protein, a 60-kDa chaperonin, matched the experimental data. This result was confirmed by subsequent N-terminal sequencing of an undigested portion of the same spot.

Advantages of this method include small sample requirements (subpicomole), applicability to N-terminally blocked proteins, and tolerance to posttranslational modifications. The protein sequence database is growing by over 20,000 sequences per year. Thus, this mass measurement/searching method will likely become the method of choice to identify proteins and obtain their sequence, and is already being used routinely by a number of groups.[102,104-106]

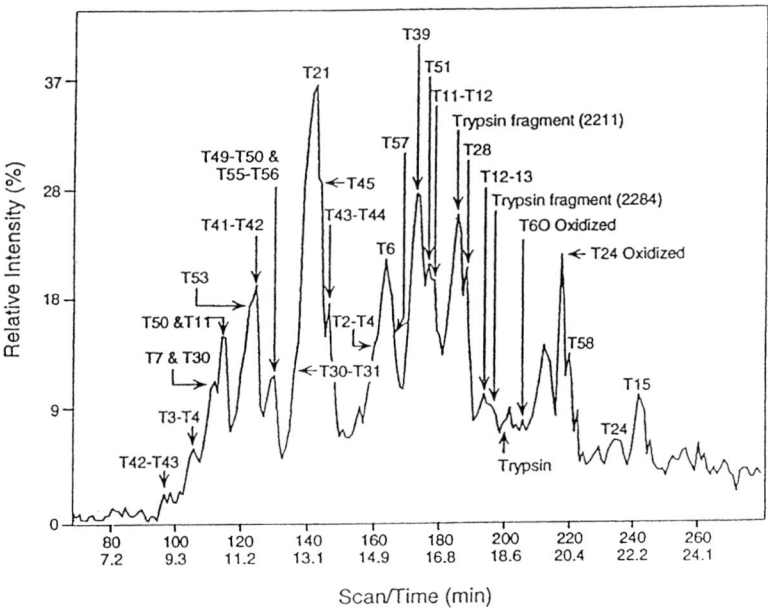

FIGURE 7
Packed capillary LC-MS of the *in situ* tryptic digestion of a protein spot from the blot (PVDF) of a 2-D gel separation of *E. coli* whole-cell lysate. The excised membrane was destained, reduced, alkylated, and digested with trypsin. Peptides were eluted from the membrane into the digestion buffer which contained 10% acetonitrile. The buffer was dried *in vacuo*, then the peptides were redissolved and separated with a 180 μm × 10 cm, C18 column (LC Packings), using a linear gradient of 0 to 70% B in 50 min. The peptide masses were used to search a protein sequence database with the program FRAGFIT. The masses identified a single *E. coli* protein, a 60-kDa chaperonin, that was subsequently confirmed by N-terminal sequencing of an identical spot from a duplicate gel. The N-terminal sequence indicated that approximately 2 pmol of protein was present in the spot. (From Henzel, W.J., Grimely, C., Bourell, J.H., Billeci, T.M., Wong, S.C., and Stults, J.T., *Methods:* Companion to *Methods Enzymol.*, 6, 239, 1994. With permission.)

2.5 PROTEIN MIXTURES

The subject of this chapter so far has been peptide analysis. LC-MS is equally useful for analysis of proteins and protein mixtures.[107] One of the chief advantages of electrospray ionization is its high mass capability. Small proteins exhibit similar ionization efficiencies to peptides, although compound to compound variations may be substantial even for molecules of similar size.

The detection limits for large proteins tend to decrease because the ion current is distributed over a larger number of charge states. The coupling of ESI with HPLC addresses a potential problem in the analysis of proteins (and peptides, for that matter): the presence of contaminating salts, buffers, and surfactants. This problem is particularly significant for many proteins for which these solution components may be required for solubility or activity. Removal of the "contaminants" must be done prior to mass spectrometry. LC-MS is a convenient method for sample desalting and removal of denaturants, but surfactants require other approaches. After loading the sample, the column is extensively washed with aqueous solvent, then a rapid gradient or step gradient is used to elute the protein. If the protein solution contains high concentrations of salt, or chaotropes (e.g., 8 M guanidine), either the column eluent must be diverted or the orifice (or capillary) entrance to the mass spectrometer must be blocked during the wash phase to prevent clogging of the interface.

One should exercise some caution in the use of LC-MS for all protein separations. Some proteins may be irreversibly absorbed to reversed phase packing, but the use of C4 or polymeric packings can reduce this problem.

3 SUMMARY: TODAY AND TOMORROW

Packed capillary HPLC with electrospray ionization mass spectrometry is a valuable technique for analyzing peptide mixtures at low picomole levels. Capillary HPLC offers high peptide concentrations in the eluent and low flow rates that are easily accommodated by the mass spectrometer. All components are available from commercial sources. The technique is used for confirming the sequences of proteins of putative sequence, for locating and identifying posttranslational modifications, for sequencing proteins of unknown sequence, and for aiding sequencing by Edman degradation. Recently, the masses of peptides have been used to identify proteins from a sequence database.

There is significant potential for greater sensitivity with trapping techniques such as Fourier transform and quadrupole ion traps, but these methods are not yet routinely available. Improvements in capillary column technology will permit greater column-to-column reproducibility and resolution. Considerable improvements in the ease of use of LC-MS instruments are expected to occur in the near future. More reliable electrospray interfaces, and dramatic improvements in the user interface due to greater computer hardware and software sophistication, are expected to make LC-MS routine in many laboratories. The mass spectrometer will transform rapidly from an "exotic" instrument to a "black box" detector that is ubiquitous on HPLC systems.

REFERENCES

1. Tsuda, T. and Novotny, M., *Anal. Chem.,* 50, 271, 1978.
2. Hewick, T.M., Hunkapiller, M.W., Hood, L.E., and Dreyer, W.H., *J. Biol. Chem.,* 256, 7990, 1981.
3. Erdjument-Bromage, H., Geromanos, S., Chodera, A., and Tempst, P., *Techniques in Protein Chemistry,* Vol. IV, Angelletti, R., Ed., Academic Press, San Diego, 1993, 419.
4. Atherton, D., Fernandez, J., DeMott, M., Andrews, L., and Mische, S.M., *Techniques in Protein Chemistry,* Vol. IV, Academic Press, San Diego, 1993, 409.
5. Caprioli, R.M., *Anal. Chem.,* 62, 477, 1990.
6. Fenn, J.B., Mann, M., Meng, C.K., Wong, S.F., and Whitehouse, C.M., *Science,* 246, 64, 1989.
7. Karas, M., Bachman, D., Bahr, U., and Hillenkamp, F., *Int. J. Mass Spectrom. Ion Processes,* 87, 53, 1987.
8. Hillenkamp, F., Karas, M., Beavis, R.C., and Chait, B.T., *Anal. Chem.,* 63, 1193A, 1991.
9. Arpino, P.J. and Guiochon, G., *J. Chromatogr.,* 185, 529, 1979.
10. Games, D.E. and Ramsey, E.D., *J. Chromatogr.,* 323, 67, 1985.
11. Kim, H.Y., Pilosof, D., Dyckes, D.F., and Vestal, M.L., *J. Am. Chem. Soc.,* 106, 7304, 1984.
12. Pilosof, D., Kim, H.Y., Dyckes, D.F., and Vestal, M.L., *Anal. Chem.,* 56, 1236, 1984.
13. Stachowiak, K., Wilder, C., Vestal, M., and Dyckes, D.F., *J. Am. Chem. Soc.,* 110, 1758, 1987.
14. Fink, S.W. and Freas, R.B., *Anal. Chem.,* 61, 2050, 1989.
15. Ito, Y., Takeuchi, T., Ishii, D., and Gotto, M., *J. Chromatogr.,* 346, 161, 1985.
16. Caprioli, R.M., Fan, T., and Cottrell, J.S., *Anal. Chem.,* 58, 2949, 1986.
17. Caprioli, R.M., DaGue, B., Fan, T., and Moore, W.T., *Biochem. Biophys. Res. Commun.,* 146, 291, 1987.
18. Caprioli, R.M., Moore, W.T., DaGue, B., and Martin, M., *J. Chromatogr.,* 443, 355, 1988.
19. Caprioli, R.M., DaGue, B.B., and Wilson, K., *J. Chromatogr. Sci.,* 26, 640, 1988.
20. Caprioli, R.M., Moore, W.T., DaGue, B., and Martin, M., *Anal. Chem.,* 62, 477A, 1990.
21. Caprioli, R.M., *J. Am. Chem. Soc.,* 27, 513, 1988.
22. Bell, D.J., Brightwell, M.D., Neville, W.A., and West, A., *Rapid Commun. Mass Spectrom.,* 4, 88, 1990.
23. Page, J.A. and Beer, M.T., *J. Chromatogr.,* 474, 51, 1989.
24. Deterding, L.J., Moseley, M.A., Tomer, K.R., and Jorgenson, J.W., *Anal. Chem.,* 61, 2504, 1989.
25. Kassel, D.B., Musselman, B.D., and Smith, J.A., *Anal. Chem.,* 63, 1091, 1991.
26. Ashcroft, A.E., Chapman, J.R., and Cottrell, J.S., *J. Chromatogr.,* 394, 15, 1987.
27. Henzel, W.J., Bourell, J.H., and Stults, J.T., *Anal. Biochem.,* 187, 228, 1990.
28. Caprioli, R.M., Ed., *Continuous-Flow Fast Atom Bombardment Mass Spectrometry,* John Wiley & Sons, Chichester, 1990.
29. Whitehouse, C.M., Dreyer, R.N., Yamashita, M., and Fenn, J.B., *Anal. Chem.,* 57, 675, 1985.
30. Fenn, J.B., Mann, M., Meng, C.K., Wong, S.F., and Whitehouse, C.M., *Mass Spectrom. Rev.,* 9, 37, 1990.
31. Smith, R.D., Loo, J.A., Loo, R.R.O., Busman, M., and Udseth, H.R., *Mass Spectrom. Rev.,* 10, 359, 1991.

32. Covey, T.R., Bonner, R.F., Shushan, B.I., and Henion, J., *Rapid Commun. Mass Spectrom.*, 2, 249, 1988.
33. Hemling, M.E., Roberts, G.D., Johnson, W., Carr, S.A., and Covey, T.R., *Biomed. Environ. Mass Spectrom.*, 19, 677, 1990.
34. Tomer, K.B., Perkins, J.R., Parker, C.E., and Deterding, L.J., *J. Biol. Mass Spectrom.*, 20, 783, 1991.
35. Suter, M.J.-F., DaGue, B.B., More, W.T., Lin, S.-N., and Caprioli, R.M., *J. Chromatogr.*, 553, 101, 1991.
36. Deterding, L.J., Parker, C.E., Perkins, J.R., Moseley, M.A., Jorgenson, J.W., and Tomer, K.B., *J. Chromatogr.*, 554, 329, 1991.
37. Bruins, A.P., Covey, T.R., and Henion, J.D., *Anal. Chem.*, 59, 2642, 1987,
38. Chowdhury, S.K., Katta, V., and Chait, B.T., *Rapid Commun. Mass Spectrom.*, 4, 81, 1990.
39. Shen, S., Whitehouse, C., Banks, F., and Fenn, J.B., 40th ASMS Conf. Mass Spectrom. and Allied Topics, Washington, D.C., May 31, 1992.
40. Meng, C.K., McEwen, C.N., and Larsen, B.S., *Rapid Commun. Mass Spectrom.*, 4, 147, 1990.
41. McEwen, C.N. and Larsen, B.S., *Rapid Commun. Mass Spectrom.*, 6, 173, 1992.
42. Chapman, J.R., Gallagher, R.T., Barton, E.C., Curtis, J.M., and Derrick, P., *J. Org. Mass Spectrom.*, 27, 195, 1992.
43. Gallagher, R.T., Chapman, J.R., and Mann, M., *Rapid Commun. Mass Spectrom.*, 4, 369, 1990.
44. Cody, R.B., Tamura, J., and Musselman, B.D., *Anal. Chem.*, 64, 1561, 1992.
45. Henry, K.D., Williams, E.R., Wang, B.H., McLafferty, F.W., Shabanowitz, J., and Hunt, D.F., *Proc. Natl. Acad. Sci. U.S.A.*, 86, 9075, 1989.
46. Beu, S.C., Senko, M.W., Quinn, J.P., Wampler, F.M.I., and McLafferty, F.W., *J. Am. Soc. Mass Spectrom.*, 4, 557, 1993.
47. Winger, B.E., Hofstadler, S.A., Bruce, J.E., Udseth, H.R., and Smith, R.D., *J. Am. Soc. Mass Spectrom.*, 4, 566, 1993.
48. McLuckey, S.A., VanBerkel, G.J., Flish, G.L., Huang, E.C., and Henion, J.D., *Anal. Chem.*, 63, 375, 1991.
49. Boyle, J.G. and Whitehouse, C.M., *Anal. Chem.*, 64, 2084, 1992.
50. Novotny, M., *Anal. Chem.*, 60, 500A, 1988.
51. Moritz, R.L. and Simpson, R.J., *J. Chromatogr.*, 599, 119, 1992.
52. Karlsson, K.E. and Novotny, M., *Anal. Chem.*, 60, 1662, 1988.
53. Balogh, M.P. and Stacey, C.C., *J. Chromatogr.*, 562, 73, 1991.
54. Davis, M.T. and Lee, T.D., *Protein Sci.*, 1, 935, 1992.
55. Kassel, D.B., Shushan, B., Sakuma, T., and Saltzmann, J.-P., *Anal. Chem.*, 66, 236, 1994.
56. Hunt, D.F., Alexander, J.E., McCormack, A.L., Martino, P.A., Michel, H., Shabanowitz, J., Sherman, N., Moseley, M.A., Jorgenson, J.W., and Tomer, K.B., *Techniques in Protein Chemistry*, Vol. II, Villafranca, J.J., Ed., Academic Press, New York, 1991, 441.
57. Kennedy, R.T. and Jorgenson, J.W., *J. Am. Chem. Soc.*, 61, 1128, 1989.
58. Shelley, D.C., Gluckman, J.C., and Novotny, M.V., *J. Am. Chem. Soc.*, 56, 2990, 1984.
59. Andreolini, F., Borra, C., and Novotny, M., *Anal. Chem.*, 59, 2428, 1987.
60. Hancock, W.S. and Sparrow, J.T., *HPLC — Advances and Perspectives*, Vol. 3, Horvath, C., Ed., Academic Press, New York, 1983, 49.
61. Geng, X. and Regnier, F.E., *J. Chromatogr.*, 296, 15, 1984.
62. Pearson, J.D., Lin, N.T., and Regnier, F.E., *Anal. Biochem.*, 124, 217, 1982.
63. Wong, S.C., Grimley, C., Padua, A., Bourell, J.H., and Henzel, W.J., *Techniques in Protein Chemistry*, Vol. IV, Angeletti, R., Ed., Academic Press, San Diego, 1993, 371.

64. Cobb, K.A. and Novotny, M., *Anal. Chem.*, 61, 2226, 1989.
65. Eckerskorn, C. and Lottspeich, F., *Chromatographia*, 28, 92, 1989.
66. Aebersold, R., Leavitt, J., Saavedra, R.A., Hood, L.E., and Kent, S.B., *Proc. Natl. Acad. Sci. U.S.A.*, 84, 6970, 1987.
67. Patterson, S.D., Hess, D., Yungwirth, T., and Aebersold, R., *Anal. Biochem.*, 202, 193, 1992.
68. Iwamatsu, A., *Electrophoresis*, 13, 142, 1992.
69. Henzel, W.J., Billeci, T.M., Stults, J.T., Wong, S.C., Grimley, C., and Watanabe, C., *Techniques in Protein Chemistry*, Vol. V, Crabb, J.W., Ed., Academic Press, San Diego, 1994, 3.
70. Zhang, W., Czernik, A.J., Yungwirth, T., Aebersold, R., and Chait, B.T., *Protein Sci.*, 3, 677, 1994.
71. Ling, V., Guzzetta, A.W., Canova-Davis, E., Stults, J.T., Hancock, W.S., Covey, T.R., and Shushan, B.I., *Anal. Chem.*, 63, 2909, 1991.
72. Carr, S.A., Hemling, M.E., Bean, M.F., and Roberts, G.D., *Anal. Chem.*, 63, 2802, 1991.
73. Mock, K., Hail, M., Mylchreest, I., Zhou, J., Johnson, K., and Jardine, I., *J. Chromatogr.*, 646, 169, 1993.
74. Watkins, P.J.F., Jardine, I., and Zhou, J.X.G., *J. Biol. Mass Spectrom.*, 19, 957, 1991.
75. Canova-Davis, E., Chloupek, R.C., Baldonado, I.P., Battersby, J.E., Spellman, M.W., Basa, L.J., O'Connor, B., Pearlman, R., Quan, C., Chakel, J.A., Stults, J.T., and Hancock, W.S., *Am. Biotech. Lab.*, 6, 8, 1988.
76. Hail, M., Lewis, S., Jardine, I., Liu, J., and Novotny, M., *J. Microcol. Sep.*, 2, 285, 1990.
77. Hunt, D.F., Henderson, R.A., Shabanowitz, J., Sakaguchi, K., Michel, H., Sevilir, N., Cox, A.L., Appella, E., and Englehard, V.H., *Science*, 255, 1261, 1992.
78. Rossomando, A.J., Wu, J., Michel, H., Shabanowitz, J., Hunt, D.F., Weber, M.J., and Sturgill, T.W., *Proc. Natl. Acad. Sci. U.S.A.*, 89, 5779, 1992.
79. Carr, S.A. and Roberts, G.D., *Anal. Biochem.*, 157, 396, 1986.
80. Yip, T.-T. and Hutchens, T.W., *Techniques in Protein Chemistry*, Vol. IV, Angeletti, R., Ed., Academic Press, San Diego, 1993, 201.
81. Conboy, J.J. and Henion, J.D., *J. Am. Soc. Mass Spectrom.*, 3, 804, 1992.
82. Carr, S.A., Huddleston, M.J., and Bean, F.M., *Protein Sci.*, 2, 183, 1993.
83. Huddleston, M.J., Bean, M.F., and Carr, S.A., *Anal. Chem.*, 65, 877, 1993.
84. Huddleston, M.J., Annan, R.S., Bean, M.F., and Carr, S.A., *J. Am. Soc. Mass Spectrom.*, 4, 710, 1993.
85. Ding, J., Burkhart, W., and Kassel, D.B., *Rapid Commun. Mass Spectrom.*, 8, 94, 1994.
86. Covey, T., Shushan, B., and Bonner, R., *Methods in Protein Sequence Analysis*, Jornvall, H., Hoog, J.-O., and Gustavsson, A.-M., Birkhauser Verlag, Basel, 1991, 249.
87. Andersson, L. and Porath, J., *Anal. Biochem.*, 154, 250, 1986.
88. Michel, H., Hunt, D.F., Shabanowitz, J., and Bennett, J., *J. Biol. Chem.*, 263, 1123, 1988.
89. Nuwaysir, L.M. and Stults, J.T., *J. Am. Soc. Mass Spectrom.*, 4, 662, 1993.
90. Hou, J., McKeehan, K., Kan, M., Carr, S.A., Huddleston, M.J., Crabb, J.W., and McKeehan, W.L., *Protein Sci.*, 2, 86, 1993.
91. Hunt, D.F., Yates, J.R., III, Shabanowitz, J., Winton, S., and Hauer, C.R., *Proc. Natl. Acad. Sci. U.S.A.*, 83, 6233, 1986.
92. Hunt, D.F., Michel, H., Dickinson, T.A., Shabanowitz, J., Cox, A.L., Sakaguchi, K., Appella, E., Grey, H.M., and Sette, A., *Science*, 256, 1817, 1992.
93. Huang, E.C. and Henion, J.D., *Anal. Chem.*, 63, 732, 1991.

94. Griffin, P.R., Coffman, J.A., Hood, L.E., and Yates, J.R.I., *Int. J. Mass Spectrom. Ion Processes,* 111, 131, 1991.
95. Covey, T.R., Huang, E.C., and Henion, J.D., *Anal. Chem.,* 63, 1193, 1991.
96. Katta, V., Chowdury, S.K., and Chait, B.T., *Anal. Chem.,* 63, 174, 1991.
97. Gillece-Castro, B.L., Barber, L., and Parham, P., Unpublished observations.
98. Henzel, W.J., Aswad, D.W., and Stults, J.T., *Techniques in Protein Chemistry,* Hugli, T.E., Ed., Academic Press, San Diego, 1989, 127.
99. Johnson, R.S. and Walsh, K.A., *Protein Sci.,* 1, 1083, 1992.
100. Geromanos, S., Casteels, P., Elicone, C., Powell, M., and Tempst, P., *Techniques in Protein Chemistry,* Vol. V, Crabb, J.W., Ed., Academic Press, San Diego, 1994.
101. Hess, D., Covey, T.C., Winz, R., Brownsey, R.W., and Aebersold, R., *Protein Sci.,* 2, 1342, 1993.
102. Henzel, W.J., Billeci, T.M., Stults, J.T., Wong, S.C., Grimely, C.G., and Watanabe, C.W., *Proc. Natl. Acad. Sci. U.S.A.,* 90, 5011, 1993.
103. Henzel, W.J., Grimely, C., Bourell, J.H., Billeci, T.M., Wong, S.C., and Stults, J.T., *METHODS: A companion to Methods in Enzymology,* 6, 239, 1994.
104. Yates, J.R.I., Speicher, S., Griffin, P.R., and Hunkapillar, T., *Anal. Biochem.,* 214, 397, 1993.
105. Mann, M., Hojrup, P., and Roepstorff, P., *Biol. Mass Spectrom.,* 22, 338, 1993.
106. Pappin, D.J.C., Hojrup, P., and Bleasby, A.J., *Curr. Biol.,* 3, 327, 1993.
107. Moseley, M.A., III and Unger, S.E., *J. Microbiol. Sep.,* 4, 393, 1992.
108. Stults, J.T., O'Connell, K.L., Garcia, C., Wong, S., Engel, A., Garbers, D.L., and Lowe, D.G., *Biochemistry,* 33, 11372, 1994.

CHAPTER 6

CAPILLARY ELECTROPHORESIS – MASS SPECTROMETRY IN PEPTIDE MAPPING

Jon H. Wahl, Harold R. Udseth, and Richard D. Smith

CONTENTS

1 Introduction 143

2 CE-MS Interfacing Methods 145
    2.1 Off-Line Methods 145
    2.2 Continuous-Flow Fast Atom Bombardment Interfaces 146
    2.3 Electrospray Ionization Interfaces 147

3 Experimental Methods and Considerations for CE-MS Analyses 152
    3.1 Separation Considerations 152
    3.2 Mass Spectrometric Considerations 153

4 CE-MS Applications 158

5 Limitations of Present CE-MS Methods and Instrumentation 164

6 Practical Approaches to Improved CE-MS Sensitivity 168
    6.1 Reduced Elution Speed Detection 168
    6.2 Effect of Capillary Inner Diameter Upon CE-MS
         Sensitivity 171

Acknowledgments 175

References 176

1 INTRODUCTION

The increasing application of capillary electrophoresis (CE) to biochemical analyses has led to a growing demand for detection methods that are sensitive, selective, or structurally informative for the separated components. Mass spectrometry (MS) has long been recognized as an extremely selective and broadly applicable detector for analytical separations, but the combination of CE with MS is relatively recent and the methodology continues to evolve. Recent developments in MS have led to continuing improvements in the sensitivity, speed, and degree of structural information obtainable from this powerful detector. Concurrent developments involving tandem-MS methods and higher-order MS methods (i.e., MS^n, where $n \geq 2$) illustrate instrumentation offering even greater selectivity and structural information that will be routinely achievable in the future. Although MS is still one of the most complex and expensive CE detectors, instrumentation costs continue to decrease while user friendliness of the instrument continues to increase. Indeed, when MS detection is chosen to resolve an analytical problem there is often no practical substitute.

The mass spectrometer provides the equivalent of up to several thousand discrete selective detectors essentially functioning in parallel and capable of providing both molecular weight information from the intact molecular ion as well as structurally related information from the dissociation of that molecular ion in the mass spectrometer. The hyphenated technique of CE-MS in one sense is a nearly ideal method: CE analyzes ions based roughly upon their mass to charge ratio, which causes differing mobilities in solution, and MS analyzes ions based upon their mass to charge ratio (m/z) in the gas phase. Conversely, these two analytical methods exploit ion movement in two very different environments; moderately conductive liquid systems are used in CE and high-vacuum conditions are used in MS. A primary role of the CE-MS interface is to couple these different environments. Chronologically, the combination of liquid separation methods with MS has long been of interest, and a variety of innovative liquid chromatographic (LC)-MS interfacing schemes have been investigated over the last 25 years. The CE-MS combination is more recent, however, and places significantly different demands on such an interface compared to the LC-MS combination. For example, CE flow rates are quite low or negligible, the buffer system is moderately conductive, electrical contact must be maintained with both ends of the analytical capillary to define the CE field gradient, and any extracapillary broadening contribution to the solute zones due to thermal heating within the capillary or from laminar flow must be avoided or minimized if the high separation efficiencies possible with CE are to be realized.

In this chapter, we review CE-MS interfacing methods, experimental considerations, and applications, with an emphasis upon peptide mapping. Although we discuss the major interfacing methods — continuous-flow fast atom bombardment (CF-FAB) and electrospray ionization (ESI) — the emphasis of this review is on the latter, which reflects the explosive growth in ESI methodology in the field of mass spectrometry. This emphasis is not meant to diminish the impressive results obtained with FAB interfaces, particularly for small-molecule applications where results comparable to ESI interfaces in terms of sensitivity have been achieved. Finally, we wish to report on the current status of CE-MS, an extremely promising marriage still somewhat limited by imperfect control of the ESI interface, sensitivity, and the scan speed constraints of current MS technology. We also describe some simple and practical approaches that may alleviate these limitations. These developments hold the promise of CE-MS/MS for routine sequencing applications, a development that would greatly expand the use of this technology. Because the MS limitations that initially inhibited broad application of CE-MS are almost sure to be overcome in the next few years, the future for CE-MS appears to hold exceptional promise.

2 CE-MS INTERFACING METHODS

As noted above, two on-line interfacing methods have been used primarily for CE-MS — continuous-flow fast atom bombardment (CF-FAB) and electrospray ionization (ESI). In general, there are two variations reported for interfacing FAB and ESI ionization for use in CE-MS. These variations are based upon either a liquid junction or a coaxial sheath flow approach, and are differentiated by the methods of establishing electrical contact at the capillary terminus. The two methods also differ in the process used to transfer ions from the liquid to the gas phase so as to allow for the subsequent m/z analysis in the MS. ESI/MS interfaces require an atmospheric pressure ionization sampling inlet, and FAB/MS methods operate with conventional probe inlets and produce ions under vacuum. In addition to these approaches, off-line methods, involving the collection of CE effluents and subsequent MS analysis have been reported and are briefly discussed below.

2.1 OFF-LINE METHODS

The advantages of off-line CE-MS include the flexibility of using any available ionization or MS method. The primary disadvantages of off-line methods include limited resolution of the separation due to sample collection constraints, reduced sensitivity, and the need for extra sample handling. The latter is particularly problematic when small sample sizes and injection volumes are used in CE (<10 nL). For these reasons, the earliest reported off-line CE-MS studies involved isotachophoresis, where sample sizes are generally much larger than encountered with CE. One of the earliest reports is from Kenndler and Kaniansky[1] and demonstrated the identification of pesticides using conventional electron ionization (EI)/MS. Later work by Kenndler and Haidl[2] demonstrated the potential for quantitative application for amenable compounds, such as the hydrogenation products of quaternary ammonium salts. A major limitation of EI methods is that ionization occurs in the gas phase, and the general problem of getting thermally labile or nonvolatile analytes into the gas phase is not solved.

A much more powerful approach for off-line CE-MS has recently been described by Takigiku et al.[3] using plasma desorption (PD)/mass spectrometry. PD/MS allows analysis of nonvolatile compounds and has been successfully applied to proteins with average molecular weights (M_r) as large as 45 kDa.[4] These workers used a porous glass joint, similar to that first described by Ewing and Wallingford,[5] that functioned similar to the liquid junction interfacing methods described below. Electrical contact with the capillary terminus was established through the porous glass capillary, and the eluents were collected on nitrocellulose-coated aluminum foil. Samples were then lyophilized, redissolved in a compatible solution, and again deposited on nitrocellulose-coated foils and washed with deionized water prior to analysis.[3] Picomole-level sensitivities were obtained for a number of peptides and proteins, and improved sensitivity and molecular weight range can be anticipated using matrix-assisted laser desorption methods.[6] Some buffer systems, however, are incompatible with PD/MS, and off-line methods reported thus far remain constrained due to the limited number of fractions generally collected, small sample sizes, and extensive sample manipulation. Recent developments have indicated that the off-line combination, with matrix-assisted laser desorption ionization (MALDI), has several attractive features including greater sensitivity and M_r range than PD.

2.2 Continuous-Flow Fast Atom Bombardment Interfaces

FAB ionization is a widely used MS method for analysis of labile and nonvolatile compounds and, as such, provides a basis for CE-MS. The liquid junction and coaxial sheath flow variations upon CE-FAB/MS interfacing have been reported, each imposing its own constraints.

The liquid junction approach couples the analytical capillary with an additional length of capillary used for transferring analytes in a "make-up" flow to the FAB source where solute ionization occurs. A variety of liquid junction FAB interfaces have been reported.[7-16] A variation on this interface for the higher flow rates of simultaneous capillary chromatography and electrophoresis, which is termed pseudoelectrochromatography, has been described.[17] The FAB method generally requires a viscous liquid component, where a 5 to 25% glycerol solution in a variety of solvent systems is commonly used with the liquid junction interface. Generally 50 to 100 μm i.d. capillaries are used for CE, and the CE flow rate (<0.1 μL/min) is much lower than the flow introduced through the liquid junction (5 to 10 µL/min). The long transfer capillary (~1 m) results in some degradation of CE separation efficiency, where about an order of magnitude decrease in the number of theoretical plates for FAB/MS detection has been noted compared with detection before the liquid junction.[14,15] Suter et al.[9] have also noted a loss in separation efficiency as well as reduced sensitivity compared to sheath flow FAB interface. They note, however, that the liquid junction interface is easier to handle and setup.

A coaxial sheath flow continuous-flow FAB interface for CE-MS was developed by Deterding and co-workers.[18-24] The sheath flow arrangement in this case is restricted to relatively small CE capillaries (10 to 15 µm i.d.), which minimizes laminar flow within the capillary due to the large pressure gradient along the capillary. As a result, sample injection volumes are quite small, often <1 nL. Because CE flow rates are extremely low, lower sheath flows (~0.5 to 2 µL/min) containing up to 25% glycerol have been used, resulting in improved detection limits compared to the liquid junction approach. In addition, the sheath flow arrangement appears to preserve CE separation efficiency better than the liquid junction arrangement. Indeed, Moseley et al. have obtained up to 4×10^5 theoretical plates using 10 to 100 fmol of sample/injection and somewhat larger samples for tandem-MS.[21] Suter et al.[9] have compared the sheath flow and liquid junction interfaces, finding the former somewhat more difficult to operate and the latter less sensitive. They note that a significant loss is experienced with both FAB interfaces, but also note that this restriction is of limited practical significance at present due to the trade-offs in MS detection. At this time, and although FAB interfaces are viable, it appears that superior performance is attainable using ESI interfaces.

2.3 ELECTROSPRAY IONIZATION INTERFACES

Although the choice of the CE-MS interface will ultimately be governed by the MS instrumentation available, the clear trend favors ESI methods. The ESI method is well suited for CE-MS interfacing because it produces ions directly from liquid solutions at atmospheric pressure.[25-27] CE-MS interfaces based upon ESI require a mass spectrometer incorporating an atmospheric pressure ionization inlet. The availability of such MS instruments is rapidly increasing with numerous commercial implementations. Considerations for interfacing generally derive from limitations upon buffer composition and the desire to position the ESI source (i.e., the point of charged droplet formation) as close as possible to the analytical capillary terminus, so as to avoid lengthy transfer lines. However, it has been demonstrated that, by preventing flow of the buffer to the ESI source, problems associated with undesirable buffer components can be reduced or eliminated. Generally, there have been two variations used on ESI interfaces. These are again based upon a liquid

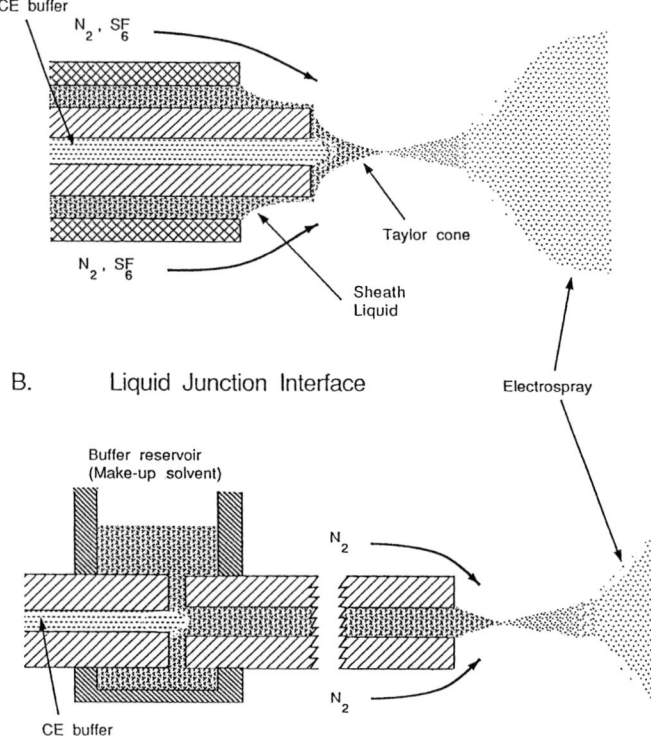

FIGURE 1
Schematic illustration of the coaxial sheath flow interface (A); and the liquid junction interface (B) used for CE-ESI/MS.

junction and coaxial sheath flow approach, which are shown schematically in Figure 1. More recently, we have also demonstrated the use of "sheath-less" interfaces, in which the capillary terminus is etched and gold deposited to create an electrical contact. Indeed, this approach is actually a refined version of the original design for CE-MS.[30,31]

ESI is an extremely soft ionization technique and will, under appropriate conditions, yield intact molecular ions without contributions due to dissociation unless induced during transport into the MS. Molecular weight measurements for large biopolymers that exceed the MS mass range can be obtained because their ESI mass spectra generally consist of a distribution of molecular ion charge states.[25,27] The envelope of molecular ion charge states for peptides and proteins, arising generally from protonation for positive ion ESI, yields a distinctive pattern of peaks due to the discrete nature of the electronic

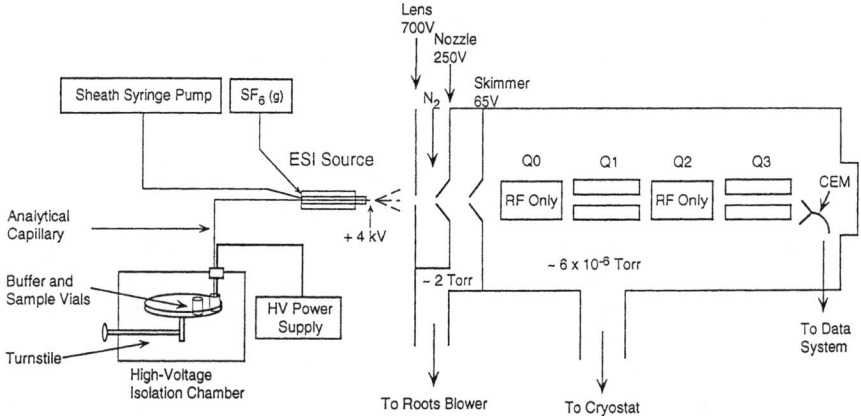

FIGURE 2
Schematic illustration of the experimental arrangement for CE-MS. Details of the sheath flow interface are given in Figure 3.

charge, where adjacent peaks vary by addition or subtraction of one charge. In the case of noncovalently bonded species, the mass spectra typically show only the individual subunits under conditions typically used. With proper care, however, it is also possible to select appropriate interface conditions so as to detect a range of noncovalently associated species by ESI/MS.

Figure 2 shows a general schematic illustration for CE-MS/MS instrumentation based upon the sheath flow interface such as is used at our laboratory. The ESI source is generally operated at an elevated voltage of 2 to 5 kV relative to the sampling orifice, where ions that are formed at atmospheric pressure enter the mass spectrometer entrained in a flow of gas.[26] Many of the practical constraints and considerations for CE-MS interfacing derive from the voltage bias between these components. The ESI source requires a high electric field that causes a charge to accumulate on the liquid surface that extends from the capillary terminus so as to disrupt and disperse the liquid flow to droplets. The ESI liquid nebulization process can be assisted mechanically by vibration at high frequency or pneumatically by a high-velocity annular gas flow at the capillary terminus. The pneumatically assisted electrospray is sometimes referred to as "ion spray" — an approach originally described for ESI by Dole and co-workers.[28,29] The use of low flow rates with the sheathless design results in a smaller droplet size that facilitates closer positioning of the capillary to the ion inlet orifice. This has the effect of reducing the voltage needed for ESI and increasing the efficiency of ion sampling/utilization (i.e., sensitivity).

The first CE-MS, based upon an electrospray ionization interface, was developed at our laboratory.[30,31] This initial approach used a metalized capillary terminus to make contact with the CE eluent under conditions of high electroosmotic flow. We later reported an interface using a flowing sheath liquid and also demonstrated the combination of capillary isotachophoresis (CITP) with MS and CE-MS/MS.[32-36] More recently, other researchers have reported similar interfaces.[37-45]

The liquid junction variation was developed by Henion and co-workers,[46-48] who have reported extensively upon its applications,[49-53] and was explored more recently by Pleasance et al.[54] who compared its performance to the sheath flow approach. As indicated in Figure 1B, electrical contact with the capillary terminus for the liquid junction interface is established through a liquid reservoir that surrounds the junction of the analytical capillary and a transfer capillary. The gap between the two capillaries is typically adjusted to 10 to 20 µm, a compromise resulting from the need for sufficient make-up liquid to be drawn into the transfer capillary while avoiding analyte loss by diffusion into the reservoir. The flow of make-up liquid arises from a combination of the gravity-driven flow, due to the height of the make-up reservoir, and flow induced in the transfer capillary due to a mild vacuum generated by the venturi effect of the nebulizing gas used at the ESI source.[54]

In the sheath flow electrospray interface (Figure 1A) a liquid, which is generally methanol, methoxyethanol, or acetonitrile, and often with 10 to 20% formic acid, acetic acid, or water, flows through the annular space between the CE capillary (~140–180 µm o.d.) and a fused silica or stainless steel capillary (~250 µm i.d.). The analytical capillary typically extends approximately 0.2 mm from the sheath capillary. A recent version of this interface developed at the authors' laboratory is shown in Figure 3. In this design a 300 µm i.d. fused silica capillary surrounds the smaller-diameter electrophoresis capillary. For the positive ion mode ESI operation, a voltage of +2 to +5 kV is applied to the sheath liquid from which a syringe pump delivers the flow of sheath liquid at 2 to 5 µl/min. Potential problems from the formation of bubbles in the connecting lines are minimized by the inclusion of a static gas trapping volume. Enhanced stability can sometimes be obtained by degassing the organic solvents used in the sheath, by minimizing heating from the ion source, and filtering all solutions prior to use. Cooling of the ESI source, or the use of less volatile sheath solvents such as methoxyethanol, have also been useful in some situations.[55,56] A number of reports from our laboratory have described work based upon the sheath flow approach.[26,34,55,57-61]

It appears that most of the distinctions in terms of performance between the sheath flow and liquid junction ESI interfaces are relatively minor and the performance in terms of spectral quality is similar. However, some differences have been noted. Pleasance et al.[54] compared the liquid junction and sheath flow approaches using a pneumatically assisted ESI interface, and noted that the latter provided a "more robust and reproducible interface with the added advantage of offering an independent means of calibration and quantitation ... through the sheath."[42] Improved signal to noise ratios and slightly improved separations were obtained with the sheath flow interface. Differences from

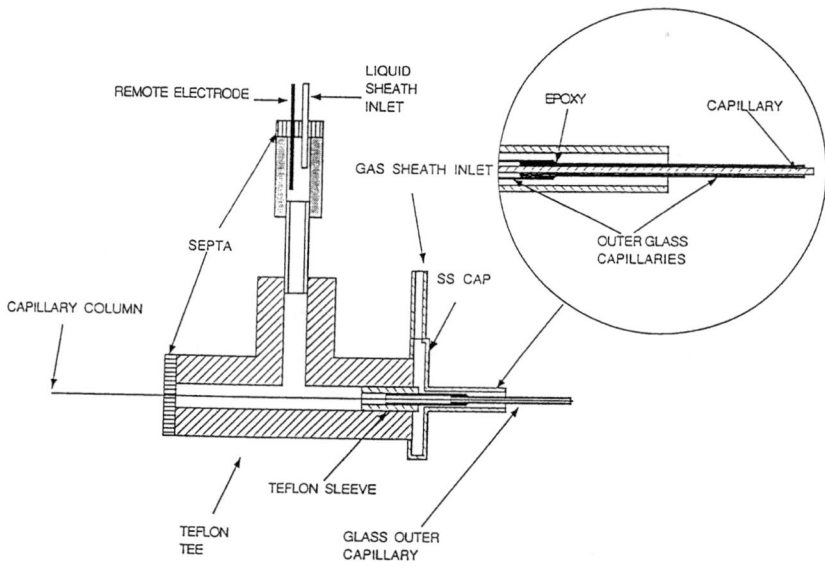

FIGURE 3
One version of a coaxial sheath flow electrospray ionization interface for CE-MS developed at the authors' laboratory. See also Figures 1 and 2.

run to run in elution times using the liquid junction interface were reported by these workers and were attributed to modification of the liquid junction buffer due to flushing of the CE capillary between analyses. At this time, the sheath flow interface appears to be the design of choice, and it has been the basis for the first commercially available interfaces.

The dependence of ESI ion current upon solution conductivity is relatively weak — generally 0.1 to 0.3 µA at atmospheric pressure — and only about 20 to 200 pA integrated ion current (i.e., the sum of all ions) is actually transmitted through the MS and detected. An electron scavenger can be used to inhibit electrical discharge at the capillary terminus, particularly for ESI of aqueous solutions. Sulfur hexafluoride has proven particularly useful for suppressing corona discharges for such applications, and improving the stability of negative ion ESI. Introduction of this gas is most effectively accomplished using a gas flow (~50 to 200 ml/min) through an annular volume surrounding the sheath liquid capillary (Figures 1A and 2). Importantly, it is now recognized that the use of small o.d. capillaries, or etched capillary "tips", is attractive since it serves to stabilize the electrospray current.

The formation of ions suitable for MS requires conditions affecting solvent evaporation from the initial droplet population produced by ESI. Droplets must shrink to the point where repulsive coulombic forces approach the level of droplet cohesive forces (e.g., surface tension). This evaporation can be accomplished at atmospheric pressure by a counter current flow of dry gas at moderate temperatures (~80°C), by heating during ion transport through the sampling region, and (particularly in the case of ion trapping MS methods) by energetic collisions at relatively low pressure. As noted above, the use of a sheathless micro-spray has several advantages that include more rapid droplet desolvation, due to both the smaller total volume of liquid and the smaller droplet size produced.

As applications for CE have developed the advantages of CE combined with mass spectrometry have been increasingly recognized, and the merits of various interfaces identified. When compared to FAB, the ESI methods appear to offer advantages in most cases due to better sensitivity, reduced background, and interface designs that do not require long transfer lines or incur a pressure drop across the capillary. Perhaps the most significant advantage of the ESI interface to CE is its applicability to higher molecular weight compounds, impractical by CF-FAB.

3 EXPERIMENTAL METHODS AND CONSIDERATIONS FOR CE-MS ANALYSES

3.1 SEPARATION CONSIDERATIONS

The success of CE–MS methods depends upon various factors that relate to the interfacing techniques; in particular, the obtainable sensitivity. In general, all the issues relevant in CE concerning sample injection, buffer composition, solute interaction with the capillary inner surface, and separation efficiency are applicable to CE-MS. Most CE-MS experiments have been performed using 50 to 100 μm i.d. capillaries; however, as we discuss later, improved detection sensitivities are obtained with smaller capillaries in CE-ESI/MS. Moreover, CE with more conventional detectors is conducted with these larger capillaries, and their selection represents a compromise among factors relating to detector sensitivity, injection method, capillary surface interactions, and reduced separation quality due to resistive heating in larger-diameter capillaries. A wide range of CE buffer systems can be used with either the liquid junction or the sheath flow interface for electrospray ionization.[26,34] Buffer concentrations of at least 0.1 M for more volatile buffers can be used for CE-MS. Due to detection sensitivity constraints, however, CE buffer concentrations are generally minimized; furthermore, sensitivity can vary significantly with buffer composition. In general, the best sensitivity is obtained using volatile buffer components at the lowest practical concentration, and by minimizing other nonvolatile, charge-carrying components. In addition, buffer components that interact strongly with the sample (e.g., denaturants) substantially degrade sensitivity, largely due to nonspecific association in the gas phase (i.e., cluster formation). The use of surfactants, e.g., sodium dodecyl sulfate (SDS), gives rise to intense background signals in both positive and negative ion ESI, and presents a major difficulty to micellar electrokinetic capillary chromatography-MS. Recent progress in this area has been achieved by eliminating the flow of such components to the ESI source.

Probably the most widely used buffers to date for CE-ESI/MS are acetic acid, ammonium acetate or bicarbonate, and formic acid systems — choices made primarily due to their volatility. Nonvolatile buffer systems may be used for CE-MS; however, different ESI interfaces may be required for their routine use. Figure 4 shows ESI mass spectra for comparison of a 0.01 M acetic acid solution (pH ~3.4) and a 0.01 M phosphate buffer (pH ~2.5) with identical sheath liquid composition. The major background ions due to the buffer systems are below m/z 150. As shown in Figure 5, however, inspection of the same mass spectra at a factor of 32.5 times greater scale magnification depicts that the constraints due to background ions for these solutions are different. The phosphate buffer produces cluster ions extending to at least m/z 1000 with sufficient abundance to substantially limit CE-MS applicability. Conversely, the acetic acid buffer produces a broad range of background ions presumably due to solution impurities, which illustrates the need to use high-purity buffer systems in CE-MS. The MS sensitivity, which is ultimately limited by the ability of the mass spectrometer to efficiently analyze ions produced at atmospheric pressure by ESI, is probably the most important factor related to MS detection and is discussed at length later. Sensitivity does not appear to depend significantly on the mechanical details of a well-designed and optimized electrospray source, although its design details can influence ESI stability and ease of operation.

Increasingly, CE-MS instrumentation benefits from automation of electrokinetic or hydrostatic injection and capabilities for various on-capillary spectrophotometric detectors, as well as capillary temperature control. The advantages of automated injection are significant for both accuracy and precision, as well as freedom from artifacts arising from less reliable manual injection methods.[61] Temperature control has also been recognized as important in obtaining good reproducibility. In the case of high CE currents, cooling of the capillary can reduce difficulties with the interface. Gas generated in either the CE capillary or the ESI sheath capillary, which often contains more volatile organic solvent, can create a region of high resistance and effectively terminate a separation. On the other hand, the ability to use higher temperatures may be essential in some cases to obtaining good separations of proteins, where higher-order structural heterogeneity can be strongly temperature dependent. In addition, higher temperatures tend to denature proteins and increase the extent of charging during ESI.[59]

3.2 Mass Spectrometric Considerations

Because the ESI interface is generally operated at atmospheric pressure, the pressure drop across the CE capillary is normally zero, and no degradation

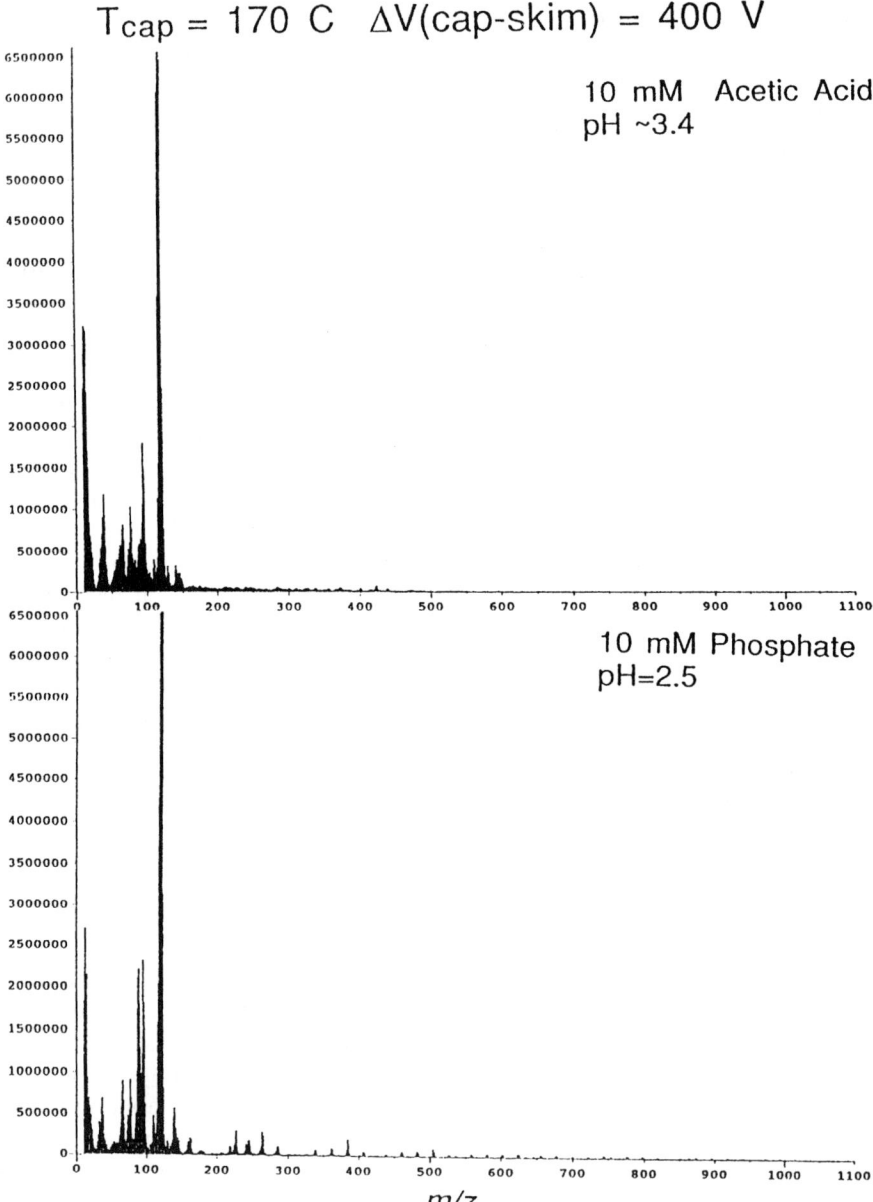

FIGURE 4
Comparison of background mass spectra from ESI of 0.01 M acetic acid and phosphate buffer solutions obtained by direct infusion, and with a sheath liquid of the same composition.

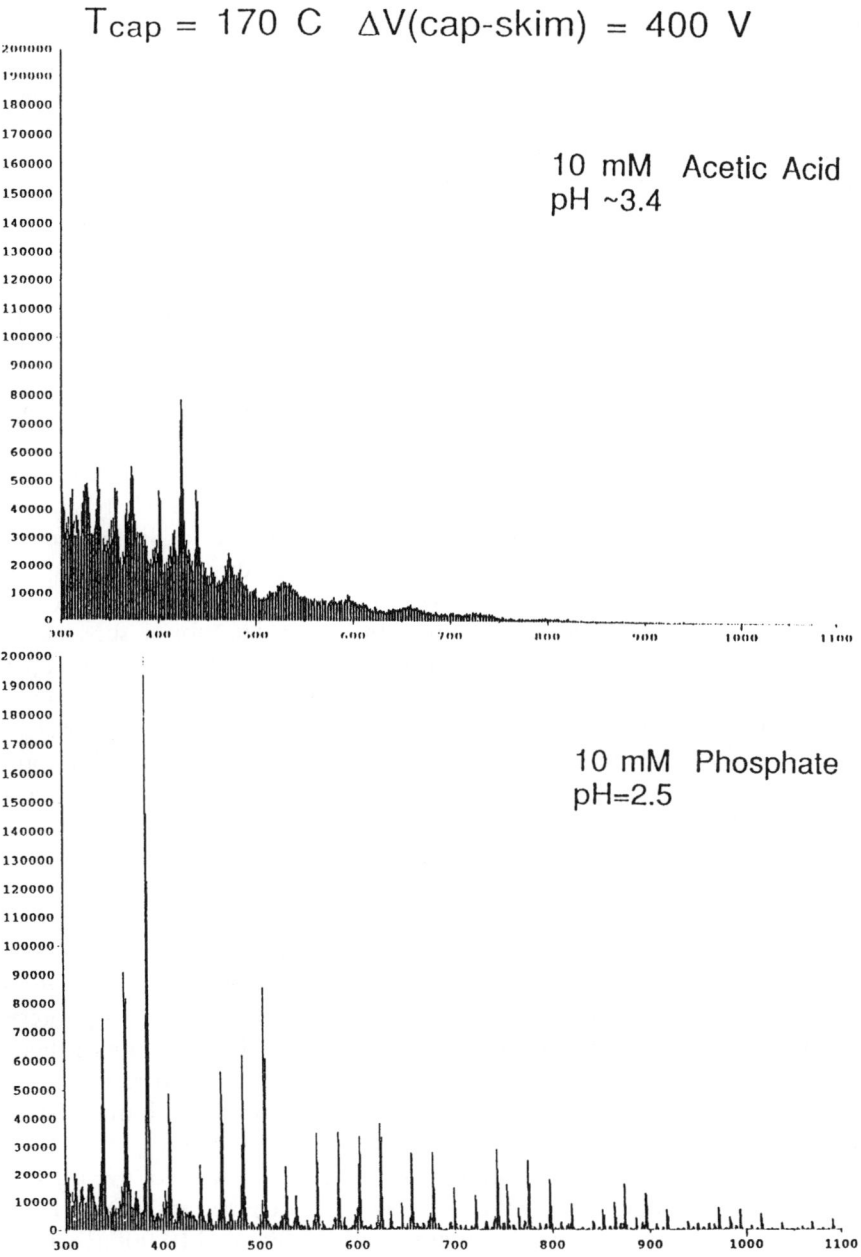

FIGURE 5
Comparison of mass spectra at high gain (× 32.5 the scale in Figure 4) for acetic acid and phosphate buffers.

of separation efficiency should arise due to laminar flow. The ESI method is expected to be amenable to essentially any charged species in solution, but this in itself does not guarantee success of the CE-MS method. Gas-phase ions must be substantially desolvated, relatively free of adducted buffer-related species, and must have an m/z within the capabilities of the MS. Moseley et al. have reported an evaluation of sheath and buffer constituents upon ESI/MS,[44] but note that variations in ESI response for different solvents could be dependent upon the details of the interface with regard to ESI droplet-evaporation and ion desolvation.

The major considerations relevant to MS detection are generally due to the nature and complexity of the particular sample and pragmatic constraints due to MS detection sensitivity, resolution, and related scan speed compromises. For quadrupole mass spectrometers, single or multiple ion monitoring, sometimes referred to as selected ion monitoring (SIM), leads to enhanced detection limits compared to scanning operation, which is due to the greater time spent (dwell time) at specific m/z values. For samples where analyte molecular weights are known and m/z values can be predicted, SIM detection is an obvious choice. If sufficient sample is available, a direct infusion experiment can be used to produce a mass spectrum of the unseparated mixture and the results used to select an ensemble of m/z values for a subsequent separation using SIM detection. This approach can be problematic, however, because relative ionization efficiencies can be different for direct infusion of the mixture compared to the separated species. In addition, noncovalent associations between mixture components may further complicate interpretation. Thus, important components may be overlooked during the direct introduction of mixtures. One of the advantages of CE-MS is that mixture components strongly discriminated against for such direct infusion experiments, such as polypeptides from enzymatic digests, often show much more uniform response after CE separation.

The realization of high efficiency CE-MS separations, leading to peak widths of only a few seconds or less, presents challenges for most mass spectrometers. The on-line combination of capillary zone electrophoresis with mass spectrometry was first demonstrated in this laboratory for synthetic mixtures of quaternary ammonium salts.[30,31] These initial separations used 100 μm i.d. by 100 cm uncoated fused silica capillaries and a buffer system of 10^{-4} M potassium chloride dissolved in water-methanol 1:1 (v/v) with a field gradient of 200 V/cm. As shown in Figure 6, a mixture of five quaternary ammonium cations (m/z from 74 to 242) were detected at subfemtomole levels, with separation efficiencies ranging from 35,000 to 140,000 theoretical plates. The use of scanning a broad m/z range with the quadrupole mass spectrometer (full scan detection) often results in either limited signal intensities due to the short dwell time at each m/z or too few scans obtained during peak elution. One option is to reduce MS resolution, which increases signal intensities and allows faster scan speeds. This is not an ideal solution because mass measurement accuracy is reduced and chemical noise due to

CAPILLARY ELECTROPHORESIS – MASS SPECTROMETRY

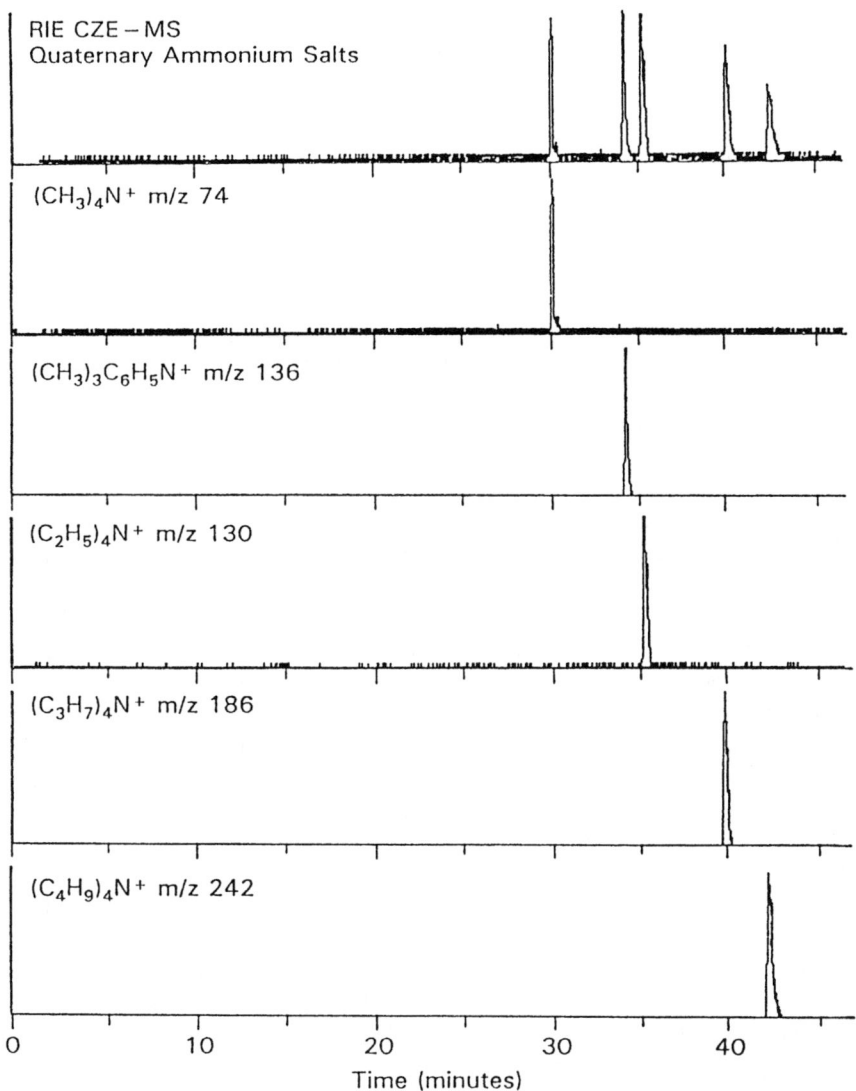

FIGURE 6
Total and reconstructed single ion electropherograms for CE-MS of a mixture of five quaternary ammonium salts. The separation was conducted on a 2 m × 100 μm i.d. bore fused-silica capillary in a 0.01 M phosphate buffer at pH 4 with an electric field of 200 V/cm.

background signals at adjacent m/z values may mitigate any gain. For these reasons, obtaining the maximum number of theoretical plates possible with combined CE-MS has not been a major concern. Thus CE conditions involving longer separations and larger injection volumes than optimum for separation quality have often been used. Similar, but magnified, constraints arise for the use of tandem-MS methods for obtaining structural information, as in polypeptide sequencing, due to the low signal intensities with conventional MS/MS instruments. These limitations are a major driving force for the implementation of improved MS instrumentation using array detection or ion trapping methods.

4 CE-MS APPLICATIONS

Many of the early CE-MS reports concerned investigation and demonstration of the methodology and comparison of these studies to other techniques, particularly microcolumn LC-MS. Because CE-MS is a less familiar technique to the analytical chemist, the initial reports generally are those arising from laboratories often more interested in new methodologies and instrumental approaches as opposed to routine applications or specific analytical problems. Gradually, however, subsequent studies have examined more realistic problems, and such studies should increase as CE becomes more common.

Complex mixtures of peptides generated from tryptic digestion of large proteins present a difficult analytical challenge because the fragments cover a large range of both pI and hydrophobicity. Because trypsin specifically cleaves peptide bonds C-terminally at lysine and arginine, the resulting peptides generally form doubly and singly charged molecular ions in positive ESI. Such doubly charged tryptic peptides generally fall within the m/z range of modern quadrupole mass spectrometers. The resolving power of these methods is illustrated in Figure 7, which shows a UV electropherogram (top) obtained in conjunction with a CE-MS analysis[55] for a separation of a tryptic digest of tuna cytochrome *c* using a 0.05 M acetate buffer system at pH 6.1 that is mixed with an equal volume of acetonitrile. Shown are single ion electropherograms for two charge states of two tryptic fragments. In this work, a commercial CE instrument (Beckman P/ACE 2000) was modified to allow UV detection of the separation at one-half of the MS detection time. The individual peaks indicate a separation efficiency corresponding to ~400,000 theoretical plates. It was noted that the apparent efficiency for MS detection is significantly greater than for UV detection, which is ~120,000 theoretical plates for the same peak, a fact largely attributed to the longer separation time and the very small detector cell-broadening contribution of the ESI interface compared to conventional UV detection.[55]

The total ion electropherograms obtained from tryptic digests of bovine, *Candida krusei*, and horse cytochrome *c* are shown in Figure 8. For each separation, a 0.01 M ammonium acetate/acetic acid buffer system, pH 4.4, was used. The separation capillary used was 10 μm i.d., which was chemically modified with 3-aminopropyltrimethoxysilane and was 50 cm in length. The mass spectrometer scanned from m/z 600 to 1200 in 2 m/z steps at 0.6 s/scan.

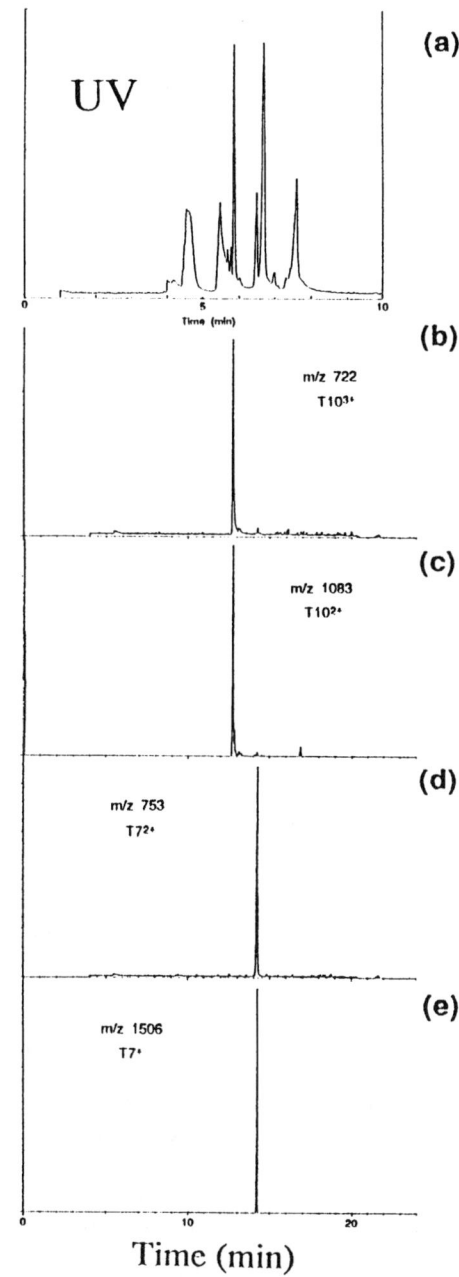

FIGURE 7
Comparison of the UV (214 nm) detection (top) and MS reconstructed ion electropherograms (bottom), with time axes adjusted for the corresponding peaks, for a CE-MS separation of a tryptic digest of tuna cytochrome c.

In each case, the separation was complete within 6 min. From Figure 8 it can be concluded that greater separation resolution could be obtained by using a longer capillary. Mass spectrometrically, however, the individual tryptic fragments may be isolated. For example, shown in Figure 9 are the extracted

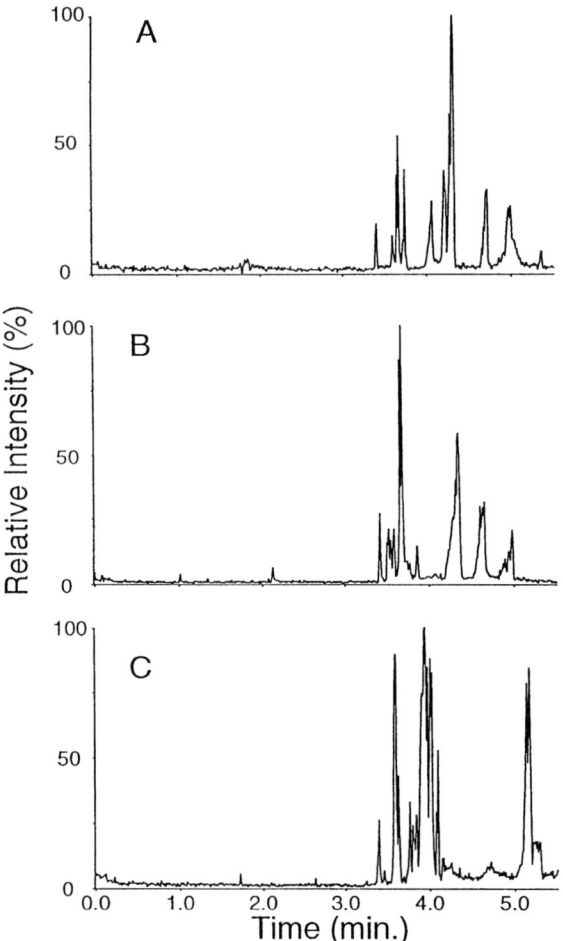

FIGURE 8
Total ion electropherograms obtained from tryptic digests of bovine (A), *Candida krusei* (B), and horse cytochrome *c* (C). Separation conditions: 0.01 M ammonium acetate/acetic acid buffer system, pH = 4.4; capillary: 10 μm i.d., 50 cm in length, and chemically modified with 3-aminopropyltrimethoxysilane; injection: 5 s at −1 kV; concentration of original proteins: ~750 μg/ml; mass spectrometer: scanning from *m/z* 600 to 1200, 2 *m/z* steps, 0.6 s/scan.

electropherograms for the individual tryptic fragments YIPGTK (*m/z* 678), which is a common tryptic fragment for all three proteins, EDLIAYLK (*m/z* 964), which is common to bovine and horse cytochrome *c*, and MAFGGLK (*m/z* 723), which is specific to *Candida krusei* cytochrome *c*. In addition, the three extracted electropherograms show additional solute zones of the same *m/z* as the tryptic fragments. These additional species may arise due to incomplete digestion, adduction of ion species, noncovalent complexes, or fragmentation occurring within the interface of the mass spectrometer.

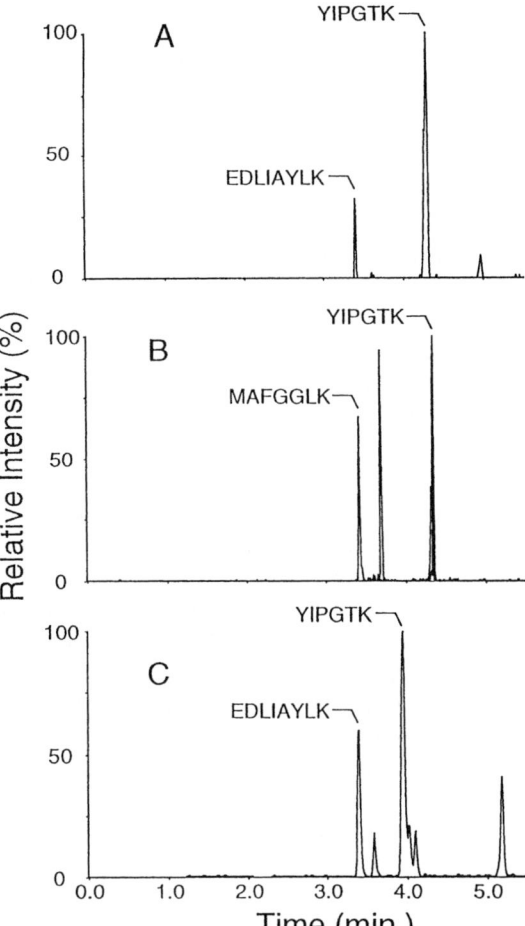

FIGURE 9
Extracted ion electropherograms obtained from Figure 8 for the individual tryptic fragments YIPGTK (*m/z* 678) which is a common tryptic fragment for all three proteins, EDLIAYLK (*m/z* 964) which is common to bovine (A) and horse (C) cytochrome c, and MAFGGLK (*m/z* 723) which is specific to *Candida krusei* (B) cytochrome c. The additional solute zones of the same *m/z* as the tryptic fragments are probably due to incomplete digestion and fragmentation occurring within the mass spectrometer.

Lee et al. showed the detection and identification of 11 components of a tryptic digest of recombinant bovine somatotropin using SIM detection.[48] Their separations were conducted in a 100 µm i.d. capillary using a 50/50 acetonitrile/0.005 M ammonium acetate buffer, and obtained efficiencies as high as 300,000 theoretical plates. In separate experiments, CE-MS/MS was demonstrated for one of the components, providing information on the amino acid sequence. Subsequently Johansson et al. reported the analysis of a hemoglobin tryptic digest.[49,50] They used a 100 µm i.d. capillary, a field strength of 260 V/cm and a 50/50 acetonitrile/0.015 M ammonium acetate buffer adjusted to pH 5.0 with acetic acid. Approximately 10 pmol were injected electrokinetically during 25 s at 10 kV injection using a 154 pmol/µL digest. The digest was expected to have 28 components, and a complex total ion electropherogram with several coeluting solute zones was observed from a scanning experiment covering a 600 *m/z* range. The authors reported that all of the expected enzymatic products were detected. In separate experiments a 10 pmol injection of the digest was used to obtain a MS/MS spectrum of the T-3 fragment. The product daughter spectrum showed fragment ions that provided information on the peptide. In another experiment these workers used a 75 μm i.d. capillary with a 0.015 M ammonium acetate buffer adjusted to pH 2.5 by the addition of trifluoroacetic acid and 15% methanol. They monitored 8 individual ions (SIM) characteristic of the expected fragments identified from their previous experiments and demonstrated efficiencies from 85,000 to 135,000 theoretical plates for an 8 pmol injection.

Thibault et al. examined a separation of the tryptic fragments of a glucagon digest.[41] The separation was done in a 50 μm i.d. capillary at 390 V/cm, using either a 0.01 M, pH 3.4, or 0.0025 M acetic acid buffer, pH 2.1. They used a commercial (Microcoat™) polymeric coating reagent; 2 pmol of digested glucagon were injected. The four expected fragments, which produced both doubly and singly charged ions, were observed when MS was scanning a 950 m/z-wide range. In addition, evidence for the partial oxidation of the sample was obtained along with a small amount of CID fragmentation that occurred in the ESI/MS interface. Higher efficiency, ~34,000 theoretical plates, but longer analysis times were obtained with the lower pH buffer.

Experience to date suggests that nearly all tryptic fragments can be effectively detected by ESI/MS methods if first separated; however, infusion of the unseparated digest often results in dramatic discrimination against some components. Fragments observed with the UV detector are almost always detected by ESI/MS. However, detection has sometimes been problematic for very small fragments, such as individual amino acids and dipeptides, due to difficulties in obtaining an optimum ESI/MS condition over a relatively wide m/z range. Excessive internal excitation or solvent-related background peaks at low m/z appear to be the origin of some of these difficulties. There is also evidence that some interface designs discriminate strongly against low m/z ions. Available evidence indicates, however, that if tryptic fragments are not detected by CE-MS, they generally are not present prior to the separation — a problem particularly evident with nanoscale sample handling. As observed in the past, full realization of more sensitive analytical methods will depend upon concurrent improvements in the earlier stages of sample handling and preparation.

An important attribute of the ESI/MS interface is the capability to obtain structurally related information on large molecules by dissociation in the interface by a collisional heating process, which is induced by a simple electric field gradient. This capability was first demonstrated for large molecules by our laboratory in 1988, and has the advantage compared to CID MS/MS methods of much greater selectivity, but the disadvantage of greater ambiguity because only one MS step is utilized and a good separation is generally required.[62] The complexity of large-molecule collisional dissociation spectra prevents one from obtaining complete sequence information with current MS quadrupole technology, but its potential for fingerprinting has been established.[27,63] As shown in Figure 10, extensive sequence-related fragmentation

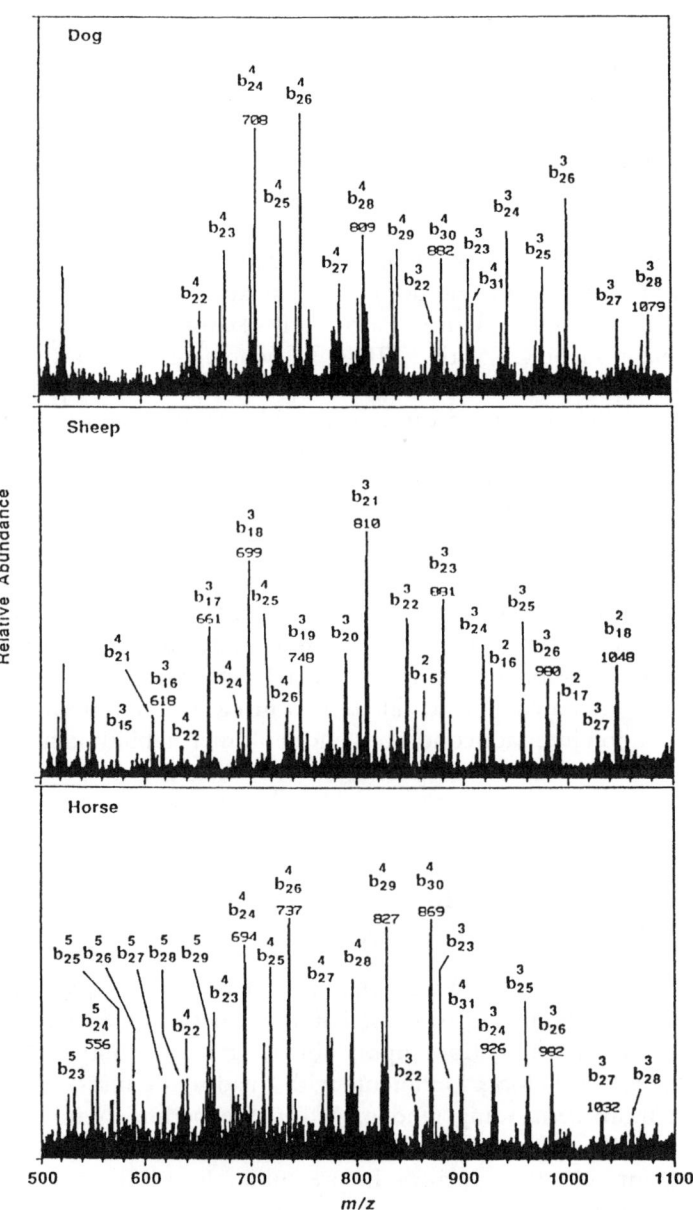

FIGURE 10
Electrospray ionization mass spectra of three serum albumins ($M_r \sim 66,000$) obtained with a nozzle-skimmer voltage (ΔNS) of 335 V using a modified Sciex TAGA 6000E triple quadrupole mass spectrometer. The high ΔNS spectra provide structurally related information with much greater sensitivity than is possible with triple quadrupole (MS/MS) methods, that is useful for fingerprinting. The labels (b_x^y) refer to ions arising due to products incorporating the amino terminus at the x peptide bond and having a charge state of y+.

can be obtained for molecules as large as serum albumins (~66 kDa), yielding spectra suitable for fingerprinting purposes. We have also shown that direct infusion experiments using low-picomole sample quantities provide spectra for which a limited amount of sequence information can be obtained for materials of substantially known sequences. This suggests the potential utility of direct CE-MS analysis of proteins for evaluation of suspected primary structural or post-translational modifications.

Recent improvements in MS instrumentation, particularly in quadrupole or Fourier transform ion-trap devices, are making structural studies of large molecules increasingly practical. Particularly exciting have been results from the combination of CE with FTICR,[72-75] where recent work has demonstrated the capability for direct studies of proteins from small populations of cells.[76] The CE-MS method can also provide information on impurities or possibly multimeric species, specifically noncovalently associated dimers, or larger complexes stable in solution.[27] For example, the CE-MS analysis of a sample of B-chain insulin in 0.01 M TRIS at pH 8.3, electric field strength of 150 V/cm, using 0.7 pmol injection of a sample, showed a second, smaller solute zone migrating using both MS and UV detection. This second solute zone was attributed to either an impurity of very similar M_r or a dimer and perhaps larger multimers that would dissociate to multiply charged monomers due to collisional heating in the ESI interface.[27] Such species can generally be separated due to differences in electrophoretic mobility and, depending upon interface conditions and multimer stability, can be detected as either the monomer or intact multimer.

5 LIMITATIONS OF PRESENT CE-MS METHODS AND INSTRUMENTATION

Most early ESI/MS instruments were based upon quadrupole mass spectrometers although instrumental improvements have rapidly changed this situation. The limitations upon obtaining high-quality full-scan mass spectra for CE-ESI/MS experiments with quadrupole mass spectrometers, and even more demanding tandem MS/MS experiments for studies that involve peptide sequencing, for example, require further discussion because the constraints arise directly from the present levels of ESI/MS performance. Consideration and understanding of the nature of these constraints is important for understanding the trade-offs between scan speed, resolution, m/z range, and sensitivity as well as for the development of improved methods and instrumentation that may lead to their resolution or circumvention.

Perhaps the greatest limitation of present ESI/MS instrumentation arises due to the small fraction of the total ESI ion current actually introduced and analyzed by the mass spectrometer. At present, a typical ESI/MS quadrupole instrument provides an ion current of approximately 10^{-11} A to the detector, which corresponds to ~6×10^7 singly charged ions per second. Although not always the case, the ionic strength of the buffer system generally exceeds the analyte by about 10^2 to realize the high efficiency possible with CE. Consequently, if the ESI process does not discriminate efficiently between buffer and analyte ions, at most only about 6×10^5 analyte ions per second will be detected. In practice, ESI interfaces generally discriminate in favor of higher molecular weight species, and somewhat higher sample ion currents can be obtained using volatile buffer systems such as acetic acid. Examination of full-scan spectra, for example from m/z 10 to 2000 and under conditions that minimize discrimination, generally show that even the most intense peaks correspond to ⩽10% of the total ion current, consistent with such expectations. The best resolved and fastest migrating solute zones for CE capillaries of conventional dimensions are 1 to 3 s full width at half height (FWHH). If 10 scans are required during solute migration, and a m/z range of 100 to 2000 is scanned at 0.2 m/z steps, then the most intense peak that can be detected would correspond to only ~20 ions. Even with pulse-counting detection, more than 3 to 5 ions are required to provide minimal confidence of detection, which suggests a dynamic range of ~5. Clearly, this quality of MS performance is impractical with present quadrupole instruments. For scans at a reduced width of 1000 m/z units and 1 m/z steps, ~200 ions can result for major solutes, giving a more reasonable dynamic range of ~50. Slower scans are also useful in practice. For a 1.4 s scan over a 500 m/z unit window, with a 1 m/z step size, a dynamic range of ~400 may be estimated, which is qualitatively consistent with the present results. Clearly, while results obtained at slower scan speeds, lower resolution, and a smaller m/z scan range are far less than ideal, simultaneous optimization of both CE and MS components is generally impractical given existing instrument performance. The situation is more severe for tandem MS/MS using triple quadrupole instrumentation, due to the much lower signal intensities after collisional dissociation, which are generally at least one to two orders of magnitude smaller.

There are several possible approaches to resolving these limitations. One is to increase the efficiency of ion transport from the ESI source into the mass spectrometer, which is now generally only ~0.01% or less in overall transport efficiency.[26] A second approach is to analyze all the ions that enter the mass spectrometer, using either ion trapping methods or array detection. Indeed, a number or workers are beginning to examine the quadrupole ion trap mass spectrometer (ITMS) for this purpose.[64,65] Because the ITMS can trap and accumulate ions over a wide m/z range, and then rapidly conduct a swept ejection/detection step, the fraction of total ions detected compared to that entering the ITMS is much greater, allowing the utilization of a much larger fraction of the total ESI current, assuming efficient ion injection and trapping. Much faster scan speeds over extended m/z ranges are thus possible. The best MS performance, in principle, would be obtained with full-range array detection, but practical instrumentation for this purpose is not yet available and would likely be much more expensive than either the ITMS or existing instrumentation. It should be noted that projections regarding ITMS performance must be made with caution, because the trapping efficiencies upon injection from external ESI sources are not yet well established, and may be on the order of only 1%. In addition, the m/z measurement accuracy currently obtainable has been reported to be significantly lower than for quadrupole instruments, a particular concern for multiply charged species and sequencing applications. However, preliminary results indicate the sensitivity obtainable with ESI is indeed better than for quadrupole instrumentation, although concerns regarding mass measurement accuracy have not been completely addressed, and capabilities for both high resolution and high m/z detection have been demonstrated.[64,65]

Other developments are promising. McLafferty and co-workers have pioneered the combination of ESI with Fourier transform-ion cyclotron resonance mass spectrometry (FTICR) and have shown that very high resolution is obtainable.[66] Laude and Hofstadler have demonstrated that substantial gains in sensitivity are obtainable by having the ESI interface in the high magnetic field of an FTICR using a superconducting magnet.[67] Based on ESI-FTICR instrumentation developed at our laboratory,[68] we have now extensively demonstrated the significant promise of CE interfacing.[72-76] As is shown in Figure 11, recent work has demonstrated the extensive potential of FTICR. This figure shows preliminary results which demonstrate the feasibility of using the CE-FTICR combination as a high performance detection scheme for the analysis of cellular proteins acquired directly from small populations of intact living cells. The human erythrocyte (red blood cell) was chosen as a model system owing to its availability, relatively homogeneous composition, and thorough documentation of contents by previous researchers. The contents of the erythrocyte are unusually homogeneous; nearly the entire volume of the cell is filled with hemoglobin, approximately 450 amol per cell, a challenging but attainable level for mass spectrometric detection with current instrumentation. Online acquisition of high resolution mass spectra (average resolution ≥45,000 FWHM) of the α and β hemoglobin chains were acquired from the injection of as few as 10 human erythrocytes[76] (this corresponds to ≈4.5 fmol of hemoglobin). Shown in Figure 11a is a total ion current trace from the injection of 20 erythrocytes derived from the detection of the electrospray ion current signal from the ion beam impinging on the front shutter of the FTICR instrument.[72] Approximately one minute after the buffer peak elutes, a broad peak due to the major cell contents is observed from which mass spectra of the α and/or β chains of hemoglobin are acquired. The noncovalent tetrameric hemoglobin complex consisting of two α chains and two β chains ($\alpha_2\beta_2$) which are dissociated due to the low pH of the 10 mM HOAc CE buffer. In this initial work the eluting peaks were broad and poorly resolved due to variations in cell lysing and injection times. The mass spectrum of the α chain shown in Figure 11b was acquired from the injection of only 10 erythrocytes, and demonstrates a nearly complete separation of the α and β chains. The MS/MS

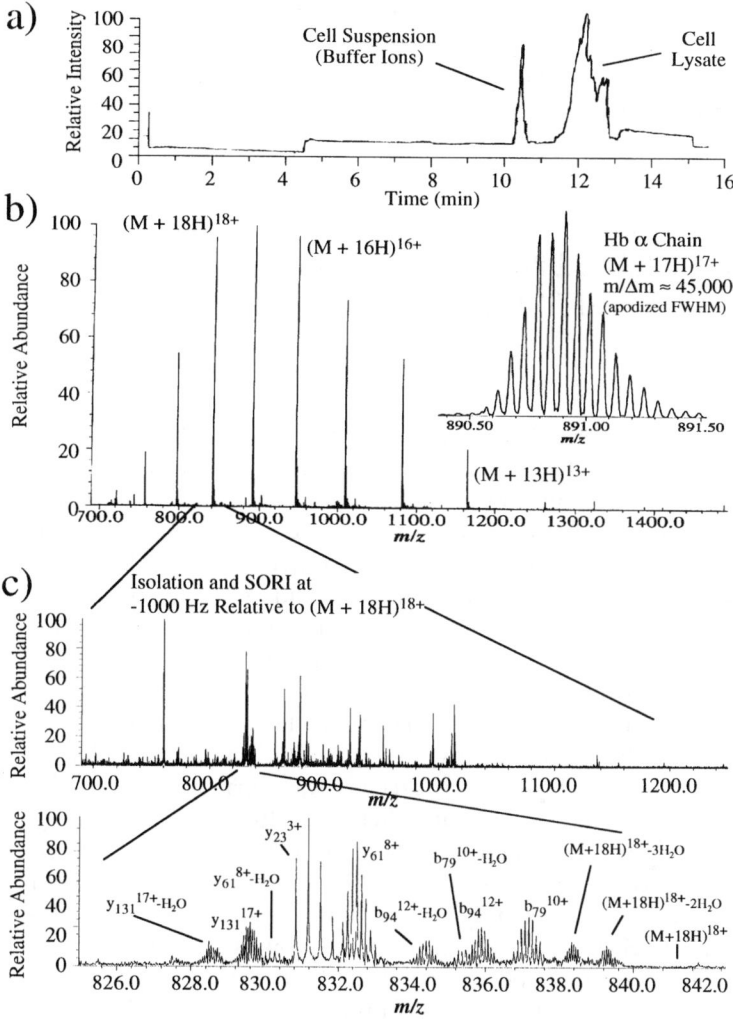

FIGURE 11
CD-FTICR results for the analysis of hemoglobin from intact red blood cells. (a) Total ion electropherogram; (b) high resolution FTICR mass spectrum for α chain hemoglobin and (c) collisional dissociation product spectrum for α chain hemoglobin from (b).

spectrum in (c) was acquired for dissociation of the 18+ charge state of the α chain from ~75 erythrocytes. Extensive dissociation producing predominantly y and b fragments is observed, yielding substantial sequence information. Future efforts are anticipated to result in the extension of these methods to the single cell level, and it appears that there is still the potential for significant improvements in sensitivity feasible with the incorporation of more efficient ion transmission, trapping, ion accumulation and ion remeasurement methods.

The potential for higher-order mass spectrometry, MS^n ($n \geq 3$), using ICR or ITMS instruments also appears promising. In the next section, we discuss the issue of MS scan speed and sensitivity in more detail, and described two recently developed approaches to improve these constraints in CE-MS using current technology.

6 PRACTICAL APPROACHES TO IMPROVED CE-MS SENSITIVITY

6.1 REDUCED ELUTION SPEED DETECTION

The small quantities of solute used in CE require highly sensitive detection methods. The low signal intensities generally produced by ESI/MS, typically resulting in maximum analyte ion detection rates of no greater than about 10^6 to 10^7 counts per second, effectively limit the maximum practical scan speeds with quadrupole mass spectrometers. Thus, depending on the desired m/z range, solute concentration, and other factors related to the nature of the solute and buffer species, maximum m/z scan speeds are often insufficient to exploit the high-quality separations feasible with CE when coupled to quadrupole or other scanning mass spectrometers.

For fundamental reasons the maximum electrospray ion current is a weak function of solution conductivity. When the amount of charge-carrying solute entering the ESI source exceeds the capability of the electrospray process for complete ionization, then the efficiency of solute ionization decreases. Figure 12 qualitatively illustrates the nature of ESI ion formation as a function of delivery rate to the ESI source. Two general regimes exist. At higher analyte concentrations or flow rates, ESI/MS signal strengths become relatively insensitive to flow rate. At low flow rates or concentrations, the ESI signal strength becomes limited by the number of charge-carrying species in solution. In this regime optimum sensitivity will be obtained.

In fact, most ESI work, and nearly all CE-MS, has been conducted in this regime where the efficiency of solute ionization is substantially limited by factors that include competition for available electrospray current between analyte and buffer components. CE separations generally incur higher currents (5 to 50 µA) than typical ESI currents (0.1 to 0.3 µA) and therefore deliver charged species to the ESI source at a rate where ionization is necessarily inefficient. Any contaminant present may also compete with the analyte for available charge, thus decreasing the solute ionization efficiency. Ideal buffer components have characteristics that allow CE separation of the analytes, are volatile, are discriminated against during ESI, and form minimal gas phase contributions to the mass spectra arising from ionized clusters of buffer constituents.

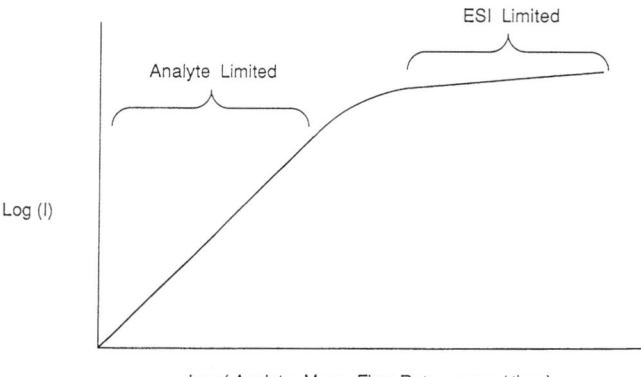

FIGURE 12
Illustration of the two regimes for ionization efficiency in ESI. Analyte signal intensities approximately directly relate to analyte concentration at low CE flow rates where the delivery of charged species to the ESI source is less than the ESI current. At higher delivery rates, signal intensities are nearly independent of CE flow rates and are related to analyte concentration.

An approach to obtaining enhanced sensitivity for CE-MS has recently been developed at our laboratory based upon reduced elution speed (RES) detection, which aids in alleviating both the sensitivity and scan speed limitations of CE-MS with scanning mass spectrometers.[69] Involving only the ability to cause step changes in the CE electric field strength, the technique is simple and readily implemented. Figure 13 illustrates the basis of RES detection. A hypothetical CE-MS separation where the electric field strength is held constant throughout the analysis is shown in Figure 13A. Alternatively, prior to elution of the first solute to the ESI source, the electrophoretic voltage is decreased in the RES CE-MS experiment, yielding the result shown in Figure 13B for the same two solutes. Elution of solutes is slowed, allowing more scans to be recorded without a significant loss in ion intensity. Under conditions where the amount of solute entering the ESI source per unit of time exceeds the current available for complete ionization of the solute, no substantial decrease in maximum ion intensity is expected when the electric field strength is decreased. As a result, a greater fraction of the analyte ions will be transferred to the gas phase during an RES CE-MS experiment than during normal constant electric field strength CE-MS experiment.

Some of the greatest challenges for CE-MS involve analysis of complex mixtures of biopolymers. An important goal is to decrease the quantity of protein required for sequencing using methods based upon an initial enzymatic digestion of a protein. Comparison of constant electric field strength and RES CE-MS for a 40 fmol injection of peptides produced by digestion of bovine serum albumin with trypsin (Figures 14A and B) shows only a small decrease (20%) in ion intensity when the electrophoretic voltage was decreased to slow elution. Single ion electropherograms extracted for the

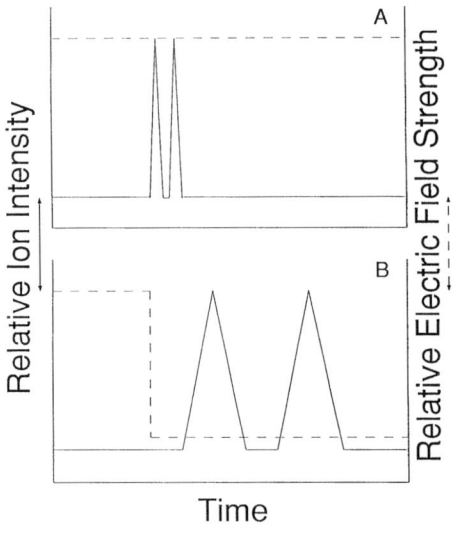

FIGURE 13
Hypothetical CE-ESI/MS experiment showing the separation of a two-component mixture by (A) constant field strength CE-ESI/MS and (B) reduced elution speed CE-ESI/MS. Relative ion intensity is denoted by (——) and relative electric field strength by (– – –).

peptide AWSVAR, produced by cleavage of bovine serum albumin with trypsin at amide bonds 32 and 37, are shown in Figure 15A for constant electric field strength CE-MS, and in Figure 15B for RES CE-MS. A total ion electropherogram (TIE) for the bovine serum albumin tryptic peptide AWSVAR, analyzed by RES CE-MS (Figure 15B), showed only a minor increase in peak width at half height when compared to the constant electric field strength CE-MS analysis (Figure 15A).

Figure 16A shows a comparison of the mass spectra obtained during elution of this peptide. Five scans from the TIE shown in Figure 14B, for the corresponding longer RES elution period, were averaged to produce the mass spectrum shown in Figure 16B. The mass spectrum recorded during one scan under constant electric field strength CE-MS for peptide AWSVAR (Figure13A) shows the presence of ~15 additional significant peaks, due to a poorer overlap of the peak elution time with the MS scan period. A reduction in the complexity of individual scans can facilitate data interpretation of complex mixtures due to the greater effective scan speed and the reduced likelihood of other components eluting during the same scan.

Reduced elution speed CE-MS provides an increase in the efficiency of mass spectrometric scanning compared to conventional CE-MS methods. The prolonged residence time of solute in the electrospray can be exploited in several ways including: (1) increasing the m/z range scanned, (2) increasing the number of scans recorded during migration of a given solute, and (3) enhancement of sensitivity, the integrated ion intensity, for a given solute. The method does not increase solute consumption, provides improved sensitivities for peptide and protein analyses extending into the low femtomole

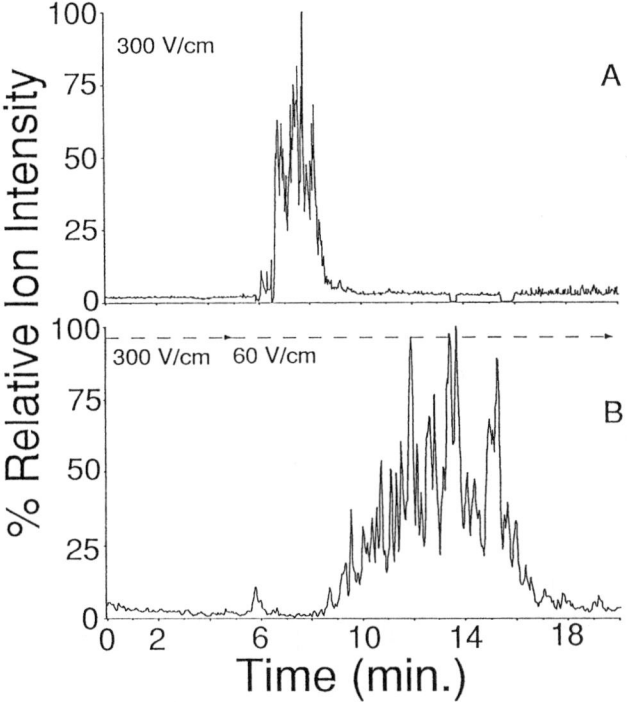

FIGURE 14
Comparison of constant field strength (A) and reduced elution speed (B) CE-ESI/MS analysis of bovine serum albumin after digestion by trypsin. The constant field strength CE-ESI/MS analysis was conducted at 300 V/cm. The reduced elution speed CE-ESI/MS analysis was conducted at 300 V/cm until 1 min prior to elution of the first peptide, when the electric field strength was reduced to 60 V/cm.

regime, and incurs very little loss in ion intensity, factors particularly important for MS/MS methods and their potential application to peptide sequencing.

6.2 Effect of Capillary Inner Diameter upon CE-MS Sensitivity

As indicated by Figure 12, optimum ESI/MS sensitivity is generally obtained at the low CE currents, where the rate of delivery of charged species to the ESI source is minimized. Consequently, an optimum capillary diameter meets several criteria: it should be (1) available commercially or readily prepared, (2) be amenable to alternative detection methods, and (3) allow optimum detector sensitivity. For the last criterion, we would expect that optimum sensitivity would be obtained at CE currents approximately equal to or less than the ESI current. Figure 17 shows the ESI and CE current for

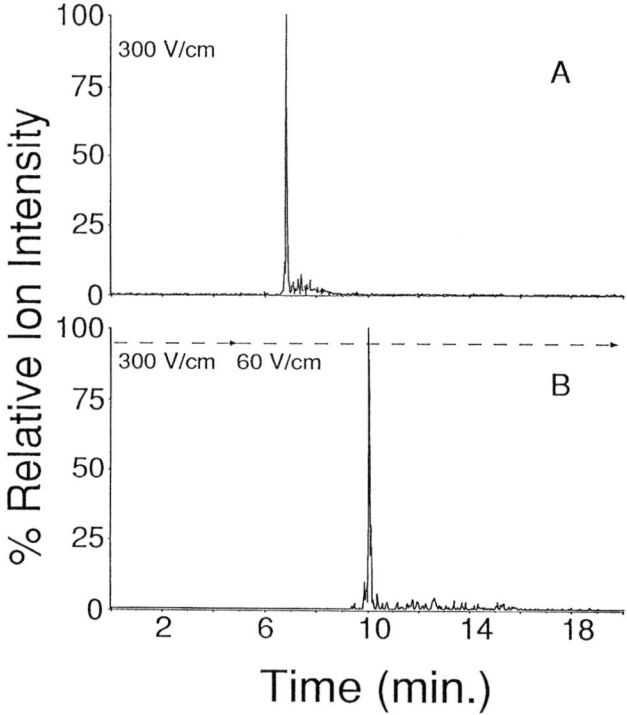

FIGURE 15
Extracted single-ion electropherograms for a single peptide (residues 32 to 37) from constant field strength (A) and reduced elution speed (B) CE-ESI/MS analysis of bovine serum albumin after digestion by trypsin (see Figure 14).

a 0.01 M acetic acid buffer, pH 3.4, with a 75/24/1 v/v/v methanol/water/acetic acid sheath at a flow rate of 2 µL/min. It can be seen that, even for this relatively low conductivity buffer system, CE capillary diameters greater than 40 µm will have currents that exceed that of the ESI source.

Wahl et al. have examined ESI/MS detection sensitivity as a function of capillary diameter using standard analyte mixtures.[70,71] Figure 18 shows the total ion current profiles for separation of a three-component mixture in capillaries of 100, 50, 25, and 10 µm diameter. Because the same electrokinetic injection technique was used for each capillary, the injection volume is approximately proportional to the square of capillary diameter. As shown in Figure 18, only relatively small decreases in the ESI/MS ion currents were observed for the smaller capillary diameters. For the 10 µm i.d. capillary, where a factor of ~100 less sample was injected, analyte signal intensities are observed to decline by only factors of 2 to 4, compared to the 100 µm diameter capillary.

Figure 19 shows the actual ESI/MS response observed for one analyte, leucine enkephalin, as a function of injection size for the four capillary

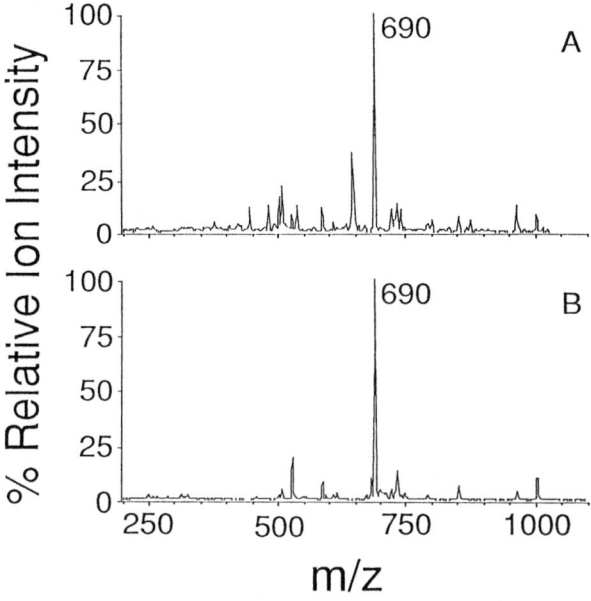

FIGURE 16
Mass spectra of a peptide (residues 32 to 37) produced during digestion of bovine serum albumin with trypsin and analyzed by constant electric field strength (A) and reduced elution speed (B) conventional CE-ESI/MS (see Figure 14).

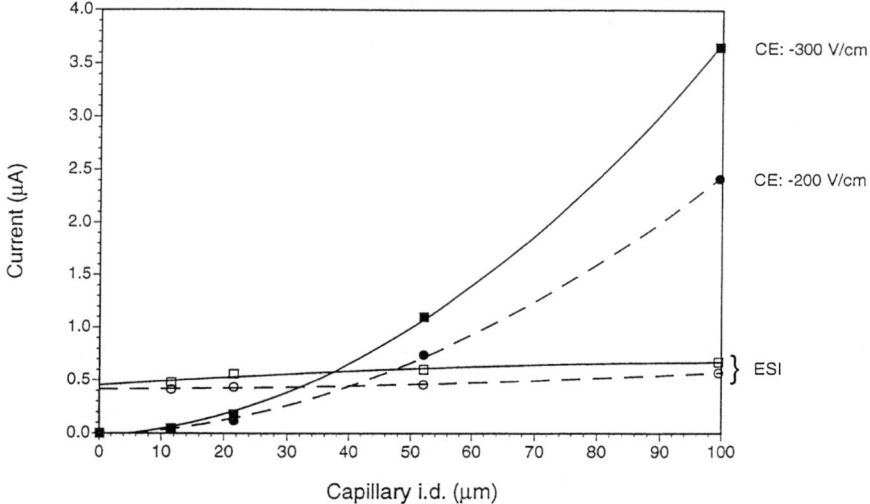

FIGURE 17
Comparison of the CE and ESI currents for a 0.01 M acetic acid buffer as a function of capillary diameter for two different CE field strengths. The CE current has the expected quadratic dependence upon capillary inner diameter.

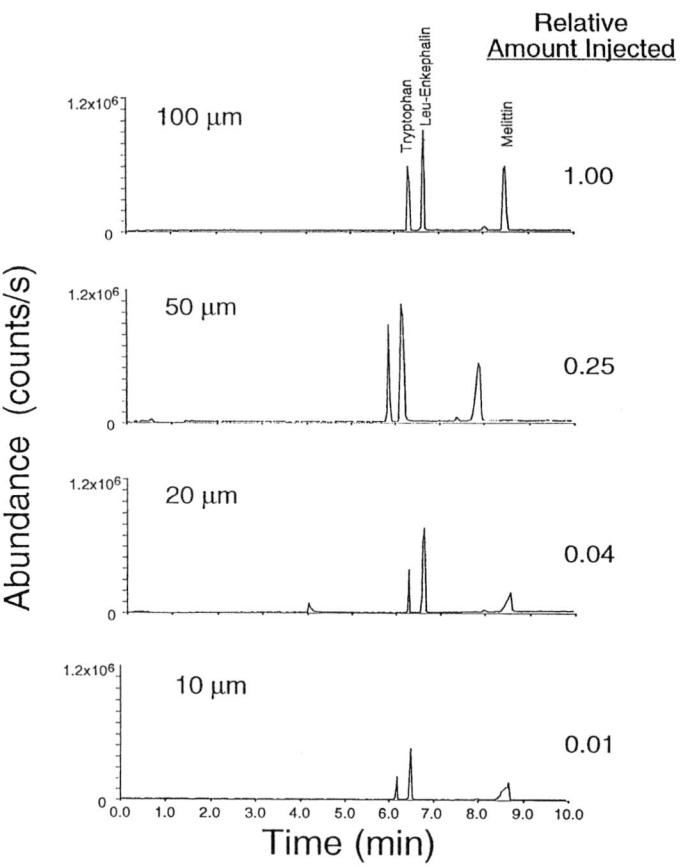

FIGURE 18
Comparison of the CE-MS total ion electropherograms for separation of tryptophan, leucine-enkephalin, and melittin for four different capillary diameters. Electromigration injection and separation conditions were identical for all experiments and signal intensities are shown on the same scale. The relative amount of analyte injected is a factor of 100 lower for the 10-μm capillary compared to the 100-μm capillary.

diameters.[71] Consistent with the above results, and qualitatively with the behavior predicted by Figure 12, sensitivity increases as capillary diameter decreases. Also, in accord with this is the observation that roughly equivalent sensitivity is obtained using the 10 μm and 20 μm diameter capillaries. Figure 20 shows the exceptional sensitivity that can be obtained using a 20 μm i.d. capillary for a simple mixture of polypeptides with selected ion monitoring of seven m/z values.

In the regime of ultrasensitive analysis, and in particular for biopolymer characterization, mass spectrometry has lagged far behind techniques such as laser-induced fluorescence and electrochemical detection. These detection

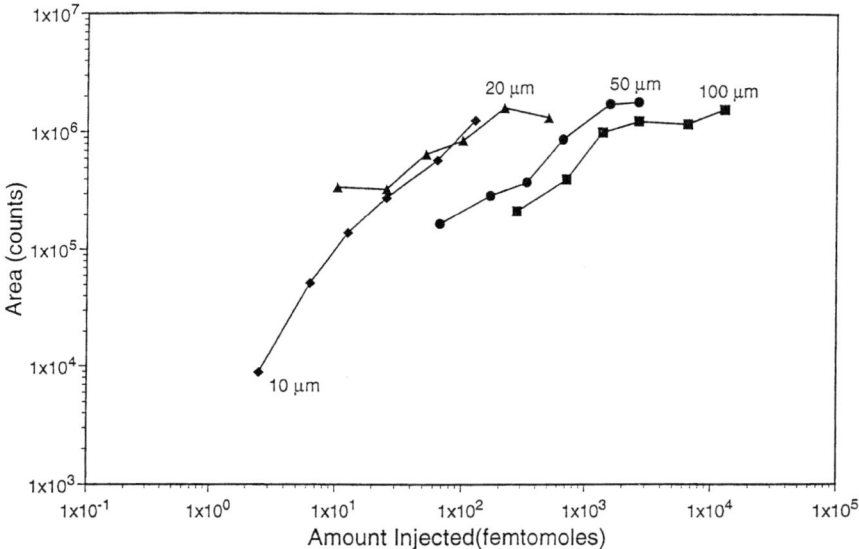

FIGURE 19
CE-MS signal intensities for leucine-enkephalin observed as a function of sample size injected using four different capillary diameters. Injections utilized similar electrokinetic conditions and a range of analyte concentrations.

methods have provided the most impressive sensitivities yet demonstrated with CE, but application is restricted and information for compound identification is generally limited due to a reliance on electrophoretic mobilities. The attraction of MS detection is that accurate molecular weight information and component identification can be performed, and techniques for obtaining structural information based upon tandem methods are currently being extended to larger molecules. These results, and on-going developments in mass spectrometric instrumentation, suggest that CE-MS has the potential to move into the regime of ultrasensitive detection. The recent development of an improved "sheathless" CE-MS interface,[77] our improving understanding of interface conditions and buffer selection,[78] and the use of alternative CE formats, such as isotachophoresis,[78] also portend much broader applications.

ACKNOWLEDGMENTS

We thank the U. S. Department of Energy, Office of Health and Environmental Research (Contract DE-AC06-76RLO 1830), for support of this research, and Drs. D.R. Goodlett and C.J. Baringa for helpful discussions and contributions to the research described.

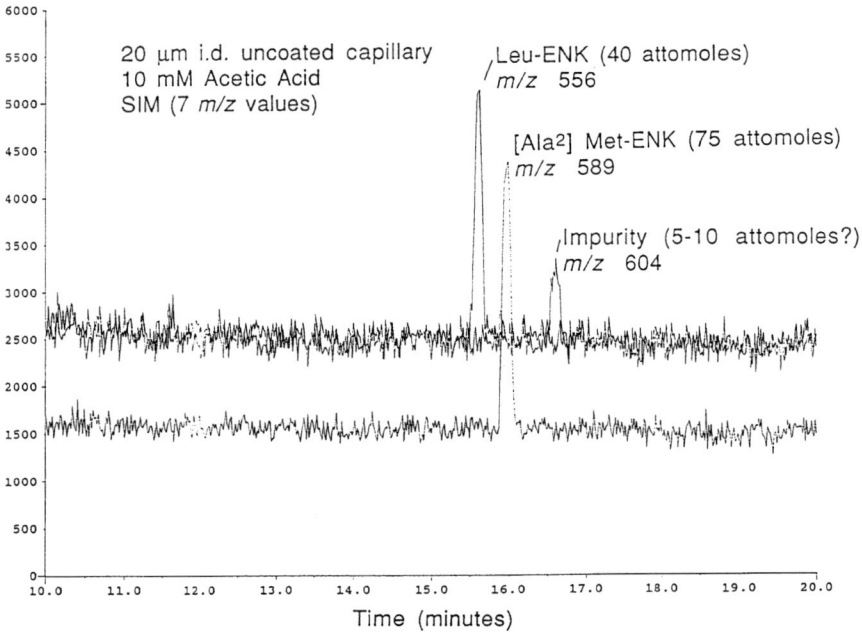

FIGURE 20
CE-MS obtained with a 20 μm i.d. capillary for high sensitivity with SIM detection. The amount of the unknown impurity is estimated.

REFERENCES

1. Kenndler, E. and Kaniansky, D., *J. Chromatogr.*, 209, 306, 1981.
2. Kenndler, E. and Haidl E., *Fresenius' Z. Anal. Chem.*, 322, 391, 1985.
3. Takigiku, R., Keough, T., Lacey, M.P., and Schneider, R.E., *Rapid Commun. Mass Spectrom.*, 4, 24, 1990.
4. Loo, J.A., Edmonds, C.G., Smith, R.D., Lacey, M.P., and Keough, T., *Biomed. Environ. Mass Spectrom.*, 19, 286, 1990.
5. Wallingford, R.A. and Ewing, A.G., *Anal. Chem.*, 59, 1762, 1987.
6. Hillenkamp, F, Karas, M., Beavis, R.C., and Chait, B.T., *Anal. Chem.*, 63, 1193, 1991.
7. Caprioli, R.M., Moore, W.T., Martin, M., and DaGue, B.B., *J. Chromatogr.*, 480, 247, 1989.
8. Caprioli, R.M., *Proc. Jpn. Soc. Biomed. Mass Spectrom*, 1991.
9. Suter, M.J.-F., DaGue, B.B., Moore, W.T., Lin, S.-N., and Caprioli, R.M., *J. Chromatogr.*, 553, 101, 1991.
10. Moore, W.T. and Caprioli, R.M., *Techniques in Protein Chemistry*, Vol. II, Academic Press, New York, 1991, 511.
11. Suter, M.J.-F. and Caprioli, R.M., *J. Am. Soc. Mass Spectrom.*, 3, 198, 1992.
12. Minard, R.D., Luckenbill, D., Curry, R., Jr., and Ewing, A.G., *Proc. 37th ASMS Conf. Mass Spectrometry and Allied Topics*, 1989, 950.
13. Wolf, S. M., Vouros, P., Norwood, C., and Jackim, E., *J. Am. Soc. Mass Spectrom.*, Miami Beach, FL, May 21–26, 3, 757, 1992.

14. Reinhoud, N.J., Niessen, W.M.A., and Tjaden, U.R., *Rapid Commun. Mass Spectrom.*, 3, 348, 1989.
15. Reinhoud, N.J., Schroder, E., Tjaden, U.R., Niessen, W.M.A., Ten Noever de Brauw, M.C., and van der Greef, J., *J. Chromatogr.*, 516, 147, 1990.
16. Tehrani, M., Macomber, R., and Day, L., *J. High Resolut. Chromatogr.*, 14, 10, 1991.
17. Verheij, E.R., Tjaden, U.R., Niessen, W.M.A., and Van Der Greef, J., *J. Chromatogr.*, 554, 339, 1991.
18. DeWit, J.S.M., Deterding, L.J., Moseley, M.A., Tomer, K.B., and Jorgenson, J.W., *Rapid Commun. Mass Spectrom.*, 2, 100, 1988.
19. Moseley, M.A., Deterding, L.J., Tomer, K.B., and Jorgenson, J.W., *J. Chromatogr.*, 480, 197, 1989.
20. Moseley, M.A., Deterding, L.J., Tomer, K.B., and Jorgenson, J.W., *Rapid Commun. Mass Spectrom.*, 3, 87, 1989.
21. Moseley, M.A., Deterding, L.J., Tomer, K.B., and Jorgenson, J.W., *Anal. Chem.*, 63, 109, 1991.
22. Deterding, L.J., Moseley, M.A., Tomer, K.B., and Jorgenson, J.W., *J. Chromatogr.*, 554, 73, 1991.
23. Deterding, L.J., Parker, C.E., and Perkins, J.R., *J. Chromatogr.*, 554, 329, 1991.
24. Deterding, L.J., Parker, C.E., Perkins, J.R., Moseley, M.A., Jorgenson, J.W., and Tomer, K.B., *J. Chromatrogr.*, 554, 329, 1991.
25. Fenn, J.B., Mann, M., Meng, C.K., and Wong, S.F., *Mass Spectrom. Rev.*, 9, 37, 1990.
26. Smith, R.D., Loo, J.A., Edmonds, C.G., Barinaga, C.J., and Udseth, H.R., *Anal. Chem.*, 62, 882, 1990.
27. Smith, R.D., Loo, J.A., Ogorzalek Loo, R.R., Busman, M., and Udseth, H.R., *Mass Spectrom. Rev.*, 10, 359, 1991.
28. Bruins, A.P., Covey, T.R., and Henion, J.D., *Anal. Chem.*, 59, 2642, 1987.
29. Mack, L.L., Kralik, P., Rheude, A., and Dole, M., *J. Chem. Phys.*, 52, 4977, 1970.
30. Olivares, J.A., Nguyen, N.T., Yonker, C.R., and Smith, R.D., *Anal. Chem.*, 59, 1230, 1987.
31. Smith, R.D., Olivares, J.A., Nguyen, N.T., and Udseth, H.R., *Anal. Chem.*, 60, 436, 1988.
32. Smith, R.D., Barinaga, C.J., and Udseth, H.R., *Anal. Chem.*, 60, 1948, 1988.
33. Udseth, H.R., Loo, J.A., and Smith, R.D., *Anal. Chem.*, 61, 228, 1989.
34. Smith, R.D., Fields, S.M., Loo, J.A., Barinaga, C.J., Udseth, H.R., and Edmonds, C.G., *Electrophoresis*, 11, 709, 1990.
35. Smith, R.D., Loo, J.A., Edmonds, C.G., Barinaga, C.J., and Udseth, H.R., *J. Chromatogr.*, 516, 157, 1990.
36. Edmonds, C.G., Loo, J.A., Fields, S.M., Barinaga, C.J., Udseth, H.R., and Smith, R.D., Capillary electrophoresis combined with electrospray ionization mass spectrometry and tandem mass spectrometry, in *Biological Mass Spectrometry*, Burlingame, A.L. and McCloskey, J.A., Eds., Elsevier, Amsterdam, 1990, 77.
37. Hail, M., Schwartz, J., Mylchreest, I., Seta, K., Lewis, S., Zhou, J., Jardine, I., Liu, J., and Novotny, M., *Proc. 38th ASMS Conf. Mass Spectrometry and Allied Topics*, Tucson, AZ, June 3–8, 1990, 353.
38. Pleasance, S., Ager, S.W., Laycock, M.V., and Thibault, P., *Rapid Commun. Mass Spectrom.*, 6, 14, 1992.
39. Parker, C.E., Perkins, J.R., and Tomer, K.B., *J. Am. Soc. Mass Spectrom.*, 3, 563, 1992.
40. Hunt, D.F., Alexander, J.E., McCormack, A.L., Martino, P.A., Michel, H., Shabanowitz, J., Moseley, M.A., Jorgenson, J.W., and Tomer, K.B., Mass spectrometric methods for protein and peptide sequence analysis, in *Techniques in Protein Chemistry*, Vol. II, Villafranca, J.J., Ed., Academic Press, San Deigo, CA, 1991, 441.

41. Thibault, P., Paris, C., and Pleasance, S., *Rapid Commun. Mass Spectrom.*, 5, 484, 1991.
42. Thibault, P., Pleasance, S., and Laycock, M.V., *J. Chromatogr.*, 542, 483, 1991.
43. Thibault, P., Pleasance, S., and Laycock, M.V., *Proc. 39th ASMS Conf. Mass Spectrometry and Allied Topics*, 1991, 593.
44. Moseley, M.A., Shabanowitz, J., Hunt, D., Tomer, K.B., and Jorgenson, J.W., *J. Am. Soc. Mass Spectrom.*, 3, 289, 1992.
45. Shida, Y., Parker, C.E., Perkins, J.R., O'Hara, K., Kono, M., and Tomer, K.B., *Proc. 39th ASMS Conf. Mass Spectrometry and Allied Topics*, Nashville, TN, May 19–24, 1989, 587.
46. Lee, E.D., Mück, W., Henion, J.D., and Covey, T.R., *J. Chromatogr.*, 458, 313, 1988.
47. Lee, E.D., Mück, W., Henion, J.D., and Covey, T.R., *Biomed. Environ. Mass Spectrom.*, 18, 253, 1989.
48. Lee, E.D., Mück, W., Henion, J.D., and Covey, T.R., *Biomed. Environ. Mass Spectrom.*, 18, 844, 1989.
49. Johansson, I.M., Huang, E.C., Henion, J.D., and Zweigenbaum, J., *J. Chromatogr.*, 554, 311, 1991.
50. Johansson, I.M., Pavelka, R., and Henion, J.D., *J. Chromatogr.*, 559, 515, 1991.
51. Mück, W.M. and Henion, J.D., *J. Chromatogr.*, 495, 41, 1989.
52. Garcia, F. and Henion, J.D., *Anal. Chem.*, 64, 985, 1992.
53. Wachs, T., Conboy, J.C., Garcia, F., and Henion, J.D., *J. Chromatogr. Sci.*, 59, 357, 1991.
54. Pleasance, S., Thibault, P., and Kelly, J., *J. Chromatogr.*, 591, 325, 1992.
55. Smith, R.D., Udseth, H.R., Barinaga, C.J., and Edmonds, C.G., *J. Chromatogr.*, 559, 197, 1991.
56. Mylchreest, I. and Hail, M., *Proc. 39th ASMS Conf. Mass Spectrometry and Allied Topics*, 1989, 316.
57. Edmonds, C.G., Loo, J.A., Barinaga, C.J., Udseth, H.R., and Smith, R.D., *J. Chromatogr.*, 474, 21, 1989.
58. Loo, J.A., Jones, H.K., Udseth, H.R., and Smith, R.D., *J. Microcol. Sep.*, 1, 223, 1989.
59. Loo, J.A., Ogorzalek Loo, R.R., Udseth, H.R., Edmonds, C.G., and Smith, R.D., *Rapid Commun. Mass Spectrom.*, 5, 101, 1991.
60. Loo, J.A., Edmonds, C.G., and Smith, R.D., *Anal. Chem.*, 63, 2488, 1991.
61. Smith, R.D., Udseth, H.R., Loo, J.A., Wright, B.W., and Ross, G.A., *Talanta*, 36, 161, 1989.
62. Loo, J.A., Udseth, H.R., and Smith, R.D., *Biomed. Environ. Mass Spectrom.*, 17, 411, 1988.
63. Smith, R.D., Loo, J.A., Barinaga, C.J., Edmonds, C.G., and Udseth, H.R., *J. Am. Soc. Mass Spectrom.*, 1, 53, 1990.
64. Van Berkel, G.J., Glish, G.L., and McLuckey, S.A., *Anal. Chem.*, 62, 1284, 1990.
65. Cooks, R.G., Glish, G.L., McLuckey, S.A., and Kaiser, R.E., *Chem. Eng. News*, March 25, 1991, 26.
66. Henry, K.D., Quin, J.P., and McLafferty, F.W., *J. Am Chem. Soc.*, 113, 5447, 1991.
67. Hofstadler, S.A. and Laude, D.A., *Anal. Chem.*, 64, 569, 1992.
68. Winger, B.E., Hofstadler, S.A., Bruce, J.E., and Smith, R.D., *J. Am. Soc. Mass Spectrom.*, 4, 566, 1993.
69. Goodlett, D.R., Wahl, J.H., Udseth, H.R., and Smith, R.D., submitted to *J. Microcol. Sep.*, 5, 57, 1993.
70. Wahl, J.H., Goodlett, D.R., Udseth, H.R., and Smith, R.D., *Anal. Chem.*, 64, 3194, 1992.

71. Wahl, J.H., Goodlett, D.R., Udseth, H.R., and Smith, R.D., *Electrophoresis,* 14, 448, 1993.
72. Smith, R.D., Wahl, J.H., Goodlett, D.R., and Hofstadler, S.A., *Anal. Chem.,* 65, 574A–584A, 1993.
73. Hofstadler, S.A., Wahl, J.H., Bruce, J.E., and Smith, R.D., *J. Am. Chem. Soc.,* 115, 6983–6984, 1993.
74. Wahl, J.H., Hofstadler, S.A., and Smith, R.D., *Anal. Chem.,* 67, 462–465, 1995.
75. Hofstadler, S.A., Wahl, J.H., Bakhtiar, R., Anderson, G.A., Bruce, J.E., and Smith, R.D., *J. Amer. Soc. Mass Spectrom.,* 5, 894–899, 1994.
76. Hofstadler, S.A., Swanek, D.D., Gale, D.C., Ewing, E.G., and Smith, R.D., *Anal. Chem.,* 67, 1477–1480, 1995.
77. Wahl, J.H., Gale, D.C., and Smith, R.D., *J. Chromatography, A,* 659, 217–222, 1994.
78. Wahl, J.H. and Smith, R.D., *J. Cap. Electrophor.,* 1, 62–67, 1994.
79. Zhao, Z., Wahl, J.H., Udseth, H.R., Hofstadler, S.A., Fuciarelli, A.F., and Smith, R.D., *Electrophoresis,* 16, 389–395, 1995.

CHAPTER 7

ANALYZING REVERSED PHASE PEPTIDE MAPS OF RECOMBINANT HUMAN GLYCOPROTEINS USING LC/ES/MS

A.W. Guzzetta and William S. Hancock

CONTENTS

1 Introduction 182

2 The Marriage of Mass Spectrometry and Liquid
   Chromatography 183
   2.1 Mass Spectrometry 183
       2.1.1 The Electrospray Ion Source 183
       2.1.2 The Quadrupole Mass Filter 187
   2.2 HPLC 189
   2.3 LC/MS 190

3 Mammalian Protein Carbohydrate Chemistry 191
   3.1 The Monosaccharide 191
       3.1.1 *N*-Linked Structures 191
       3.1.2 *O*-Linked Structures 194

4 How to Choose a Method of Glycoprotein Cleavage 194
   4.1 The Protein 194
       4.1.1 Primary Amino Acid Sequence and Peptide
             Mapping 195
   4.2 Predicting *N*-Linked Glycosylation Sites and Choosing
       a Method of Protein Cleavage 195

5 Generating and Analyzing LC/MS Data From Glycoprotein

Digests 196
5.1 Locating Glycopeptides in LC/MS Peptide Maps 196
    5.1.1 CID in the High-Pressure Region Before Q1 197
    5.1.2 Locating Glycopeptides in a Contour Plot 201
5.2 Analyzing a Complex Series of Ions 203
    5.2.1 What's Missing? 203
    5.2.2 Using Sugar Residue Differences and Calculated
          Compositions to Propose Carbohydrate Structures 203
5.3 Collision-Induced Dissociation (CID) of Glycopeptides 207
5.4 Artifacts in Electrospray 209
    5.4.1 Using LC Separation and Extracted Ion Plotting
          to Uncover Artifacts 209

References 211

1 INTRODUCTION

Carbohydrate chemistry is a study of immense complexity and importance to the biotechnology industry. The nature of the monosaccharides gives rise to this complexity. Unlike amino acids that are linked through an amide bond, usually in a linear unbranched fashion, monosaccharides can be linked through a variety of hydroxyl groups present on the sugar to form a glycosidic linkage.[1] Hellerqvist and Sweetman give an extraordinary example of the linkage possibilities of a hexasaccharide yielding a possible 4.76×10^9 structures.[2] This number of structures, however, is arrived at without using the constraints and conventions of mammalian carbohydrates that exhibit common core structures and anomericity. Even with such constraints, we are left with a large degree of compositional and linkage variety at any one glycosylation site on the protein. Determining the carbohydrate composition, type, and branching pattern is an important characterization step in the development of a recombinant DNA-derived glycoprotein as a pharmaceutical.[3,4] For a broad scope of the role of oligosaccharides in biology the reader is referred to a review by Ajit Varki.[5]

The complexity of carbohydrate structures mandates that a variety of analytical methods be used for the study of these forms. A promising new procedure is described in this chapter that uses HPLC in conjunction with electrospray ionization mass spectrometry (ESI/MS) as a tool to identify the sites of glycosylation and the general nature of the glycosylation. ESI/MS can detect whether an oligosaccharide is O-linked or N-linked. It can also differentiate between complex, high mannose, or hybrid forms. While this approach is a useful addition to the battery of techniques used to characterize carbohydrates, it supplements and not replaces other important methods such as NMR techniques used to determine linkage positions of anomeric configurations. We will examine the techniques used to gain limited linkage order information using collision-induced dissociation (CID) with a triple quadrupole mass spectrometer. This chapter is intended to be an introduction to LC/MS, electrospray quadrupole mass spectrometry, and the MS techniques used to explore mammalian protein glycosylation.

2 THE MARRIAGE OF MASS SPECTROMETRY AND LIQUID CHROMATOGRAPHY

2.1 MASS SPECTROMETRY

All mass spectrometers conform to these basic principles: An ion must be formed in the gas phase and the charged species is then manipulated by the mass analyzer to gain mass information. The two primary components of a mass spectrometer are the ion source and the mass analyzer.[6] Until recently, mass spectrometry was not widely available to biotechnology due to the high cost of the technology and mass range limitations. Most institutions were limited to GC-MS systems and the low mass range capabilities of this instrument.

The FAB technique was first introduced in the early 1980s by Barber et al.[7] The strengths of this technique are its high resolution and accuracy capabilities. FAB uses a glycerol matrix that prolongs the duration of the signal, and gives it an advantage over secondary ion mass spectrometry. In FAB the sample is sputtered (ionized) into the mass spectrometer with a beam of neutral atoms. The ion source for the FAB instrument is located within the high-vacuum environment of the mass spectrometer, making high flow LC/MS difficult. To the contrary, the electrospray ion source coupled with the quadrupole mass filter has produced a mass spectrometer that is easily compatible with HPLC analysis and measurement of biological samples. Electrospray MS has put mass analysis of biological samples within the reach of most laboratories.

2.1.1 The Electrospray Ion Source

Recently, an ion source was developed that produces ions at atmospheric pressures. Dole et al. are generally given credit for the creation of the electrospray source in the late 1960s,[8] though the process of producing charged droplets can be traced back much further.[9,10] The source can ionize a large range of molecules at atmospheric pressure into the gas phase. Notably it can produce multiply charged ions of large biomolecules, making this area of MS development extremely popular. Fenn and co-workers at Yale University are given credit for resurrecting the electrospray ion source in the mid

1980s.[11,12] Their experiments showed an application for LC/MS and suggested a possible application for biopolymers. Thomson, and co-workers in the mid to late 1970s developed a similar technique and termed it ion spray.[13,14] However, the real electrospray explosion can be traced to publications in 1988 in which several laboratories published spectra of proteins up to 40 kDa.[15,16]

There are two basic designs for this new ion source: type A, ionspray, introduces a gas at the probe tip that nebulizes the LC flow (see Figure 1, upper schematic). A nebulization gas allows LC flow rates up to a practical limit of ~40 µL/min. A high voltage is applied at this source to charge the nebulized droplets. The charged droplets are sprayed toward an orifice and heated interface through a curtain gas. The droplets evaporate and the ions are ejected via coloumbic repulsion. The ions are drawn through the mass analyzer via a series of voltage gradients. This ion spray ion source was developed at the University of Toronto.[13,14] Bruins and co-workers at Cornell University exploited this technique and developed it as an LC/MS tool.[17,18] Prior to this work, Simons et al. in 1974 demonstrated the promise of a similar technique and looked at smaller compounds such as sucrose, proline, and other amino acids and suggested that the method would be applicable to LC/MS.[19]

The second type of ion source, electrospray, is shown in the central schematic of Figure 1, and is noted as type B. In electrospray, the probe tip is at ground potential and involves an organic sheath flow to aid in the ionization. The LC eluent is brought into a charged cylinder and a heated bath gas is washed over the droplets as they approach a charged capillary interface.[11,12] The sample is introduced from the LC up to a flow rate of 5 µL/min. Recent advances have improved the flow rate limitation to put it on an equal footing with ionspray.[20,21]

One advantage of having the probe tip at ground potential is that capillary electrophoresis may be easier to perform, since the power supply of the CE instrument does not have to be floated on the high voltages encountered in the ionspray instrument. Most of the CE experiments in the published literature use the probe tip at ground.[22,23]

The ions that are generated in electrospray are conducted through the mass spectrometer in much the same manner as the ions generated in ionspray, i.e., through a series of voltage gradients, skimmers, and lenses. The two types of ion sources are continually being improved and are beginning to yield similar results, and from this point on in the text they will be referred to as electrospray. A detailed evolution of the electrospray technique can be found in the review by Hamdan and Curcuruto.[24] Ikonomou and co-workers give a comparison of electrospray and ionspray in terms of sensitivity and the charging process.[25] Again, for a depiction of these two ion sources and for a detailed view of the probe tip, see Figure 1.

A major advantage of the electrospray ion source is its ability to produce a population of multiply charged ions during the ionization process.[26-28] This

ANALYZING REVERSED PHASE PEPTIDE MAPS 185

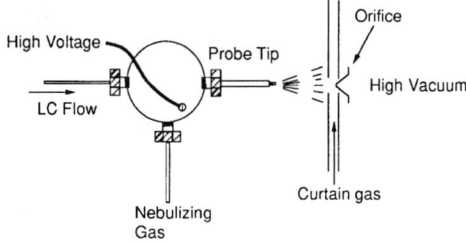

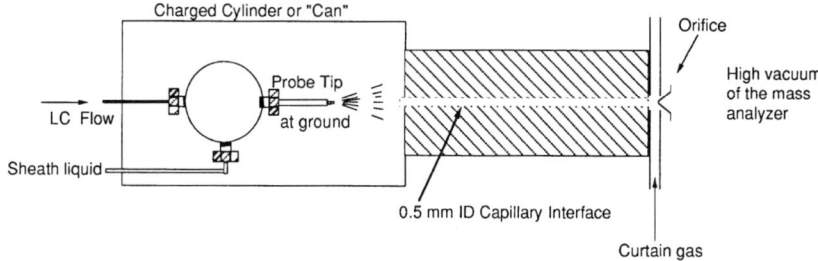

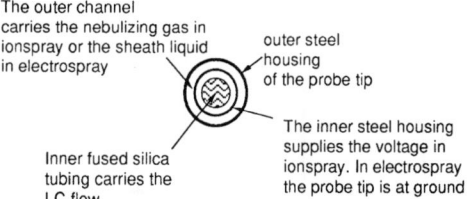

FIGURE 1
The upper diagram illustrates the basic components of an "ionspray" ion source. The middle diagram illustrates a simplified version of an electrospray ion source. The bottom figure shows a typical end view of the probe tip where the liquid stream from the LC exits the fused silica tubing to become charged droplets.

population is presented as an array of ions observed at varying mass to charge ratios in the final mass scan. An illustration of this process is made in Figure 2. In Figure 2 a population of multiply charged species is created by the ionization of a single peptide. Figure 2 shows how this population may appear in an ES mass spectrum.

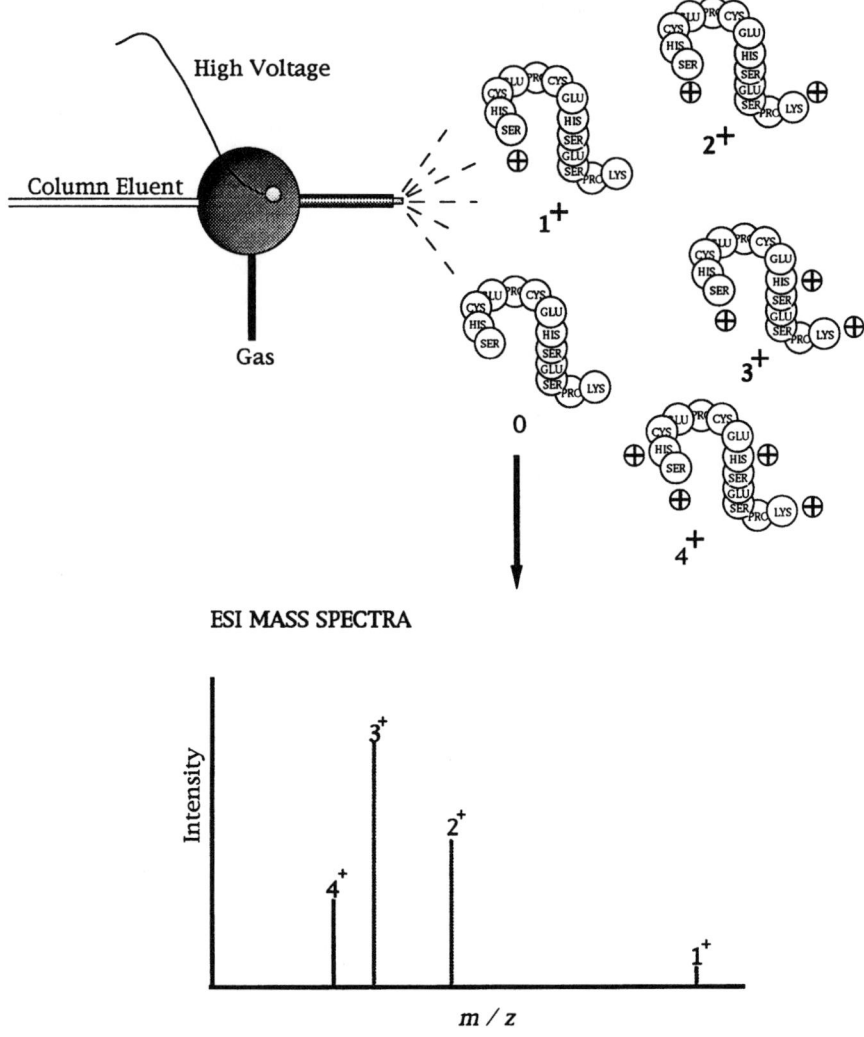

FIGURE 2
The upper portion of this figure illustrates the formation of multiply-charged species. The lower portion of this figure demonstrates how this population of multiply-charged molecules translates to an ESI mass spectra.

The advantages of the multiple charging phenomenon are twofold. Each related ion in a series can be thought of as an independent measurement of the parent mass. These multiple assessments of the parent mass are used to obtain highly accurate mass information. In most cases, the mass to charge ratio is such that the ions are in the mass range of the quadrupole mass analyzer, making mass measurement of large biomolecules possible.[29] For example, the m/z ratio of N-glycanase-treated, reduced, and S-carboxymethylated r-tPA, (M_r 61221.6 Da) with 40 proton adducts is 1531.5. This mass to charge value is well within the mass range of the analysis.[30] Using most of the automated mass analysis algorithms at least two ions are needed to make a determination of the parent M_r or charge state. Once the charge state is known, a parent mass calculation can be made using the formula: the uncharged parent mass is equal to the m/z value multiplied by the charged state of the ion minus the number of charges (see formula, Figure 2).

If only one ion is present there are a number of tricks that can be employed to deduce the charge state and thus the parent molecular weight. If the M_r of the molecule is low enough (<4000 Da), and the resolution of the instrument is high enough, it is possible to observe some resolution of the isotopes. A singly charged ion will have isotopes that differ by 1 Da (see plot A, Figure 3). If the ion is doubly charged the isotopes will differ by 0.5 Da (see plot B, Figure 3).

If the molecular weight of the molecule is higher and isotopic resolution is difficult to obtain, it may be possible to use adducts of sodium or ammonia, which are common in most samples even if the sample has been desalted. The charge state determination from sodium or ammonia adducts is accomplished in much the same way as in the previous example. The sodium and ammonium adducts are viewed as multiples of the charge state. In Figure 3, plot C shows an ion of a tryptic peptide that was digested in ammonium bicarbonate and separated by RP-HPLC. The 1054.6 ion in Figure 3 plot C has an NH_{3+} adduct that differs in mass by 8.5 Da and thus it is doubly charged. With the charge state information a parent mass is easily calculated by $(2 \times 1054.6) - 2$.

Senko et al. demonstrated the strategy of using cation adduction to determine charge state in a recent communication.[31] Charged states of ions can also be determined by their relationship to other ions that are derived from related molecular species, this method will be dealt with in the section on glycopeptide ions. Several groups have published a number of papers that describe the relationship of the ions observed in electrospray.[32-34] Fenn and co-workers have described the deconvolution of electrospray-generated ions exceeding four charges.[35,36] Lee and Vemuri have published a computer program that is helpful in deciphering MS/MS spectra and determining protein and peptide masses.[37]

2.1.2 The Quadrupole Mass Filter

The quadrupole mass filter has been used for some time. It was first introduced by Paul et al. in 1958.[38] The most commercially utilized quadrupole mass filter is found in GC/MS instrumentation. The first commercially available GC/MS quadrupole instrument was on the market in 1965 (the LKB 6000).[39] GC/MS has progressed far in the last 25 years in both technology and cost. The relatively high accuracy of the quadrupole and its low cost made it a perfect match for the electrospray ion source. The quadrupole, as the

FIGURE 3
All samples were infused into the API III mass spectrometer at a concentration of 20 pmol/μL at 5 μL/min. All scans were acquired in Q3. Plot A shows the isotopic distribution of a singly charged peptide ion. The ions differ by 1 Da, obviously making the ion singly charged. There were 41 scans acquired in the MCA mode. The starting mass for the scan was 924.5, and the ending mass was 930.5. The scan step was set at 0.2 Da. A 25-ms dwell time was set per step. The resolution was set 1 point above the Q3 calibration resolution of 132. Calibration was accomplished with a mixture of polyethylene glycols at isotopic resolution. Plot B is the isotopic distribution of a doubly charged peptide ion. The ions differ by 0.5 Da, obviously making the parent ion doubly charged. There were 37 scans acquired in the MCA mode. The starting mass for the scan was 1052.5, and the ending mass was 1057.5. The scan step was set at 0.1 Da. A 20-ms dwell time was set per step. The Q3 resolution was set 1 point above the calibration resolution of 132. Plot C is a low resolution scan of a doubly charged peptide ion at 1054.6 showing an ammonium ion adduct at 1063.1. The ammonium adduct ion differs from the protonated adduct ion by 8.5 Da, thus making the parent ion doubly charged. Nine scans were acquired in the MCA mode. The starting mass for the scan was 300, and the ending mass was 2000. The scan step was set at 0.5 Da. A 1-ms dwell time was set per step. The scans were acquired in Q3 and the Q3 resolution was set 2 point below the calibration resolution of 132 to simulate the resolution conditions during an LC/MS run.

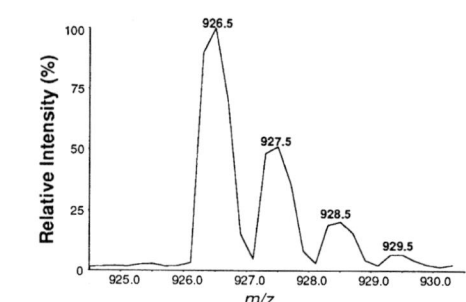

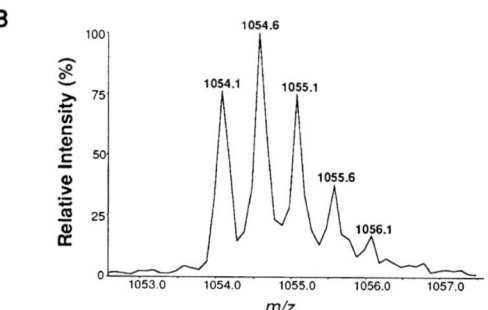

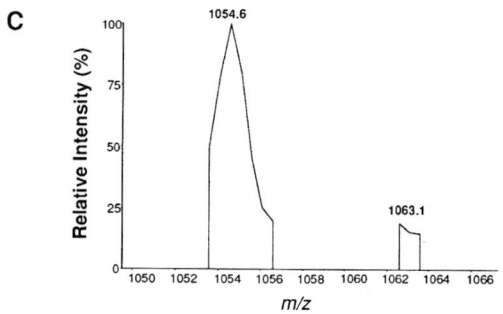

name suggests, consists of four round or hyperbolic rods. In the quadrupole, DC and RF voltages are applied to the rods. The ions enter the electric field of the quadrupoles and are filtered selectively by varying these potentials.[6] This systematic filtering results in a mass scan usually within the mass range of 100 to 3000 Da. Usual scan times can range from 1 to 8 s, depending on how the parameters are set.

2.2 HPLC

The liquid chromatograph, in this case an HPLC, has been the standard in establishing protein pharmaceutical integrity and is relied upon by the quality control departments within the biotechnology industry.[40,41] The HPLC is used to establish purity for the intact protein.[42] In a more detailed manner peptide maps are used to look for degradation products.[43] Since most MS ion sources require flow rates at less than common chromatographic levels, a split of the HPLC eluent stream must be accomplished. Most often, this split is made before the HPLC UV detector to guard against possible dilution effects caused by the HPLC flow cell.[44] The flow rate capability of the mass spectrometer is presently less of an issue with the increasing popularity of microbore (micro) and capillary HPLC. Microbore HPLC is a term used to refer to columns with an internal diameter of 1 to 2 mm utilizing flow rates of 40 and 200 µL/min, respectively.

Capillary 50 HPLC columns have an internal diameter from 500 µm to a present lower limit of 50 µm. In general, the diameter where "micro" changes to "capillary" is somewhat arbitrary. The 300 µm capillary column requires a flow rate of 3 to 5 µL/min, which makes it compatible with either electrospray ion source.[45-47] Figure 4 shows a 0.3×100 mm packed capillary column reversed phase separation of an r-tPA tryptic digest. The analysis required 3 µg of digest (about 50 pmol) to generate a total ion current chromatogram on an API III electrospray triple quadrupole mass spectrometer. From this LC/MS map it was possible to assign the peptides and glycopeptides in the separation. This capillary column technique gives equal if not better response for the peptides and glycopeptides than a conventional reversed phase column, due to the low flow rates involved and the reduced dilution effects. The low flow rate allows a direct feed of the column eluent to the electrospray tip without a connecting junction.

The low-pH solvent systems used in reversed phase chromatography help to positively charge biomolecules in electrospray. Some of the more common organic modifiers like TFA and formic acid work well. It is important to use the purest forms of these acids and solvents to eliminate signal background. TFA can be purchased in 1 g ampoules for improved purity. Several researchers have reported increased sensitivities using acetic acid or other organic acid as the modifier.[48,49] An increase in sensitivity at the expense of the chromatographic separation can be necessary when sample availability is limited.

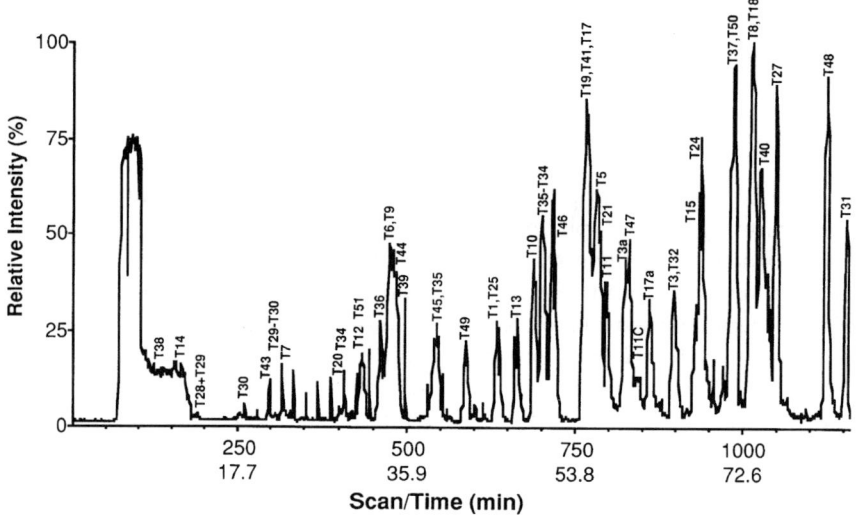

FIGURE 4
Shown above is a total ion current map generated with 30 pmol (ca. 1.8 µg) of a tryptic digest of r-tPA separated using a 0.3 × 150 mm Nucleosil C18, 5 µm capillary reversed phase column. The separation conditions were 0 to 60% B in 90 min. Solvent A was water with 0.05% TFA. Solvent B was acetonitrile with 0.05% TFA. The gradient was developed with an HP1090 HPLC at 150 µL/min and split down to 5 µL/min using a microflow processor from LC Packings Inc. The flow from the column was fed directly into the spray tip of an API III mass spectrometer. Quadrupole three was scanned from 300 to 2000 Da using a 0.5 Da step size. The dwell time per step was set at 1.2 ms for a scan duration of 4.39 s. The peptides were identified in the map and designated with a tryptic map number. The tryptic peptides are numbered sequentially from the N-terminus as T1 through T51, with T51 being the C-terminal peptide. The peptides were identified in the map by their expected ions and the LC/MS map was labeled with the tryptic map numbers.

2.3 LC/MS

The term LC/MS represents the marriage of two powerful analytical techniques. On-line LC/MS was attempted in 1970 at the Karolinska Institute when an LC eluent was deposited on a metal band and brought through to the ion source of the LKB 9000.[39] An early application of LC/MS came with the advent of LC/FAB mass spectrometry. LC/FAB mass spectrometry had been accomplished by introducing HPLC eluents at a slow flow rate on a moving belt or by a technique known as frit FAB and flow FAB. This technique dominated LC/MS through much of the 1980s and early 1990s.[50-54] These experiments were successful and ground-breaking at the time. While commercially available FAB-based LC/MS systems are still in use, they have not enjoyed the popularity of electrospray LC/MS in the 1990s. The FAB technique suffers from a sample introduction limitation of 10 µL/min and a practical mass limitation of 300 to 3000 Da. The mass limitation of FAB makes it difficult to detect intact glycopeptides. FAB/LC/MS requires that a matrix (e.g., glycerol) be used in the mobile phase.

Another characteristic of FAB is that ionization of the more hydrophilic peptides is diminished. In the 1980s chromatographers were developing HPLC assays that employed flow rates of 200 to 2000 µL/min. FAB employed a split to handle these flow rates, but this is just one of the many compromises from which the technique suffered. The LC/MS revolution began in the mid to late 1980s, when an ion source that generated gas phase ions from a liquid stream was coupled with a mass filter that had a broad mass range. The commercial availability and the reduced cost of this mass analyzer has made it a popular analytical tool. Thus, HPLC column technologies and mass spectrometry ion source capabilities, along with the lower cost and broad mass range of quadrupole mass filters, had come together to make LC/MS an exciting new analytical technique available to the biochemist.

3 MAMMALIAN PROTEIN CARBOHYDRATE CHEMISTRY

3.1 THE MONOSACCHARIDE

The most commonly occurring monosaccharides found in oligosaccharide attachments to mammalian proteins are D-mannose (man or hex), D-galactose (Gal or Hex), D-glucose (Glc or Hex), L-fucose (Fuc or dHex), N-acetylglucosamine (GlcNAc or HexNAc), N-acetylgalactosamine (GalNAc or HexNAc), and N-acetylneuraminic acid, (sialic acid or NANA). Fucose is different from the other hexose (hex) sugars mentioned because it is missing a hydroxyl group at the six-carbon position, and is thus referred to as deoxyhexose or simply abbreviated dHex. Since many of the structures have isobaric mass they are often noted with these generic terms. See Figure 5 for the monosaccharide structures and abbreviated nomenclature.[55] These coincident masses would appear to simplify and at the same time confound the world of carbohydrate chemistry for the mass spectrometrist, with only four common distinct monosaccharide masses observable. There are a number of approaches that can be made to distinguish between these isobaric forms, such as the assumption that N-acetylglucosamine is the monosaccharide that links all structures that are attached to asparagine, or "N-linked". There are further assumptions that can be made when examining N-linked oligosaccharide structures and these will be discussed later.

3.1.1 N-Linked Structures

There are three types of carbohydrate moieties commonly recognized as asparagine linked, "N-linked", structures: complex, high mannose, and hybrid. The schematic forms of these structures are noted in Figure 6. It is

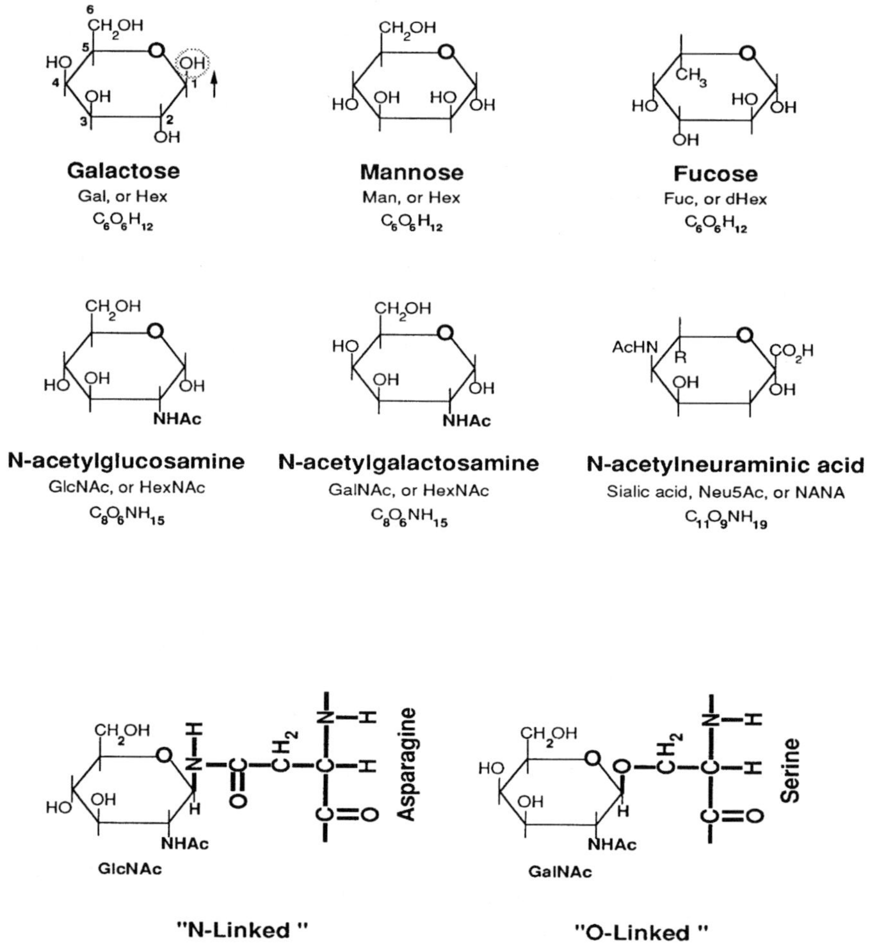

FIGURE 5
The Haworth formulas of the most commonly encountered pyranose ring forms of monosaccharides found in mammalian glycoprotein pharmaceuticals are shown in the upper portion of the figure. The less encountered sugars were omitted: glucose (hex, $C_6O_6H_{12}$), xylose (pentose, $C_5O_5H_{10}$), and NGNA. In the ring the carbons are numbered 1 through 6 in a clockwise fashion, starting with the anomeric carbon. Galactose is used to demonstrate the numbering convention. Anomericity is defined by the position of the hydroxyl group on carbon 1. Using the Haworth structure, the beta anomer has the hydroxyl group in the up position as demonstrated with galactose, and the alpha anomer has the hydroxyl group in the down position. The pyranose sugars are known to exist in the "chair conformation". In the chair ring form the beta anomer hydroxyl group would appear equatorial. The N- and O-linkage of N-acetylglucosamine and N-acetylgalactosamine to the side chains of asparagine and serine are demonstrated in the lower portion of the figure.

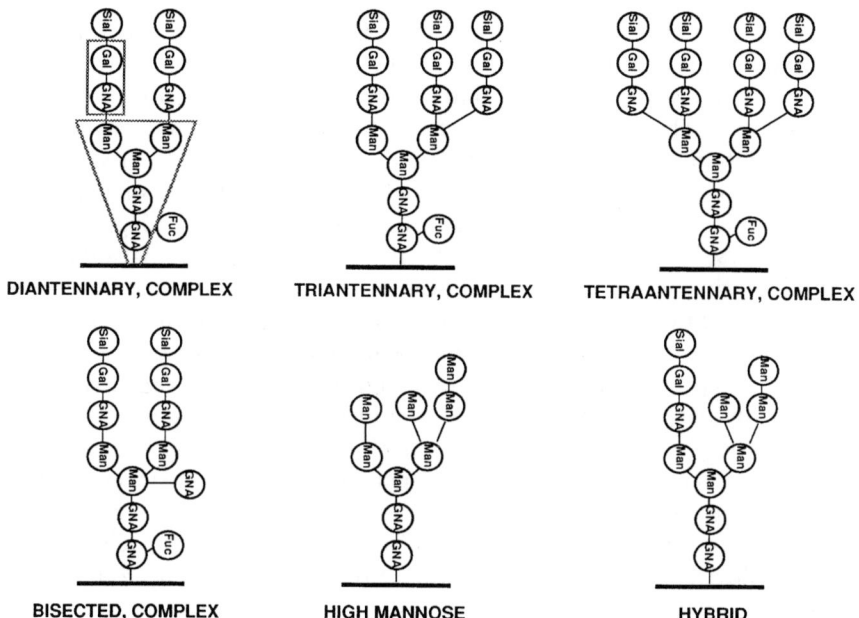

FIGURE 6
The basic forms of the common N-linked oligosaccharides are shown in this figure. All N-linked oligosaccharides possess a common core of two N-acetylglucosamine residues and three mannose residues; this core is highlighted with a shaded triangle on the complex diantennary structure. N-linked structures can be classified into three common motifs: complex, high mannose, and hybrid. The complex structures can have from two to four antennae, see top row of generalized structures. The antennae can terminate in galactose or sialic acid, usually. Sialylation adds the greatest degree of microheterogeneity to a glycoprotein. Other sources of heterogeneity can arise from variable core fucosylation. The common antenna unit is N-acetylglucosamine-galactose, and is shown in the shaded rectangle on the diantennary complex glycoform. The high mannose form is shown in the middle of the bottom row of structures. High mannose structures are termed 4 Man, 5 Man, etc.... with the three core mannose residues counted in this nomenclature. The hybrid glycoform shown in the bottom row, far right, is as its name implies; it has a branch that appears complex and others that appear as high mannose. Another variation that can occur is the bisecting N-acetylglucosamine; an example is shown on a complex structure, bottom row, far left.

helpful to know that all N-linked structure share a common core of two GlcNAc and three mannose sugar residues. This common core structure is noted in Figure 6 as a shaded triangle around the diantennary complex glycoform. When linked to a peptide, this core adds 893 Da to the peptide's mass. The core can have a fucose attached to the GlcNAc that is linked to the asparagine in the polypeptide, and it can also possess a bisecting GlcNAc residue attached to the central mannose of the core structure. Core fucosylation and bisecting GlcNAc residues are just a few of the many variations that can add to the complexity of the oligosaccharide. Complex-type structures usually have from two to four branches attached to the two outer core mannose residues. The branches are distributed over the two terminating core mannose residues. The complex structures are termed diantennary, triantennary, and tetraantennary, referring to the number of antennae. The basic branch structures are in most instances composed of one GlcNAc and one galactose residue, highlighted by the rectangle on the diantennary arm in Figure 6. This branched residue unit weighs 365 Da. If ions are observed that differ by a multiple of this unit it may be an indication of a complex-type glycosylation. To complicate matters further each of these arms can terminate in sialic acid. For example, it is possible to have a tetraantennary complex glycoform with zero to four terminating sialic acid residues, including variable core fucosylation. To limit the heterogeneity measured in a given peptide map it is important to isolate the glycosylation sites onto separate peptides through enzymatic or chemical cleavage. This will keep the dispersion of ions to a manageable level for mass analysis. The second type of *N*-linked structure is called high mannose. It contains the common core of two GlcNAc and three mannose sugar residues and has additional mannose residues emanating from the two terminating mannose residues of the core. The different forms are termed 4 Man, 5 Man, 6 Man, etc., with the three core mannose residues being counted in this terminology (see Figure 6). A series of high mannose glycoforms in a mass spectra will differ by multiples of 162 Da, which is the residue mass for a hexose such as mannose. Figure 6 shows a third example, a hybrid form with one branch of the high mannose variety and the other resembling the complex structure.

3.1.2 O-Linked Structures

O-linked structures are more difficult to define. They are linked through the oxygen present on the side chain hydroxyl of serine and threonine, but a consensus sequence for their attachment has not been reported. Recent reports include a single fucose attached to threonine 61 in recombinant tissue plasminogen activator[56] and single *O*-linked *Glc*NAc residues on various proteins.[57]

4 HOW TO CHOOSE A METHOD OF GLYCOPROTEIN CLEAVAGE

4.1 THE PROTEIN

The primary objective in choosing a cleavage scheme for a glycoprotein is to isolate the glycosylation sites onto separate peptides. The second priority should be that the glycopeptides do not coelute in the reversed phase separation. In the r-tPA reversed phase tryptic map two of the glycopeptides have nearly coincident elution.[40,58,59] While a coelution is difficult to predict and often difficult to remedy within the limitation of mobile phases compatible with the mass spectrometer, it can often be helped by trying a different protein cleavage method. This example will be reviewed in Section 5. These suggestions will hold down the heterogeneity of the spectra to a manageable level and make the data easier to interpret. The ES spectra of an intact glycoprotein will often appear as a broad hump of ions, and in most cases interpretation is nearly impossible. The large number of glycoforms divides the ion current and thus reduces the ion intensity of any single form. Lower resolution techniques such as matrix assisted, laser desorption, time of flight mass spectrometry (MALDI),[60] or even SDS-PAGE[61] will often yield better general mass information about such heterogeneous samples.

4.1.1 *Primary Amino Acid Sequence and Peptide Mapping*

As described in the earlier chapters, there are a number of methods for protein fragmentation. The methods used such as cyanogen bromide cleavage(chemical) and trypsin(enzymatic)[62] should be chosen by considering the goal of subsequent MS experiments. For MS/MS sequence analysis,[64] for example when examining the C- or N-terminal peptide of a protein or isolating an oxidized methionine on a peptide, one should aim for a peptide of a reasonable length.[63] The amino acid sequence of a protein, often available from the cDNA sequence, can be a valuable asset in deciding on the appropriate cleavage or digestion. Knowledge of the amino acid sequence can be helpful in predicting possible sites of *N*-linked glycosylation before any analysis is done.

4.2 PREDICTING *N*-LINKED GLYCOSYLATION SITES AND CHOOSING A METHOD OF PROTEIN CLEAVAGE

Possible *N*-linked glycosylation sites can be predicted if the primary amino acid sequence is known. This postulation can be made based on the *N*-linked consensus sequence where the first amino acid is asparagine, the second amino acid is not defined but is most likely not proline, and the third amino acid is threonine or serine.[3,4] The most widespread method of cleavage is with the enzyme trypsin.[40,63,65,66] Trypsin enjoys this popularity for a number of reasons: it cleaves the protein reproducibly, it is inexpensive, and it gives peptides of reasonable length (often 10 amino acids). Tryptic digestions are carried out with enzyme to protein ratios ranging from 1:25 to 1:100. Complete digestion of the protein is often dependent upon the individual protein. Most tryptic digestions are carried out slightly above pH 8, either at ambient temperature for 24 h, or at 37°C for 2 to 4 h. Digestions often involve addition of enzyme at two time points to ensure complete digestion. Due to the specificity of the enzyme tryptic fragments are almost guaranteed of having two sites that will charge at low pH in the ES source. With a *N*-terminal amino group and a *C*-terminal arginine or lysine residue the peptides will invariably protonate at two sites at low pH. *C*-terminal peptides may not have a basic residue at their *C*-terminus and thus may only have one charge. This fact can be used to find the *C*-terminal peptide. Tryptic peptides are almost certainly an advantage when MS/MS sequence analysis is performed, because CID fragmentation is enhanced near chargeable residues.[67]

5 GENERATING AND ANALYZING LC/MS DATA FROM GLYCOPROTEIN DIGESTS

Generating LC/MS data from peptide maps can be a simple task. Interpreting the data from an LC/MS run and being confident of the conclusions is not as simple. An LC/MS run can consist of 1000 single mass scans and the interpretation of a data set can be a time-consuming task. However, there are patterns that can be observed and techniques that can be used for identifying sites of glycosylation in this complex data set.

Determining whether a protein is glycosylated can be an easy task. Glycoproteins appear as broad, fuzzy bands on SDS-PAGE gels that will collapse to a tighter band when treated with glycosidases.[68] This is an approach that can be used for determination of *N*- or *O*-linkage with the help of glycosidases.[69] In LC/ES/MS there are also glycopeptide signatures that can be used to note possible sites of glycosylation. These include the observation of clusters of ions in the contour plots of m/z vs. scan data, and the observation of oxonium ions in CID data.[79]

5.1 Locating Glycopeptides in LC/MS Peptide Maps

Collision-induced dissociation in mass spectrometry is a technique that has been used for some time to gain structural information on peptides and other molecules. Traditionally, the analysis is performed using a tandem MS configuration. The ion to be analyzed is selected in the first sector and then collided with a gas (argon, xenon, etc.) in the second sector (collision cell), to produce fragments. These fragments are analyzed in the terminal mass analyzer. Several laboratories have led the field in MS/MS spectral analysis of peptides.[70-72] Complete sequences for peptides as large as 2800 Da have been reported.[73] Likewise, the analyses of oligosaccharides and glycopeptides have been carried out in a similar manner.[74] Collision-induced dissociation analysis has also been reported using a single-stage mass spectrometer. It had been observed that peptides would fragment under certain conditions before the

the first quadrupole, Q1. This phenomenon was first reported for peptides and proteins. Recently CID in the declustering region of electrospray instruments was reported for carbohydrates and glycopeptides by several researchers.

5.1.1 CID In the High-Pressure Region Before Q1

The first reports of peptide and protein fragment ions generated in the high-pressure region of electrospray ion sources were beginning to appear in the literature between 1988 and 1991.[75-76] Chait and co-workers were among the first to investigate this phenomenon using synthetic peptides.[77] The significance of this result was immediately obvious, as limited primary sequence information could be gathered using a single quadrupole electrospray mass analyzer. Furthermore, this approach allowed information to be gathered by a wide range of laboratories due to the lower cost of single quadrupole instruments. Chait's laboratory noted that the more highly charged peptides would fragment easily, and they also found that tryptic fragments, due to their doubly charged nature were natural candidates for this technique. There were other interesting observations made by Chait's group. It appeared that conformation may play a role in how a peptides fragments. Three of the peptides that were analyzed were bradykinin (RP*P*GFSP*FR), angiotensin II (DRVYIH*P*F), and synthetic renin inhibitor (P*H*P*FH*FFVYK). These peptides contain important features that make them ideal candidates for fragmentation. Noted * are prolines and also internal histidine residues. There are two general observations made by mass spectrometrists about CID: peptides tend to fragment near charged residues, and prolines in a sequence appear to enhance fragmentation. There ia a caveat, however, when the pre-Q1 CID technique is used — the discriminating nature of the tandem instrument is lost. In contrast, when a triple quadrupole instrument is used the ion to be sequenced is selected in the first quadrupole and all others are filtered out. The selected ion is then fragmented in a collision cell (Q2) with a collision gas and the fragment ions are analyzed in Q3. When the collisions are performed in the high-pressure region preceding Q1, all selectivity is lost. All ions are subject to collisions, and the analyst can only use pure peptides in the analysis or the spectra will be very difficult to interpret. Secondly, since fragmentation of the more highly charged species are favored it is possible to get fragments of pure peptides that are singly, doubly, and triply charged, which further complicates the spectra.

The studies described previously involved peptides of low molecular weight (1000 to 2000 Da). What might not appear to be obvious is that this technique is quite amenable to analyzing glycopeptides that may have molecular weights ranging up to 8000 Da and possibly larger. Henion and Conboy were the first to publish results showing that oxonium ion fragments could be obtained through collisions produced in the declustering region, pre-Q1, and demonstrated that they could be used as markers for glycopeptides in LC/ES/MS maps.[78] They showed that oxonium ions, prevalent at higher orifice potentials, could also be observed at regular operating orifice potentials but at a lower intensity, which could complicate the assessment of glycopeptide spectra. Carr et al. reported fragmentations resulting from glycopeptide CID generated in the high-pressure region of the electrospray ion source.[79,80] These authors also demonstrated that the oxonium ion fragments produced in these experiments could be used as markers for glycosylations of both N- and O-linked glycoforms in LC/MS runs of glycoprotein digests. The Carr et al. method involves increasing the orifice potential to induce fragmentation during the low mass range of the mass scan (150 to 500 Da) to observe the carbohydrate oxonium ion fragments, and then changing the orifice potential back to a setting that does not cause fragmentation for the mass range of 500 to 2000 Da. This method allows for both the observation of the carbohydrate oxonium ion marker fragments and, in the later portion of the scan, the parent glycopeptides.

Carr et al. report the observation of the following fragments at the following m/z values: Hex-HexNAc$^+$, m/z 366; NeuAc$^+$, m/z 292; HexNAc$^+$, m/z 204; Hex$^+$, m/z 163; and dHex$^+$ (fucose or xylose) at m/z 147. As a marker for both N- and O-linked oligosaccharide, they have suggested the HexNAc$^+$ oxonium ion as the best marker because a majority of the reported glycosylations have this residue as the linkage to the peptide (see Figures 5 and 6). In part, we have reproduced the pre-Q1 CID experiments with a tryptic and Asp-N digest of recombinant tissue plasminogen activator (r-tPA). This technique was very successful in locating the three glycosylation sites found on r-tPA: Asn 448, Asn 117, and Asn 184 (see Figure 7). In the trypsin digest, the upper set of extracted ion plots in Figure 7, Asn 117 and Asn 184 coelute. An Asp-N digest and LC/MS was also performed and the three glycosylation sites no longer coelute, as shown in the lower set of extracted ion plots in Figure 7. This is one example where another method of protein cleavage helped in separating carbohydrate sites into discrete regions of the map. These sites have been detailed previously by traditional carbohydrate techniques[81,82] and mass spectrometric methods.[58,59] There is also an O-linked fucose linked to Thr 61 on peptide T8,[56] but it was not detectable by the pre-Q1 fragmentation method. In our experiment, the orifice potential of the SCIEX API III was maintained at 170 V for the mass range 125 to 475 Da and at 70 V for the range 500 to 2000 Da. Regions around the T45 glycopeptide (Asn 448) were averaged, as were the scans around the T17 (Asn 184) and T11 glycopeptides (Asn 117) for the trypsin digest, LC/MS. These averages are shown in Figure 8. The Hex-HexNAc$^+$ m/z 366 and NeuAc$^+$ m/z 292 and 274 oxonium ions are more abundant than the HexNAc$^+$ ion at m/z 204. The 204 marker is obviously the smallest ion observed by this technique. However, this technique will prove to be valuable in locating the glycosylation sites in LC/MS glycopeptide maps and some measure of carbohydrate composition can be gained using this technique. One of the complicating factors that some researchers, including

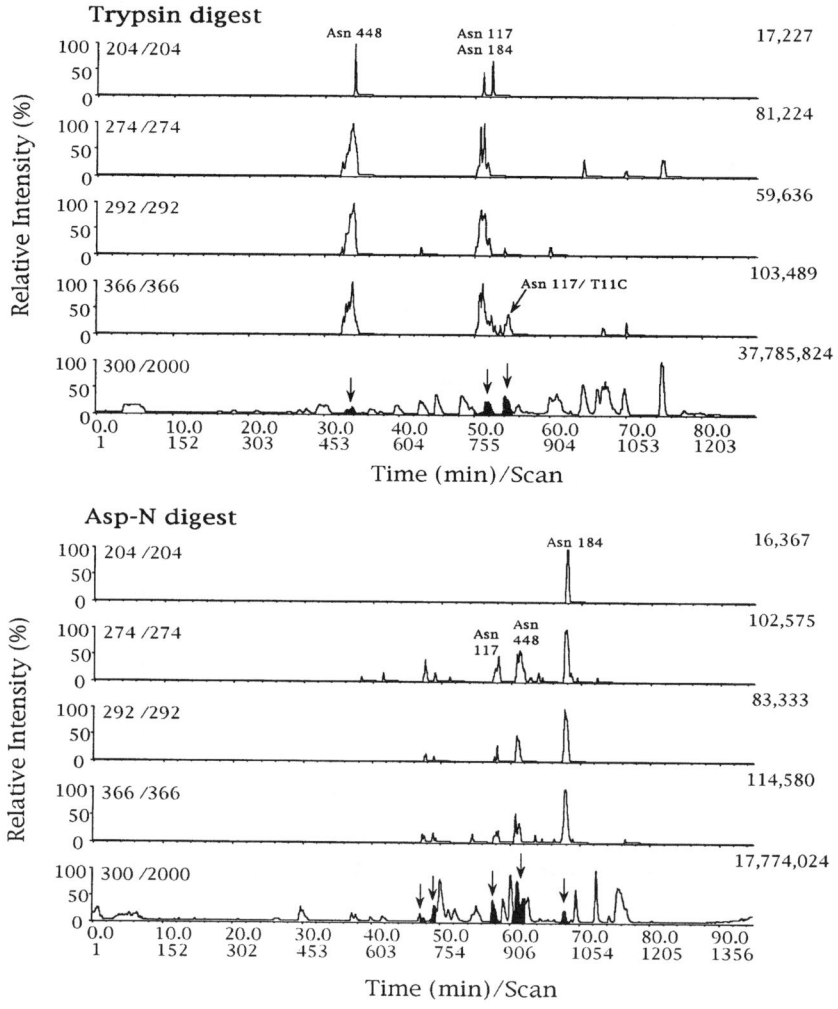

FIGURE 7
Trypsin and Asp-N digests of r-tPA were separated using a 0.3 × 150 mm Nucleosil C18, 5 μm capillary reversed phase column. The gradient and solvent systems were the same as those used in Figure 4. The LC/MS maps were generated in much the same way as the one in Figure 4. The digests were performed at ambient temperature for 24 h with two additions of enzyme resulting in a final enzyme to substrate ratio of 1:50. Oxonium ions were generated by scanning the mass range 140 to 450 Da at an orifice potential of 170V and then at 70V for the mass range of 500 to 2000 Da. This was accomplished using the multiple ion scanning mode of the SCIEX API III. The oxonium ions 204, 274, 292, and 366 were extracted from the two runs, and where multiple hits were recorded the total ion current was shaded to indicate possible glycopeptides. In the upper set of extracted ion plots (trypsin digest) Asn 117 and Asn 184 coelute; also noted in this plot is the T11C chymotryptic-like clip of the high mannose peptide. In the lower set of extracted ion plots (Asp-N digest) it was possible to separate and locate all three of the r-tPA glycosylation sites. Also noted in the lower set of extracted ion plots are several peaks at about 47 to 49 min that appear to be generating oxonium ions. These peaks have not been fully investigated yet.

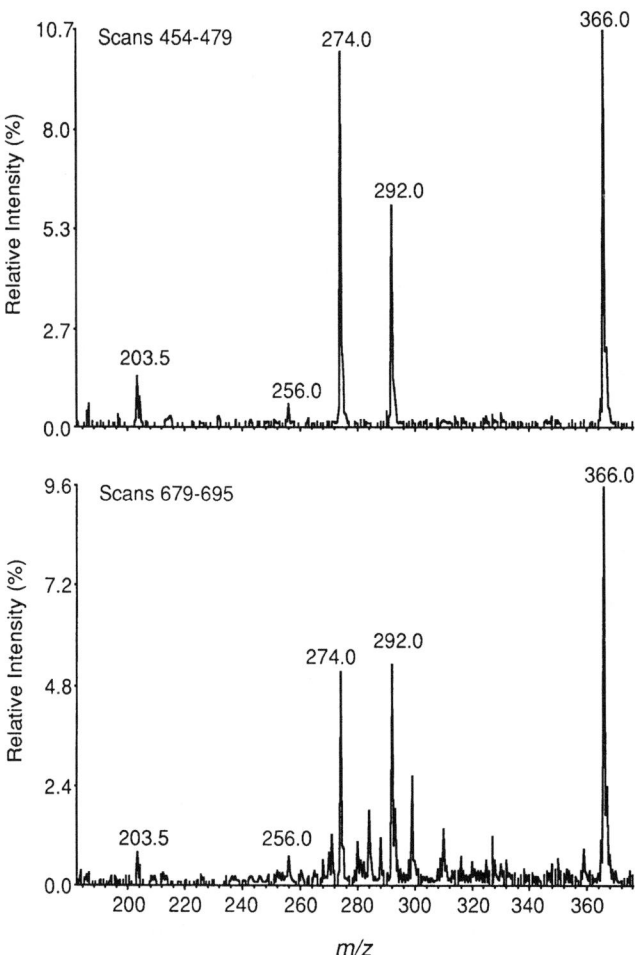

FIGURE 8
The upper plot shows the oxonium ions in the average of scans 454 to 479, the T45 peptide of r-tPA. The lower plot is the average of scans over the T17 and T11 r-tPA glycopeptides, showing just the oxonium ion region. We are observing the oxonium ion for GlcNAc at mass 203.5. The corresponding ions for sialic acid and GlcNAc-galactose are observed at masses 292 and 366, respectively. In the lower spectra many more fragment ions are observed. These additional ions are believed to be related to the coeluting nonglycosylated tryptic peptides T19 and T21. The ions at 274 and 256 are related to the sialic acid ion at 292 and are 18 and 36 mass units smaller. Each ion shows a sequential loss of water. The abundance of these ions can be shifted to 292 by lowering the orifice potential.

ourselves, have noted is that when a glycopeptide coelutes with a nonglycosylated peptide other fragment ions are generated — as an example see the lower trace of Figure 8. The glycopeptides T17 and T11 coelute with the nonglycosylated peptides T21 and T19, and we believe that this is

ANALYZING REVERSED PHASE PEPTIDE MAPS

giving rise to the additional ions observed in this spectra. Carr et al. also mentions parent ion scanning as a method of improving the selectivity of the technique.[79] A triple quadrupole mass spectrometer is used. The first mass analyzer (Q1) collects the mass spectra, collisions with argon are accomplished in Q2, and Q3 is set to select for the oxonium ion loss. The result is a very specific plot of the parent ions that have given rise to the oxonium ions.

5.1.2 Locating Glycopeptides in a Contour Plot

One of the most obvious signatures of a glycopeptide in an LC/MS reversed phase map is a cluster or cloud of ions that appear to have a negative slope in the plot of mass-to-charge vs. time and scan, or a "contour plot". The contour plot allows one to visualize all of the ions in an entire LC/MS run. The first impression is that of a 2-D gel, with spots in almost every scan.[58] With a closer inspection some very intriguing patterns begin to emerge. The duration of ions (peptides) becomes apparent, some will last ten scans while others will only endure for three or less. To some degree this is related to their concentration or the characteristics of a particular reversed phase column or to the elution behavior of the peptide. The population of ions in the plot shows a gradual positive slope over the course of the reversed phase run. This is attributed to the fact that as the organic gradient develops the more hydrophobic and generally larger peptides elute near the end of the run, while the hydrophilic and generally smaller peptides elute earlier. Often, what upsets this pattern is the appearance of a glycopeptide. Since a glycosylation can make a peptide more hydrophilic while greatly increasing its mass, the glycopeptides tend not to follow the general pattern of elution.

The observations described above led us to the identification of the glycopeptides in the analysis of a monoclonal antibody.[63] The antibody had only two major glycoforms, but they were on a peptide that eluted early in the gradient, thus pushing the ions out of the characteristic trend of nonglycosylated peptides. A signature of a heterogeneous glycopeptide is the appearance of a cloud of ions in the contour plot, as previously mentioned. This cloud may have 20 glycoforms, with ions at several charge states, all within a 1-min time frame in the LC/MS run. Figure 9 is an example from our recent study.[59] The upper trace of Figure 9 shows a contour plot of an LC/MS tryptic map of a glycosylation mutant of tissue plasminogen activator in which an extra glycosylation site occurs on tryptic peptide T11. With the presence of an extra carbohydrate moiety the peptide elutes earlier. In the wild-type contour plot, shown in the lower trace of Figure 9, T17 and T11 coelute. For mass analysis the duration of the ion dispersion is defined and the scans are averaged. The deconvolution of the T45 complex series of glycopeptide ions shown in the upper trace of Figure 9 will be demonstrated in Section 5.3.

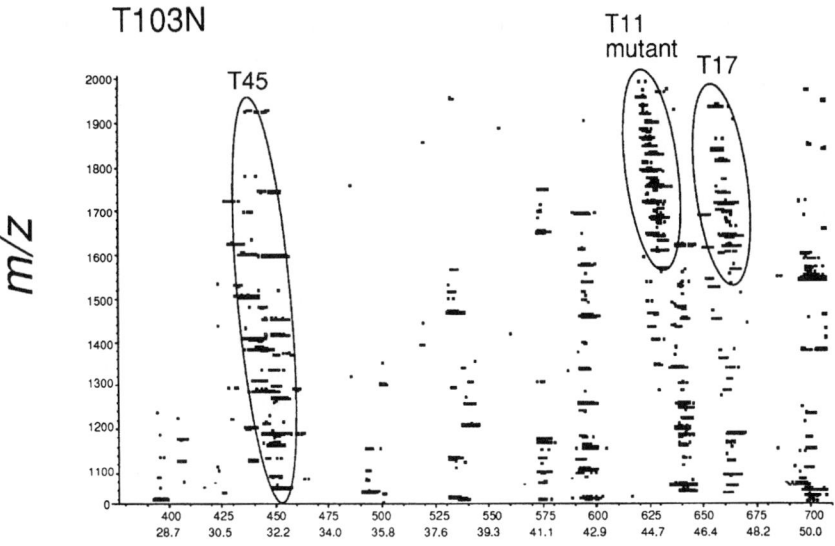
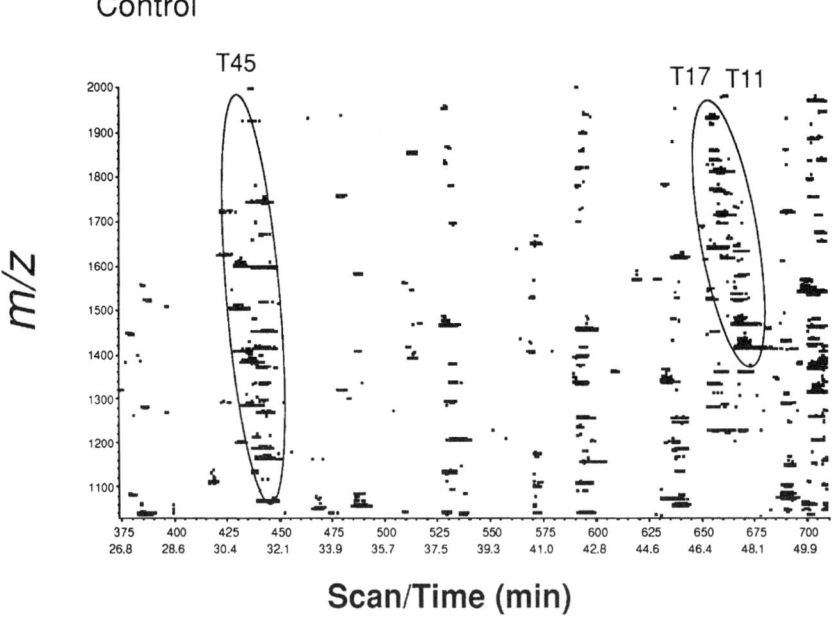

FIGURE 9
A comparison of the contour plot of a trypsin digest of the T103N mutant (upper trace), and the contour plot of the control r-tPA (bottom trace). The glycopeptide "streaks" are circled. The streak corresponding to the T11 tryptic glycopeptides of the T103N variant is noted as a cluster of ions eluting earlier than the T17 peptide (upper trace). In the control the T11 and T17 glycopeptides coelute (bottom trace).

5.2 Analyzing a Complex Series of Ions

Once the glycopeptides are identified in an LC/MS run, either through the observation of the glycosylation heterogeneity observed in a m/z vs. time contour plot or by the observation of oxonium ions produced by CID, the next step is characterization of the ions.

5.2.1 What's Missing?

Often the question of glycosylation characterization is secondary in the characterization of an LC/MS tryptic map. The researcher will try to identify as many of the predicted tryptic peptides as possible and will invariably find some masses missing. Sometimes, the missing peptides can be accounted for by an unusual cleavage or a posttranslational modification of the protein unrelated to glycosylation.[83,84] In some cases the missing peptide will possess an N-linked consensus sequence, and in other cases the possibility of an O-linked attachment. This was one of the clues that could be used when the mass for peptide T8 in r-tPA could not be found in an LC/MS map. While the location of fucose on this peptide was reported by other approaches,[86] this approach has often been useful in the mass spectral analysis of peptide maps.

5.2.2 Using Sugar Residue Differences and Calculated Compositions to Propose Carbohydrate Structures

After location of the glycopeptide by observation of an ion cloud in the contour plot or by the oxonium ion method, the next step is to average the scans defined by these glycopeptide signatures. If heterogeneity is present, the glycopeptide can be characterized initially by the monosaccharide mass differences. For example, a high mannose glycopeptide may exhibit a series of ions that differ by the residue mass of mannose at a particular charged state (see Table 1). It may then be possible to go to a different charge state in this same average of scans and observe another cluster of ions that differ by mannose residues. With this evidence one can be reasonably confident that one has found a high mannose type glycopeptide.

The next step is to link this observation with a potential candidate peptide. With the charge state established through the relationship of the ions to one another the parent mass can be calculated, even if another charge state can not be found on which to base the calculation. With the assumption that this is an N-linked high mannose glycopeptide, the core mass of two GlcNAc and three mannose residues can be subtracted from the established observed parent mass. Mannose residues can then be sequentially subtracted until the mass of one of the candidate peptides is matched. This method will give carbohydrate compositional data and general structural information.

An example of this method is demonstrated using the r-tPA T45 tryptic peptide. This peptide contains Asn 448 and has a sialylated set of complex

TABLE 1
Masses of Monosaccharide Units[a]

Sugar[b]	Mass	Charged state			
		+2	+3	+4	+5
Fucose	146.1	73.0	48.7	36.5	29.2
Hexose	162.1	81.0	54.0	40.4	32.4
HexNAc	203.2	101.6	67.7	50.8	40.6
NANA	291.3	145.6	97.1	72.8	58.2
HexNAc+Hex	365.3	182.6	121.8	91.3	73.1

[a] The residue mass of a sugar as linked in a polysaccharide.
[b] HexNAc = N-acetylhexosamine, NANA = N-acetylneuraminic acid. HexNAc+Hex is the branch unit that differentiates diantennary from triantennary and tetraantennary in complex-type carbohydrate structures in general.

carbohydrates.[81] These ions were located in the LC/MS map by their glycosylation signature in the contour plot, and then the scans that contained the ions were averaged. The T45 glycopeptide signature is located in the upper plot of Figure 9, and in Figure 10 the upper plot shows the average of the scans. The lower plot in Figure 10 is a schematic of the most obvious series of carbohydrate ions noted in the upper plot. It should be noted that the averaged mass spectrum in the upper plot of Figure 10 only shows the mass range of 1000 to 1600 Da, and shows mainly the triply charged set of ions for this peptide. Some of the ions that are not shown in the lower schematic are ions from the doubly and quadruply charged sets that overlap with this series. In addition there are also several less abundant ions that we did not attempt to identify.

In this analysis the initial observation is that there are three distinct ion groups that differ by approximately 97 Da. The 97-Da difference indicates sialic acid at the 3+ charge state (see Table 1). With this approach it was possible to give a structure to the parent mass of all the related ions at this charge state (see Table 2). Several obvious conclusions can be made about the three series noted in Figure 10. The glycosylation most likely is of the complex variety, due to the pattern of the sialic acid differences. Secondly, series A is smaller than series B and, likewise, series B is smaller than series C, which might suggest diantennary, triantennary, and tetraantennary structures since we have already made an assumption of a complex-type glycosylation. This information is summarized in Table 2. In Table 2 the parent mass is calculated. Based on the fact that series A < B < C, the assumption of di, tri, and tetraantennary is made. Assuming that the smallest member of each group is the asialo form, each of the forms is reduced to the asialo form. If our assumptions are correct, the groups should differ by 365 Da or the mass

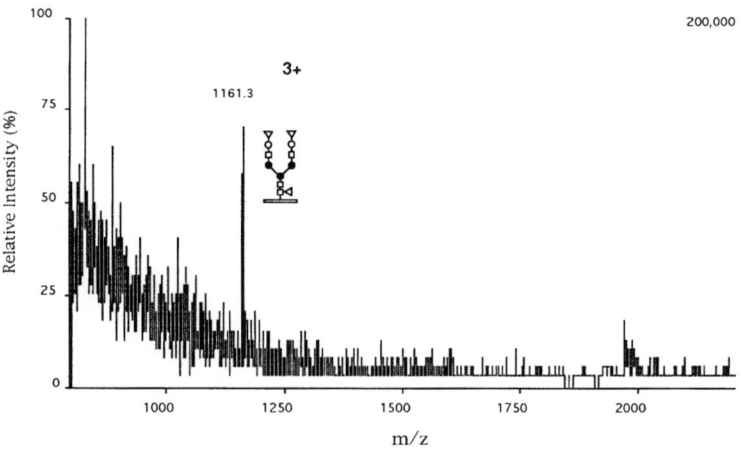

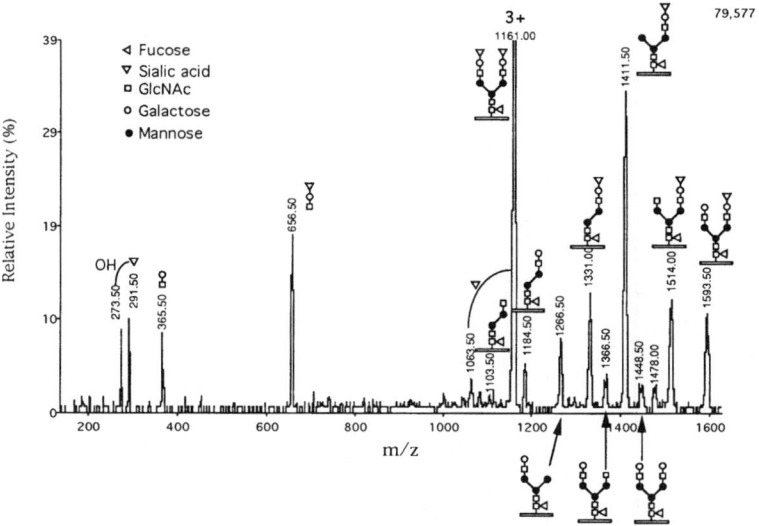

FIGURE 10
The top plot shows the triply charged parent ion of the fully sialylated diantennary glycoform of peptide T45 from an r-tPA tryptic digest as observed in the first quadrupole. The lower plot shows the product ions produced and observed in Q3 after collision with argon in Q2 of a triple quadrupole mass spectrometer.

of the common branched unit, HexNAc+Hex. This is true for series A and B, but B and C differ by a greater mass. If 365 is subtracted from this number, the analyst is left with the residue mass for sialic acid, which indicates that the first member in the tetraantennary series is the monosialylated form and not the asialo form as first assumed. True linkage information must be obtained

TABLE 2
Glycopeptide Series Deductions

Series	Calculated M_r	Assume	Subtract to asialo state	Average	Delta mass between series
A	2897.7	Diantennary, 0 NANA	2897.7		
	3189.0	Diantennary, 1 NANA	2897.7		
	3480.0	Diantennary, 2 NANA	2897.4	2897.6	
					365.0 Da = HexNAc + Hex
B	3263.4	Triantennary, 0 NANA	3263.4		
	3554.1	Triantennary, 1 NANA	3262.8		
	3844.8	Triantennary, 2 NANA	3262.2		
	4135.8	Triantennary, 3 NANA	3261.9	3262.6	
					656.9 Da = HexNAc + Hex + NANA
C	3919.8	Tetraantennary, 0 NANA	3919.8		
	4210.8	Tetraantennary, 1 NANA	3919.5		
	4502.1	Tetraantennary, 2 NANA	3919.5		
	4793.1	Tetraantennary, 3 NANA	3919.2	3919.5	

through MS/MS studies or traditional carbohydrate chemistry methods and NMR.[87-92]

5.3 COLLISION-INDUCED DISSOCIATION (CID) OF GLYCOPEPTIDES

Structural analysis of oligosaccharides, and glycopeptides by MS/MS techniques have proven to be a valuable tool in elucidating carbohydrate structures.[93-96] Certainly, compositional data can be gained easily with little ambiguity due to the mass accuracy of the electrospray instruments and the redundant nature of monosaccharide masses. A general composition for a core fucosylated, fully sialylated diantennary glycoform may be presented as $Fuc_1HexNAc_4Hex_5NANA_2$. The mass spectrometrist is constantly aware that the mass result is only a number, and not conclusive structural data. With the help of MS/MS fragmentation, however, our confidence level can be improved. For example, if the glycoform is attached to a peptide that has only one N-linked consensus sequence and no possible O-linked sites, the common core of two GlcNAc and three mannose residues can be safely assumed. From there, core fucosylation and other linkage orders can be assumed through compositional analysis based on the masses of the sugar residues. There are other rules or dogmas that can be invoked concerning linkage order and typical mammalian anomeric configurations to help one elucidate a structure, especially when very little material is available for further studies. Our general rule is to present our result as a best guess for a structure and then support it up with as much data as possible. The ocean of carbohydrate chemistry is vast enough to yield more than a few surprises and hard-and-fast rules begin to have exceptions. One nonabsolute rule is that all N- and O-linked carbohydrates are either linked through GlcNAc or GalNAc.[97] As previously mentioned, solitary O-linked fucose and GlcNAc residues have been observed.

Once a glycopeptide is found in an LC/MS reversed phase map of an enzyme digest and its possible composition is calculated, it can be subjected to MS/MS analysis. This experiment can confirm that this is a glycopeptide. Characteristic oxonium ions fragments will be observed at the low mass end of the spectra; also, larger glycopeptide fragments will show incremental losses of NANA, hexose, and HexNAc units. For an example of this technique see Figure 11. Figure 11, top frame, shows the 3+ carbohydrate ion for the fully sialylated, N-linked diantennary glycoform of r-tPA which is attached to tryptic peptide T45 at the linkage site Asn 448. The T45 glycopeptide mixture was collected from a tryptic map. Due to the dilution effects of the collection from the reversed phase peptide map only the most abundant form was observed. The ion at 1161.3 was selected for collision in Q2 and the fragments were observed in Q3, shown in the bottom frame of Figure 11. Only the fragment products from the 1161.3 ion are observed, demonstrating the selectivity of this approach. This simple MS/MS experiment still

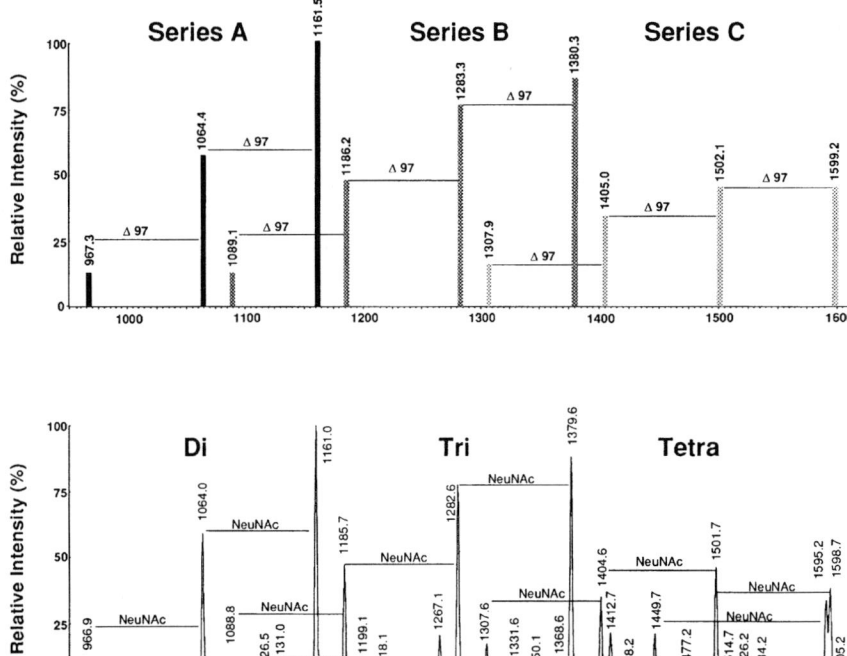

FIGURE 11
The upper plot is a schematic of the triply charged ions of the r-tPA T45 glycopeptide observed in the spectra in the lower plot. The upper plot schematic demonstrates that there are three series present, and within each of the series the ions differ by 97 Da, or the mass of sialic acid at the triply charged state. The lower spectra shows all ions in this mass region. Some of the additional ions are related to T45 at the doubly and quadruply charged state and some are left to be identified.

leaves us with a few ambiguities. For example, when a sialic acid is lost we can not tell from which branch it was taken, but we can tell that it was a terminal residue. The observation of the compositions peptide+HexNAc$_4$Fuc$_1$Hex$_5$ NANA$_1$, and then peptide+HexNAc$_4$Fuc$_1$Hex$_5$ distinguish the terminal residue. We have then confirmed this result with neuraminidase treatment of the parent molecule. The loss of the core fucose is not observed in this experiment. Often, one is able to get down to the peptide fragments peptide+GlcNAc$_2$ Fuc, peptide+GlcNAcFuc, and peptide+GlcNAc, thus demonstrating that the fucose was attached to the GlcNAc linked to the peptide (data not shown). The MS/MS experiment has proven that the lone mass at 1161.3 is a carbohydrate structure. Furthermore, through careful observations and a little common logic, a limited sequence order can be deduced and proof can be acquired that we are dealing with branched structures. Methylation analysis can further help us to decipher linkage position and structure.[98,99] The methylation analysis converts the hydroxyl groups to methyoxyl groups, HexNAc residues are N-methylated and sialic acid residues are converted to methyl esters. The use of methylation in conjunction with MS/MS analysis can confirm whether sialic acid residues are terminal or polysialic in nature, and it can also distinguish a hexose that has three linkages vs. one in a terminal branch that has two, or one. Examples of other types of chemical analyses are available in References 100 through 103.

5.4 Artifacts in Electrospray

Electrospray has been sold as a soft ionization method that creates few artifacts. In reality, a number of artifacts have been observed in electrospray. We have exploited them to our benefit in some instances. The most common artifacts are oxidation of methionines,[104] peptide fragmentation,[105] carbohydrate fragmentation,[106] metal adducts Na+ and Fe_{2+},[19] and the observation of dimeric peptides.[107] The artifacts of fragmentation and oxidation can be reduced in most cases by lowering the ionization and orifice potentials. The artifactual formation of dimers seems to be concentration related, but also may be a reflection of how the peptides exist in the final stages of droplet evaporation before ion ejection.[108,109] Another artifact, as previously mentioned, is oxonium ions produced from carbohydrate peptides under standard declustering potentials using an API SCIEX mass spectrometer (orifice voltage of 60).[78] Such observations are not common under normal operating conditions, but the analyst must be aware of such possibilities. Many of these artifacts can be tuned out, but the fact that they can exist cannot. The researcher is left with the question of what is real?

5.4.1 Using LC Separation and Extracted Ion Plotting to Uncover Artifacts

The reality that artifacts exist strengthens the marriage of liquid chromatography and mass spectrometry. An electrospray artifact should mimic the elution profile of the parent species because electrospray artifacts are created at the electrospray source. The artifactual ion in most cases should peak at the same point as the parent ion. Oxidation, which has been shown to be an artifact of electrospray, is easily differentiated from oxidation that occurs preionization. In a reversed phase separation the oxidized peptide will usually elute 5 to 15 min earlier than the nonoxidized peptide.[110] Families of carbohydrate peptides should show a logical elution profile with the larger glycopeptides being more hydrophilic and eluting earlier. An example of this elution of a series of closely related glycopeptides is shown in Figure 12. This figure is obtained by extracting the ion (usually at the most abundant charge

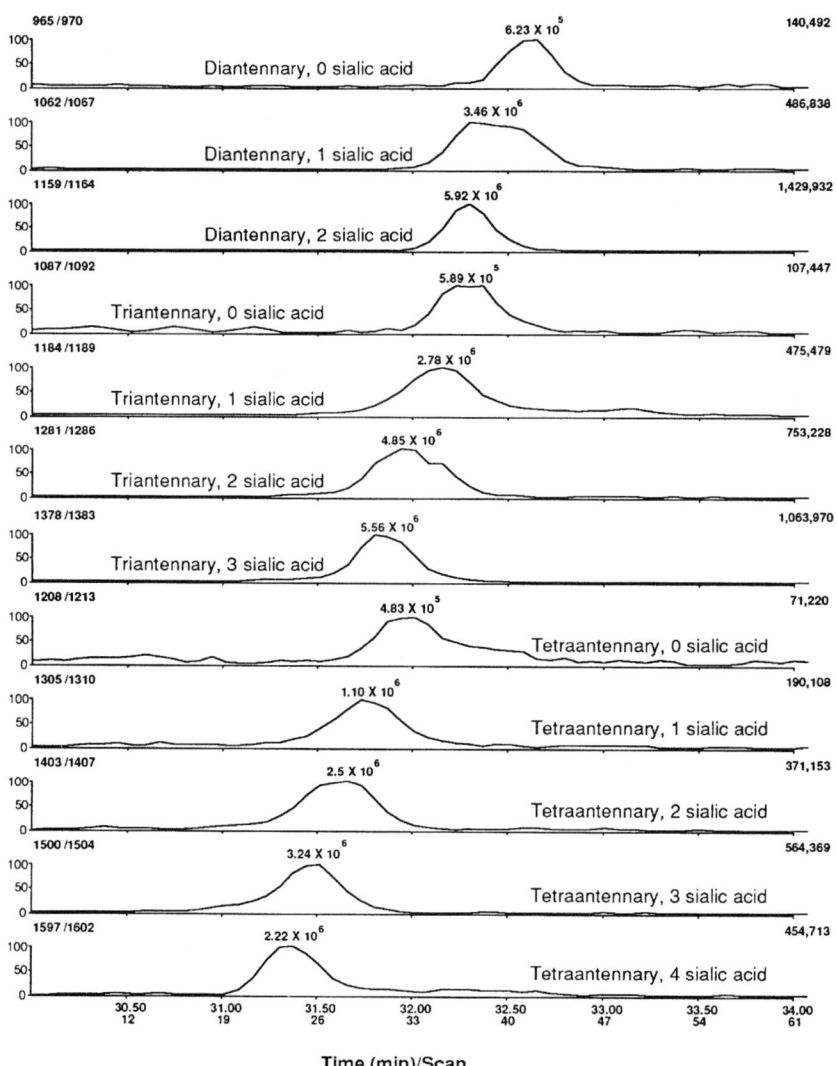

FIGURE 12
The triply charged ions of the T45 glycopeptide were extracted from an LC/MS run and plotted in separate plots. The total extracted ion current, or the area under each peak is labeled above each peak. Note that the larger glycopeptides elute earlier in the reversed phase separation.

state) for each of the glycopeptides and then plotting each individual elution profile. This method was utilized previously for ES-MS followed by extracted ion plots with ES in the extracted ion plotting of glycopeptides.[59,111] This method can be used to prove that an observed glycoform is not an artifact of the ionization, because it elutes with a different retention time. One challenge is to try and improve the chromatographic separation while maintaining the ionization potential of electrospray. Some of the best reversed phase chromatographic separations involve the addition of nonvolatile salts or neutral pH conditions. Artifacts will invariably remain a factor in electrospray mass spectrometry just as they will with all analytical techniques. For this reason, analytical chemistry must continue to maintain a multidimensional approach to any analysis.

REFERENCES

1. Hellerqvist, C. G., Linkage method using Lindberg method, *Methods in Enzymology*, Academic Press, New York, 1990, chap. 30.
2. Hellerqvist, C. G. and Sweetman, B. J., *Biomedical Applications of Mass Spectrometry*, Vol. 34, John Wiley & Sons, New York, 1990, 97.
3. Liu, D. T., Glycoprotein pharmaceuticals: scientific and regulatory considerations, and the U.S. orphan drug act, *Tibtech*, 10, 114, 1992.
4. Rudd, P., Oligosaccharides in human biology, *Glyco News*, 3, 1, 1993.
5. Varki, A., Biological roles of oligosaccharides: all theories are correct, *Glycobiology*, 3(2), 97, 1993.
6. Anderegg, R. J., *Biomedical Applications of Mass Spectrometry*, Vol. 34, John Wiley & Sons, 1990, 1.
7. Barber, M., Bordoli, R. S., Sedgwick, R. D., and Tyler, A. N., *J. Chem. Soc. (London) Chem. Commun.*, 7, 325, 1981.
8. Dole, M., Mack, L. L., Hines, R. L., Mobley, R.C., Ferguson, L. D., and Alice, M. B., Molecular beams of macroions, *J. Chem. Phys.*, 49, 2240, 1968.
9. Lenard, P., Über Wasserfallelektrizität und über die oberflächenbeschaffenheit der flüssigkeiten, *Ann. Phys.*, 47, 463, 1915.
10. Zeleny, J., Instability of electrified liquid surfaces, *Phys. Rev.*, 10, 1, 1917.
11. Yamashita, M. and Fenn, J. B., Electrospray ion source. Another variation on the free-jet theme, *J. Phys. Chem.*, 88, 4451, 1984.
12. Whitehouse, C. M., Dreyer, R. N., Yamashita, M., and Fenn, J. B., Electrospray interface for liquid chromatographs and mass spectrometers, *Anal. Chem.*, 57, 675, 1985.
13. Irbarne, J. V. and Thomson, B. A., On the evaporation of small ions from charged droplets, *J. Chem. Phys.*, 64, 2287, 1976.
14. Thomson, B. A. and Iribarne, J. V., Field induced ion evaporation from liquid surfaces at atmospheric pressure, *J. Chem Phys.*, 71, 4451, 1979.
15. Meng, C. K., Mann, M., and Fenn, J. B., Of protons or proteins, *Z. Phys. C. Atoms Molecules Clusters*, 10, 361, 1988.
16. Loo, J. A., Udseth, H. R., and Smith, R. D., Solvent effects on the charge distribution observed with electrospray ionization-mass spectrometry of large molecules, *Biomed Environ. Mass Spectrom.*, 17, 411, 1988.
17. Bruins, A. P., Covey, T. R., and Henion, J. D., IonSpray interface for combined liquid chromatography/atmospheric pressure ionization mass spectrometry, *Anal. Chem.*, 59, 2624, 1987.
18. Bruins, A. P., Weidolf, L. O. G., Henion, J. D., and Budde, W. L., Determination of sulfonated Azo dyes by liquid chromatography/atmospheric pressure ionization mass spectrometry, *Anal. Chem.*, 59, 2647, 1987.
19. Simons, D. S., Colby, B. N., and Evans, C. A., Jr., Electrohydrodynamic ioization mass spectrometry — the ionization of liquid glycerol and nonvolatile organic solutes, *Int. J. Mass Spectrom. Ion Phys.*, 15, 291, 1974.

20. Mirza, U. A., Cohen, S. L., and Chait, B. T., Heat-induced conformational changes in proteins studied by electrospray ionization mass spectrometry, *Anal. Chem.*, 65, 1, 1993.
21. Whitehouse, C. M., Andrien, B., Banks, F., Quinn, J., and Shen, S., Performance improvements of an ultrasonic nebulizer assisted electrospray ion source, (Ultaraspray™), *Proc. 41st ASMS Conf. Mass Spectrometry and Allied Topics*, San Francisco, CA, May 30th to June 4th, 1993.
22. Loo, J. A., Udseth, H. R., and Smith, R. D., Peptide and protein analysis by electrospray ionization-mass spectrometry and capillary electrophoresis-mass spectrometry, *Anal. Biochem.*, 179, 404, 1989.
23. Suter, M. J.-F., DaGue, B. B., Moore, W. T., Lin, S., and Caprioli, R. M., Recent advances in liquid chromatography-mass spectrometry and capillary zone electrophoresis-mass spectrometry for protein analysis, *J. Chromatogr.*, 553, 101, 1991.
24. Hamdan, M. and Curcuruto, O., Development of the electrospray ionization technique, *Int. J. Mass Spectrom. Ion Processes*, 108, 93, 1991.
25. Ikonomou, M. G., Blades, A. T., and Kebarle, P., Electrospray-ion spray: a comparison of mechanisms and performance, *Anal Chem.*, 63, 1989, 1991.
26. Smith, R. D., Loo, J. A., Barinaga, C. J., Edmonds, C. G., and Udseth, H. R., New development in biochemical mass spectrometry: electrospray ionization, *Anal. Chem.*, 62, 882, 1990.
27. Wong, S. F., Meng, C. K., and Fenn, J. B., Multiple charging in electrospray ionization of poly(ethylene) glycols, *J. Phys. Chem.*, 92, 546, 1988.
28. Smith, R. D., Loo, J. A., Barinaga, C. J., Edmonds, C. G., and Udseth, H. R., New development in biochemical mass spectrometry: electrospray ionization, *Anal. Chem.*, 62, 882, 1990.
29. Chait, B. T. and Kent, B. H., Weighing naked proteins: practical, high-accuraccy mass measurement of peptides and proteins, *Science*, 257, 1885, 1992.
30. Guzzetta, A. W., Keyt, B. A., Bennet, W. F., and Hancock, W. S., Strategies for analyzing glycosylated proteins using electrospray mass spectrometry, Proceedings of the 42nd ASMS Conference on Mass Spectrometry and Allied Topics, Chicago, Illinois, May 29–June 3, 1994; pp 1155.
31. Senko, M. W., Beu, S. C., and McLafferty, F. W., Mass and charge assignment for electrospray ions by cation adduction, *J. Am. Soc. Mass Spectrom.*, 4(10), 828, 1993.
32. Mann, M., Meng, C. K., and Fenn, J. B., Interpreting mass spectra of multiply charged ions, *Anal. Chem.*, 61, 1702, 1989.
33. Labowsky, M., Whitehouse, C., and Fenn, J. B., Three-dimensional deconvolution of multiply charged spectra, *Rapid Commun. Mass. Spectrom.*, 7, 71, 1993.
34. Covey, T. R., Bonner, R. F., Shushan, B. I., and Henion, J., The determination of protein, oligonucleotide and peptide molecular weights by ion-spray mass spectrometry, *Rapid Commun. Mass Spectrom.*, 2(11), 1988.
35. Fenn, J. B., Mann M., and Meng, C. K., Method of Producing Multiply Charged Ions and for Determining Molecular Weights of Molecules by use of the Multiply Charged Ions of Molecules, U.S. Patent 5,130,538, 1990.
36. Fenn, J. B., Mann M., and Meng, C. K., Multiply Charged Ions and a Method for Determining the Molecular Weight of Large Molecules, International patent BO1D 59/44, HO1J 49/00, International publication number WO 90/14148, 1990.
37. Lee, T. D. and Vemuri, S., MacProMass: a computer program to correlate mass spectral data to peptide and protein structures, *Biomed. Environ. Mass Spectrom.*, 19, 639, 1990.
38. Paul, W., Reinhard, H. P., and von Zahn, U., Das elekrische massenfilter als massenspectrometer und isotopetrenner, *Z. Phys.*, 152, 143, 1958.
39. Ryhage, R., The mass spectrometry laboratory at the Karolinska Institute 1944-1987, *Mass Spectrom. Rev.*, 12(1), 1, 1993.

40. Chloupek, R. C., Harris, R. J., Leonard, C. K., Keck, R. G., Keyt, B. A., Spellman, W. M., Jones, A. J. S., and Hancock, W. S., Study of the primary structure of recombinant tissue plasminogen activator by reversed-phase high-performance liquid chromatographic tryptic mapping, *J. Chromatogr.*, 463, 375, 1989.
41. Frenz, J., Hancock, W. S., Henzel, W. J., and Horváth, C., Reversed phase chromatography in analytical biotechnology of proteins, *HPLC of Biological Macromolecules, Methods and Applications*, Gooding, K. M. and Regnier, F. E., Eds., Marcel Dekker, New York, 1990, 145.
42. Welinder, B. S., Sorensen, H. H., and Hansen, B., Reversed-phase high-performance liquid chromatography of human growth hormone, *J. Chromatogr.*, 398, 309, 1987.
43. Johnson, B. A., Shirokawa. J. M., Hancock, W. S., Spellman, M. W., Basa, L. J., and Aswad, D. W., *J. Biol. Chem.*, 264(24), 14262, 1989.
44. Stults, J. T., Bourell, J. H., Canova Davis, E., Ling, V. T., Laramee, G. R., Winslow, J. W., Griffin, P. R., Rinderknecht, E., and Vandlen, R. L., Structural characterization by mass spectrometry of native and recombinant human relaxin, *Biomedical and Environmental Mass Spectrometry*, 19, 655, 1990.
45. Moseley, M. A., Deterding, L. J., Tomer K. B., and Jorgenson, J. W., Nanoscale packed-capillary liquid chromatography coupled with mass spectrometry using a coaxial continuous-flow fast atom bombardment interface, *Anal. Chem.*, 63, 1467, 1991.
46. Moritz, R. L. and Simpson, R. J., Application of capillary reversed-phase high-performance liquid chromatography to high-sensitivity protein sequence analysis, *J. Chromatogr.*, 599, 119, 1992.
47. Davis, M. T. and Lee, T. D., Analysis of peptide mixtures by capillary high performance liquid chromatography: a practical guide to small-scale separations, *Protein Sci.*, 1, 1992.
48. Hiraoka, K. and Kudaka, I., Formation of multiply charged ions of the oligopeptide arg-arg-arg by electrospray ionization, *Anal. Chem.*, 64, 75, 1992.
49. Fisher, S., manuscript in preparation.
50. Caprioli, R. M., Moore, W. T., DaGue, B., and Martin, M. J., Microbore high-perfomance liquid chromatography-mass spectrometry for the analysis of proteolytic digests by continuous-flow fast-atom bombardment mass spectrometry, *J. Chromatogr.*, 443, 355, 1988.
51. Henzel, W. J., Bourell, J. H., and Stults, J. T., Analysis of protein digests by capillary high performance liquid chromatography and on-line fast atom bombardment mass spectrometry, *Anal. Biochem.*, 187, 228, 1990.
52. Caprioli, R. M., DaGue, B., Fan, T., and Moore, W. T., Microbore HPLC/mass spectrometry for the analysis of peptide mixtures using a continuous flow interface, *Biochem. Biophys. Res. Commun.*, 146, 291, 1987.
53. Caprioli, R. M., Fan, T., and Cottrell J. S., Continuous-flow sample probe for fast atom bombardment mass spectrometry, *Anal. Chem.*, 58, 2949, 1986.
54. Frenz, J., Bourell, J., and Hancock, W. S., High performance displacement chromatography-mass spectrometry of tryptic peptides of recombinant human growth hormone, *J. Chromatogr.*, 512, 299, 1990.
55. Sturgeon, R. J., The glycoprotein and glycogen, in *Carbohydrate Chemistry*, Kennedy, J. F., Ed., Oxford University Press, New York, 1988, chap. 7.
56. Harris, R. J., Leonard, C. K., Guzzetta, A. W., and Spellman, M. W., Tissue plasminogen activator has an O-linked fucose attached to threonine-61 in the epidermal growth factor domain, *Biochemistry*, 30, 2311, 1991.
57. Turner, J. R., Tartakoff, A. M., and Greenspan, N. S., Cytologic assessment of nuclear and cytoplasmic O-linked N-acetylglucosamine distribution by using antistreptococcal monoclonal antibodies, *Proc. Natl. Acad. Sci. U.S.A.*, 87, 5608, 1990.

58. Ling, V., Guzzetta, A. W., Canova-Davis, E., Stults, J. T., Covey, T. R., Shushan, B. I., and Hancock, W. S., Characterization of the tryptic map of recombinant DNA derived tissue plasminogen activator by high-performance liquid chromatography-electrospray ionization mass spectrometry, *Anal. Chem.*, 63, 2909, 1991.
59. Guzzetta, A. W., Basa, L. J., Hancock, W. S., Keyt, B. A., and Bennett, W. F., Identification of carbohydrate structures in glycoprotein peptide maps by the use of LC/MS with selected ion extraction with special reference to tissue plasminogen activator and a glycosylation variant produced by site directed mutagenesis, *Anal. Chem.*, 65, 2953, 1993.
60. Chait, B. T. and Kent, B. H., Weighing naked proteins: practical, high-accuraccy mass measurement of peptides and proteins, *Science*, 257, 1885, 1992.
61. Chen, W. and Bahl, O. P., A recombinant carbohydrate variant of human gonadotropin β-subunit (hcgβ), descarboxyl terminus (115-145), *J. Biol. Chem.*, 266(10), 6246, 1991.
62. Lee, T. D. and Shively J. E., Enzymatic and chemical digestion of proteins for mass spectrometry, *Methods in Enzymology*, Academic Press, New York, 1990, chap. 19.
63. Lewis, D. A., Guzzetta, A. W., and Hancock, W. S., The characterization of humanized Anti-Tac, an antibody directed against the interleukin 2 receptor, using electrospray ionization mass spectrometry by direct infusion, LC/MS, and MS/MS, *Anal. Chem.*, 66(5), 585, 1994.
64. Hunt, D. F., Shabanowitz, J., Yates, J. R., Zhu, N.-Z., Russell, D. H., and Castro, M. E., Tandem quadrupole Fourier-transform mass spectrometry of oligopeptides and small proteins, *Proc. Natl. Acad. Sci. U.S.A.*, 84, 620, 1987.
65. Johnson, B. A., Shirokawa, J. M., Hancock, W. S., Spellman, M. W., Basa, L. J., and Aswad, D. W., Formation of isoaspartate at two distinct sites during in vitro aging of human growth hormone, *J. Biol. Chem.*, 264(24), 14262, 1989.
66. Paranandi, M. V., Guzzetta, A. W., Hancock, W. S., and Aswad, D. W., Deamidation and isoaspartate formation during in vitro aging of recombinant tissue plasminogen activator, *J. Biol. Chem.*, 269(1), 243, 1994.
67. Biemann, K. and Martin, S.A., Mass spectrometric determination of the amino acid sequence of peptides and proteins, *Mass Spectrom. Rev.*, 6, 1, 1987.
68. Petra, P. H., Griffin, P. R., Yates, J. R., III, Moore, K., and Zhang, W., Complete enzymatic deglycosylation of native sex steroid-binding protein (SBP or SHBG) of human and rabbit plasma. Effect on steroid-binding activity, *Prot. Sci.*, 1, 902, 1992.
69. Araki, Y., Orgebin-Crist, M. C., and Tulsiani, D. R., Qualitative characterization of oligosaccharide chains present on the rat zona pellucida glycoconjugates, *Biol. Reprod.*, 46(5), 912, 1992.
70. Hunt, D. F., Yates, J. R., III, Shabanowitz, J., Winston, S., and Hauer, C. R., Protein sequencing by tandem mass spectrometry, *Proc. Natl. Acad. Sci. U.S.A.*, 83, 6233, 1986.
71. Hunt, D. F., Henderson, R. A., Shabanowitz, J., Sakaguchi, K., Michel, H., Sevilir, N., Cox, A. L., Appella, E., and Engelhard V. H., Characterization of peptides bound to the class I MHC molecule HLA-A2.1 by mass spectrometry, *Science*, 255, 1261, 1992.
72. Covey, T. R., Huang, E. C., and Henion, J. D., Structural characterization of protein tryptic peptides via liquid chromatography/mass spectrometry and collision-induced dissociation of their doubly charged molecular ions, *Anal. Chem.*, 63, 1193, 1991.
73. Hunt, D. F., Zhu, N. Z., and Shabanowitz, J., Oligopeptide sequence analysis by collision-activated dissociation of multiply charged ions, *Rapid Commun. Mass Spectrom.*, 3(4), 122, 1989.

74. Gillece-Castro, B. and Burlingame, A. L., Oligosaccharide chacterization with high-energy collision-induced dissociation mass spectrometry, *Meth. Enzymol.*, 193, 689, 1990.
75. Loo, J. A., Udseth, H. R., and Smith, R. D., Solvent effects on the charge distribution observed with electrospray ionization-mass spectrometry of large molecules, *Biomed Environ. Mass Spectrom.*, 17, 411, 1988.
76. Smith, R. D., Loo, J. A., Barinaga, C. J., Edmonds, C. G., and Udseth, H. R., Collisional activation and collision-activated dissociation of large multiply charged polypeptides and proteins produced by electrospray ionization, *J. Am. Soc. Mass Spectrom.*, 1, 53, 1990.
77. Katta, V., Chowdhury, S. K., and Chait, B. T., Use of a single-quadrupole mass spectrometer for collision-induced dissociation studies of multiply charged peptide ions produced by electrospray ionization, *Anal. Chem.*, 63, 174, 1991.
78. Conboy, J. J. and Henion, J. D., The determination of glycopeptides by liquid chromatography/mass spectrometry with collision-induced dissociation, *J. Am. Soc. Mass Spectrom.*, 3, 804, 1992.
79. Carr, S. A., Huddleston, M. J., and Bean, M. F., Selective identification and differentiation of N- and O-linked oligosaccharides in glycoproteins by liquid chromatography-mass spectrometry, *Prot. Sci.*, 2, 183, 1993.
80. Huddleston, M. J., Bean, M. F., and Carr, S. A., Collisional fragmentation of glycopeptides by electrospray ionization LC/MS and LC/MS/MS. Methods for selective detection of glycopeptides in protein digests, *Anal. Chem.*, 65, 877, 1993.
81. Spellman, M. W., Basa, L. J., Leonard, C. K., Chakel, J. A., O'Connor, J. V., Wilson, S. W., and van Halbeek, H., Carbohydrate structures of human tissue plasminogen activator expressed in Chinese hamster ovary cells, *J. Biol. Chem.*, 264, 14100, 1989.
82. Parekh, R. B., Dwek, R. A., Thomas, J. R., Opdenakker, G., Rademacher, T. W., Witter, A. J., Howard S. C., Nelson, R., Siegel, N. R., Jennings, M. G., Harakas, N. K., and Feder, J., Cell-type specific and site-specific N-glycosylation of type I and type II human tissue plasminogen activator, *J. Biochem.*, 28, 7644, 1989.
83. Carr, S. A., Hemling, M. E., Bean, M. F., and Roberts, G. D., Integration of mass spectrometry in analytical biotechnology, *Anal. Chem.*, 63, 2802, 1991.
84. Huddleston, M. J., Annan, R. S., Bean M. F., and Carr, S. A., Selective detection of phosphopeptides in complex mixtures by electrospray liquid chromatography/mass spectrometry, *J. Am. Soc. Mass Spectrom*, 4, 710, 1993.
85. Harris, R. J., Leonard, C. K., Guzzetta, A. W., and Spellman, M. W., Tissue plasminogen activator has an O-linked fucose attached to threonine-61 in the epidermal growth factor domain, *Biochemistry*, 30, 2311, 1991.
86. Chloupek, R. C., Harris, R. J., Leonard, C. K., Keck, R. G., Keyt, B. A., Spellman, W. M., Jones, A. J. S., and Hancock, W. S., Study of the primary structure of recombinant tissue plasminogen activator by reversed-phase high-performance liquid chromatographic tryptic mapping, *J. Chromatogr.*, 463, 375, 1989.
87. Townsend, R. R., Hardy, M. R., Hindsgaul, O., and Lee, Y. C., High-perfomance anion-exchange chromatography of oligosaccharides using pellicular resins and pulsed amperometric detection, *Anal. Biochem.*, 174, 459, 1988.
88. Basa, L. J. and Spellman, M. W., Analysis of glycoprotein-derived oligosaccharides by high-pH anion-exchange chromatography, *J. Chromatogr.*, 499, 205, 1990.
89. Maley, F., Trimble, R. B., Tarentino, A. L., and Plummer, T. H., Characterization of glycoproteins and their associated oligosaccharides through the use of endoglycosidases, *Anal. Biochem.*, 180, 195, 1989.
90. Vliegenthart, J. F. G., Dorland, L., and Van Halbeek, H., High-resolution, ^{1}H-nuclear magnetic resonance spectroscopy as a tool in the structural analysis of carbohydrates related to glycoproteins, *Adv. Carbohydr. Chem.*, 41, 209, 1983.

91. Barker, R., Nunez, H. A., Rosevear, P., and Serianni, A. S., ^{13}C NMR analysis of complex carbohydrates, *Methods in Enzymology, 83,* Academic Press, New York, 1982, chap. 3.
92. Waard, P., Leeflang, B. R., Vliegenthart, J. F. G., Boelens, R., Vuister, G. W., and Kaptein, R., Application of 2D and 3D NMR experiments to the conformational study of a diantennary oligosaccharide, *J. Biomol. NMR,* 2, 211, 1992.
93. Angel, A. and Nilson, B., Linkage positions in glycoconjugates by periodate oxidation and fast atom bombardment mass spectrometry, *Methods in Enzymology, 193,* Academic Press, New York, 1990, chap. 32.
94. Richter, W. J., Müller, D. R., and Domon, B., Tandem mass spectrometry in structural characterization of oligosaccharide residues in glycoconjugates, *Methods in Enzymology, 193,* Academic Press, New York, 1990, chap. 33.
95. Webb, J. W., Jiang, K., Gillece-Castro, B. L., Tarentino, A. L., Plummer, T. H., Byrd, J. C., Fisher, S. J., and Burlingame, A. L., Structural characterization of intact, branched oligosaccharides by high performance liquid chromatography and liquid secondary ion mass spectrometry, *Anal. Biochem.,* 169, 337, 1988.
96. Gillece-Castro, B. and Burlingame, A. L., Oligosaccharide chacterization with high-energy collision-induced dissociation mass spectrometry, *Methods in Enzymology, 193,* Academic Press, New York, 1990, 689.
97. Spellman, M. W., Basa, L. J., Leonard, C. K., Chakel, J. A., O'Connor, J. V., Wilson, S., and Van Halbeek, H., Carbohydrate structures of human tissue plaminogen activator expressed in Chinese Hamster ovary cells, *J. Biol. Chem.,* 264, 14100, 1989.
98. Shibata, S., Midura, R. J., and Hascall, V. C., Structural analysis of the linkage region oligosaccharides and unsaturated disaccharides From Chondroitin sulfate using carbopac pa1, *J. Biol. Chem.,* 267(10), 6548, 1992.
99. Knepper, T. P., Arbogast, B., Schreurs, J., and Deinzer, M. L., Determination of the glycosylation patterns, disulfide linkages, and protein heterogeneities of baculovirus-expressed mouse interleukin-3 by mass spectrometry, *Biochemistry,* 31, 11651, 1992.
100. Nilsson, B. and Zopf, D., Gas chromatography and mass spectrometry of hexosamine-containing oligosaccharide alditols as their permethylated, *N*-trifluoracetyl derivitives, *Methods in Enzymology, 83,* Academic Press, New York, 1982, chap. 2.
101. Angel, A. and Nilson, B., Linkage positions in glycoconjugates by periodate oxidation and fast atom bombardment mass spectrometry, *Methods in Enzymology, 193,* Academic Press, New York, 1990, chap. 32.
102. Gray, G. R., Linkage analysis using reductive cleavage method, *Methods in Enzymology, 193,* Academic Press, New York, 1990, chap. 31.
103. Hellerqvist, C. G., Linkage method using Lindberg method, *Methods in Enzymology, 193,* Academic Press, New York, 1990, chap. 30.
104. Morand, K., Talbo, G., and Mann, M., Oxidation of peptides during electrospray ionization, *Rapid Commun. Mass Spectrom.,* 7, 1993.
105. Loo, J. A., Udseth, H. R., and Smith, R. D., Collisional effects on the charge distribution of ions from large molecules, formed by electrospray-ionization mass spectrometry, *Rapid Commun. Mass Spectrom.,* 2, 207, 1988.
106. Henion, J. and Conboy, J. J., High-performance anion exchange chromatography coupled with mass spectrometry for the determination of carbohydrates, *Biol. Mass Spectrom.,* 21, 397, 1992.
107. Smith, R. D., Light-Wahl, K. J., Winger, B. E., and Loo, J. A., *Org. Mass Spectrom.,* 27, 811, 1992.
108. Winger, B. E., Light-Wahl, K. J., Ogorzalek Loo, R. R., Udseth, H. R., and Smith, R. D., Observation and implications of high mass-to-charge ratio ions from electrospray ionization mass spectrometry, *J. Am. Soc. Mass Spectrom.,* 4, 536, 1993.

109. Smith, R. D. and Light-Wahl, K. J., The observation of noncovalent interactions in solution by electrospray ionization mass spectrometry: promises, pitfalls and prognosis, *Biol. Mass Spectrom.*, 22, 493, 1993.
110. Teshima, G. and Canova-Davis, E., Separation of oxidized human growth hormone variants by reversed-phase high-performance liquid chromatography. Effect of mobile phase pH and organic modifier, *J. Chromatogr.*, 625, 207, 1992.
111. Hemling, M. E., Roberts, G. D., Johnson, W., Covey, T. R., and Carr, S. A., Analysis of proteins and glycoproteins at the picomole level by on-line coupling of microbore high-performance liquid chromatography with flow fast atom bombardment and electrospray mass spectrometry: a comparative evaluation, *Biomed. Environ. Mass Spectrom.*, 19, 677, 1990.

CHAPTER 8

Matrix-Assisted Laser Desorption Mass Spectrometry for Peptide Mapping

Paolo Lecchi and Richard M. Caprioli

CONTENTS

1 Introduction 220

2 Instrumentation 221

3 Matrix 223

4 Sample Preparation 227

5 Instrument Calibration 229

6 Resolution 231

7 Peptide Mapping 233

8 Sequence-Ordered Peptide Maps 235

9 Summary and Outlook 239

Acknowledgments 240

References 240

1 INTRODUCTION

Mass spectrometry has been used for many years as an analytical technique for peptide mapping of proteins and glycoproteins. Basically, two approaches have been described, (1) direct analysis of the proteolytic digest, and (2) sample cleanup and fractionation of the peptide mixture using LC either on-line with or prior to MS analysis. The first approach is most attractive because of its speed and ease. However, a major shortcoming is that often all peptide ions are not observed even though they are present in the sample. These ion suppression effects commonly occur when complex mixtures of compounds and salts enter the ion source at the same time, i.e., some peptides do not form ions of sufficient intensity to be recorded in the presence of salts or other peptides. The use of LC prior to MS analysis, either off-line or on-line (LC/MS), minimizes and often eliminates this effect, but at the cost of more complex instrumentation and additional time in achieving the analysis.

The introduction of fast atom bombardment mass spectrometry (FAB MS) in the early 1980s[1,2] provided the biochemist with a technique for the mass specific analysis of proteolytic digests where chemical derivatization and purification were not a mandatory part of the analysis procedure. FAB mapping[3] of proteins was introduced and continues to be used quite successfully today. Two major limitations of this technique include a relatively poor sensitivity (typically, 10 to 50 pmol or more of protein is needed) and a limited mass range of about 4000 to 5000 Da. In the latter case, although many tryptic fragments lie below mass 3000, large unhydrolyzed peptides and also those produced from other proteolytic enzymes and chemical cleavage methods, such as cyanogen bromide, are not amenable to FAB mapping. The introduction of continuous-flow FAB (CF-FAB)[4,5] significantly improved the sensitivity and decreased ion suppression effects observed relative to standard FAB conditions, although it operates over the same mass range. CF-FAB also introduced the capability of FAB ionization for LC/MS applications.[6] Plasma desorption mass spectrometry (PDMS), first described by MacFarlane et al. in 1982,[7] provided the means to extend the mass range to 20,000 Da or more and could be used for the analysis of mixtures. The introduction of the nitrocellulose surface for the PDMS technique provided a methodology to bind peptides to the target while allowing salts and other nonbound substances to be washed off.[8] More recently, electrospray ionization, as described by Dole and co-workers[9] and Fenn and Yamashita,[10] also has been used for the analysis of proteolytic digests and has been quite successful in this regard, particularly when used as the ionization method for LC/MS analysis of the digests. This technique is described in detail in Chapter 5 of this volume.

Matrix-assisted laser desorption ionization (MALDI) mass spectrometry was first described by Karas and Hillenkamp in 1988[11] and has since been shown to be extremely effective in the analysis of biomolecules, particularly peptides and proteins, and can produce ions beyond m/z 300,000 in some cases. In addition, the technique has advantages in having a relatively simple sample preparation, a very fast analysis time, and simple instrumentation. The latter is quite attractive to the biological community, who generally are inexperienced in mass spectrometry. Initial experiments with MALDI also indicated that it was effective in the analysis of mixtures including those produced from protease digests.

This chapter will describe the use of MALDI for peptide mapping. Although the technique is still quite new and changes and innovations designed to improve its capabilities are being introduced continuously, a number of papers and investigations involving the analysis of peptide mixtures have been reported. This chapter will describe work done thus far, and draw some general conclusions about the applicability of the technique to analysis of peptides. Unless otherwise specified, the nitrogen laser (337 nm) system will be described since it is most common, especially in commercial instruments. Presently, MALDI is a static technique, i.e., sample aliquots are independently spotted on a target and are subsequently analyzed, on-line LC/MS methods involving proteolytic digests have not yet been described. However, some observations concerning the effectiveness of using off-line LC cleanup and fractionation are made later in this chapter.

2 INSTRUMENTATION

MALDI is a pulsed ionization technique that is capable of producing ions of high mass and therefore is normally coupled with a time-of-flight (TOF) mass analyzer. This analyzer measures the time ions take to traverse the analyzer tube; low mass ions travel faster than high mass ions of the same mass to charge ratio. Under typical conditions, ions at m/z 50,000 would require about 100 µs to traverse a 1-m flight tube, assuming an accelerating voltage of about 25 kV. The technique is based on the use of a solid organic matrix which absorbs light energy at the wavelength produced by the laser. This matrix is mixed in great molar excess with the sample being analyzed. For proteins, approximately 1 µL of a sample containing 0.5 to 1 pmol of protein in water or dilute buffer is mixed with 1 µL of a saturated solution of the matrix in an appropriate solvent, e.g., mixtures of water and acetonitrile. The mixture is then gently dried on the target for subsequent analysis. For an instrument using a nitrogen laser (337 nm), matrix compounds commonly employed are cinnamic acid analogs[12] such as sinapinic acid (SA), caffeic acid (CA), ferulic acid (FA), α-cyano-4-hydroxy cinnamic acid (CCA), and dihydroxybenzoic acid (DHBA), as well as others. The Nd:YAG laser (266 nm)

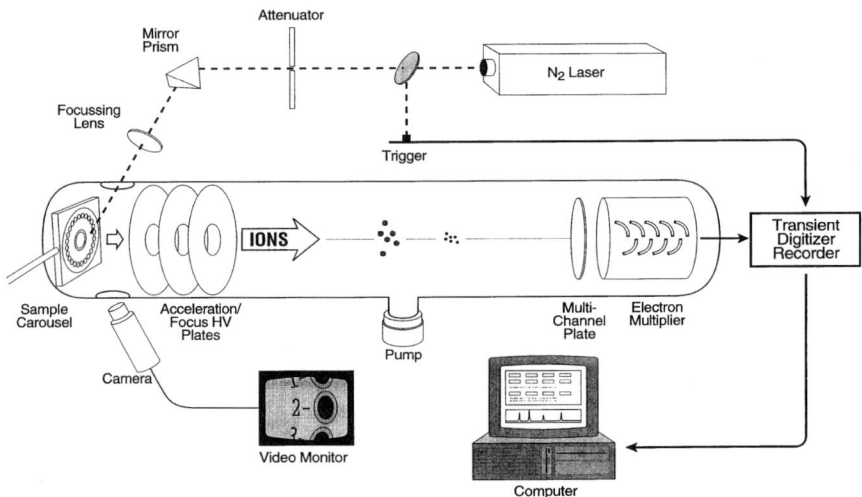

FIGURE 1
Schematic diagram of a MALDI/TOF system of benchtop design (Vestec Corp.).

which is sometimes used employs other matrices, such as nicotinic acid analogs. The general instrumental setup is shown in Figure 1. When the laser beam irradiates the target, energy is absorbed principally by the matrix and, in what is thought to be an explosive expansion and ejection of material from that part of the target surface, ions are desorbed. These ions are accelerated into the TOF analyzer where the flight times needed to reach the detector, in some instruments, a multi-channel plate and an electron multiplier, are recorded by a transient digitizer. Individual spectra are usually summed to achieve better signal-to-noise before being transferred to the data system. In some commercially available instruments, a camera provides a view of the target, a useful aid in viewing the area of the target being irradiated.

TOF analyzers are most often used with MALDI sources because of their high mass range and their compatibility with the pulsed laser process. An additional advantage is their ease of fabrication and thus lower cost of manufacture. Perhaps the greatest disadvantage, at least with respect to current technology, is the limited resolution commonly obtained, e.g., about 500 to 2000 for most linear TOF designs. The use of a reflectron device, first introduced by Mamyrin et al.,[13] allows TOF analyzers to achieve resolutions of 5000 or more, but it is most effective in the low mass range (below about m/z 10000).

The mass measurement accuracy, one of the most important parameters, can vary considerably depending on various analysis conditions. Under the best conditions, a mass measurement accuracy of ±0.01% can be achieved up to moderately high mass (about m/z 20,000), but this measurement can be much poorer (±0.1 to ±0.5%) under less than optimal conditions. This subject is dealt with in more detail later in this chapter as it applies to peptide mixtures.

Although a MALDI/TOF instrument is a relatively simple device — with commercial instruments usually having only computer controlled operation with no manually operated switches, valves, or knobs — there are several analysis parameters that can be changed by the operator and these can affect the performance of the technique for a given analysis. Thus, factors such as the choice of matrix, calibration method, concentration of sample, laser power, instrument resolution, complexity of the sample, presence of salts and/or detergents, etc., can significantly affect both the qualitative and quantitative aspects of the analysis.

3 MATRIX

The choice of the matrix used for the analysis of peptide digests is important, although no single matrix compound is best for all compounds or mixtures. Sinapinic acid and α-cyano-4-hydroxy cinnamic acid[14] are commonly used and generally give high quality spectra for the nitrogen laser system. A number of workers have successfully used other matrices, including caffeic acid, ferulic acid, 2,5-dihydroxybenzoic acid (DHBA), 6-aza-2-thiothymine (ATT), and mixtures of these. Others have used a mixture of dihydroxybenzoic acid and co-matrices such as 5-methoxysalicylic acid or carbohydrates, e.g., glucose and fructose[15] and fucose.[16]

Whatever the particular matrix used, it is clear that as yet no one matrix will give all of the information normally required for the detailed analysis of the digest. Two major difficulties are observed, (1) all matrices produce a background which potentially interferes with small peptides in the mixture between m/z 50 to 600, and (2) different matrices can give different spectra with respect to relative peak heights and even the presence or absence of a particular ion. The backgrounds produced by two different matrices are shown in Figure 2. The top panel for each matrix example was recorded at threshold laser power, sufficient to just produce the major ion species for that matrix. The lower panel represents the laser power required to produce an ion from a peptide added to the matrix, and more closely represents the background one would get during a peptide analysis. Here it is seen that the multiplicity of peaks would prevent any high sensitivity analysis below m/z 600. Dihydroxybenzoic acid (DHBA) perhaps has the lowest intensity background for the matrices commonly used in MALDI analysis of peptide mixtures, but also generally does not give the best spectrum in that often some of the peptides present in the mixture are not observed in the spectrum.

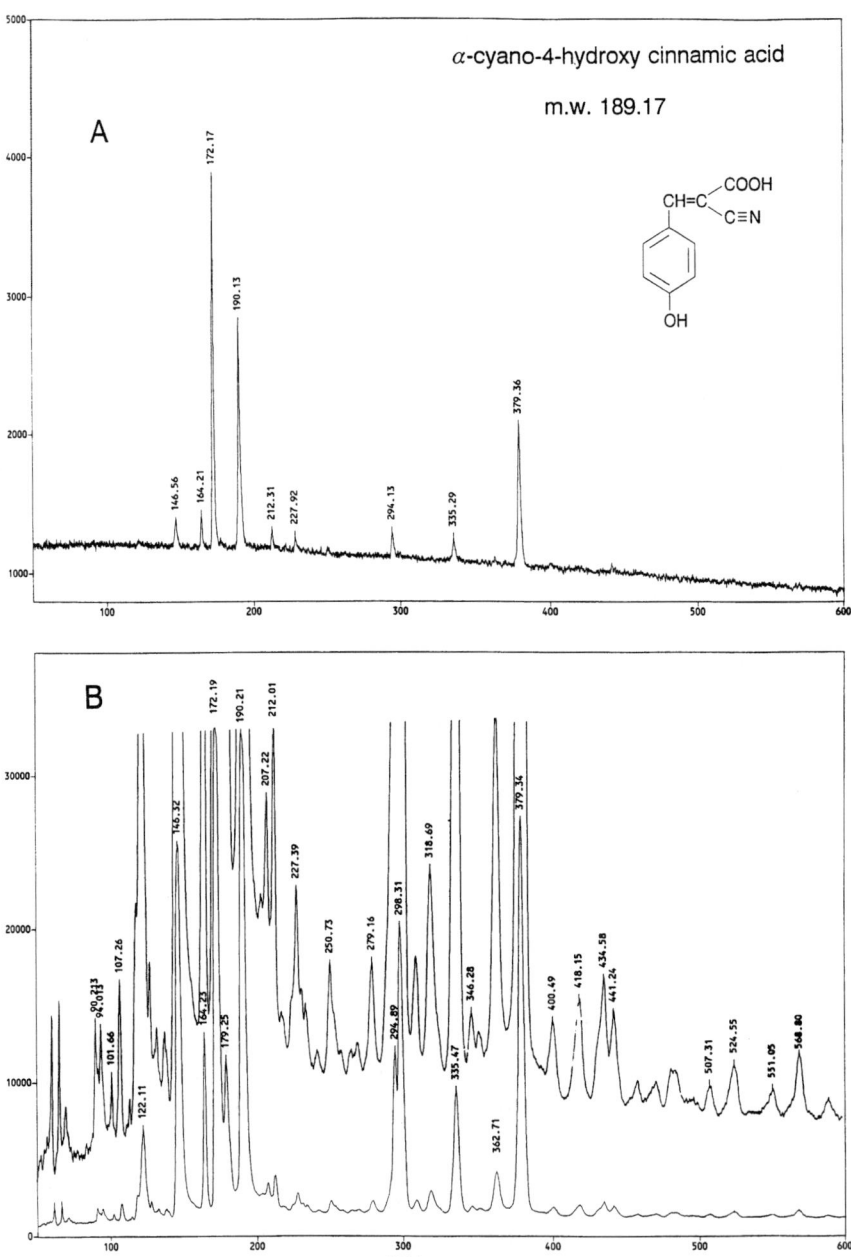

FIGURE 2
Backgrounds produced by two matrices commonly used in MALDI MS. The spectra were produced in panels (A) using minimum laser power to generate low mass ions, and (B) using laser power necessary to record a protein signal (cytochrome c) co-mixed with the matrix.

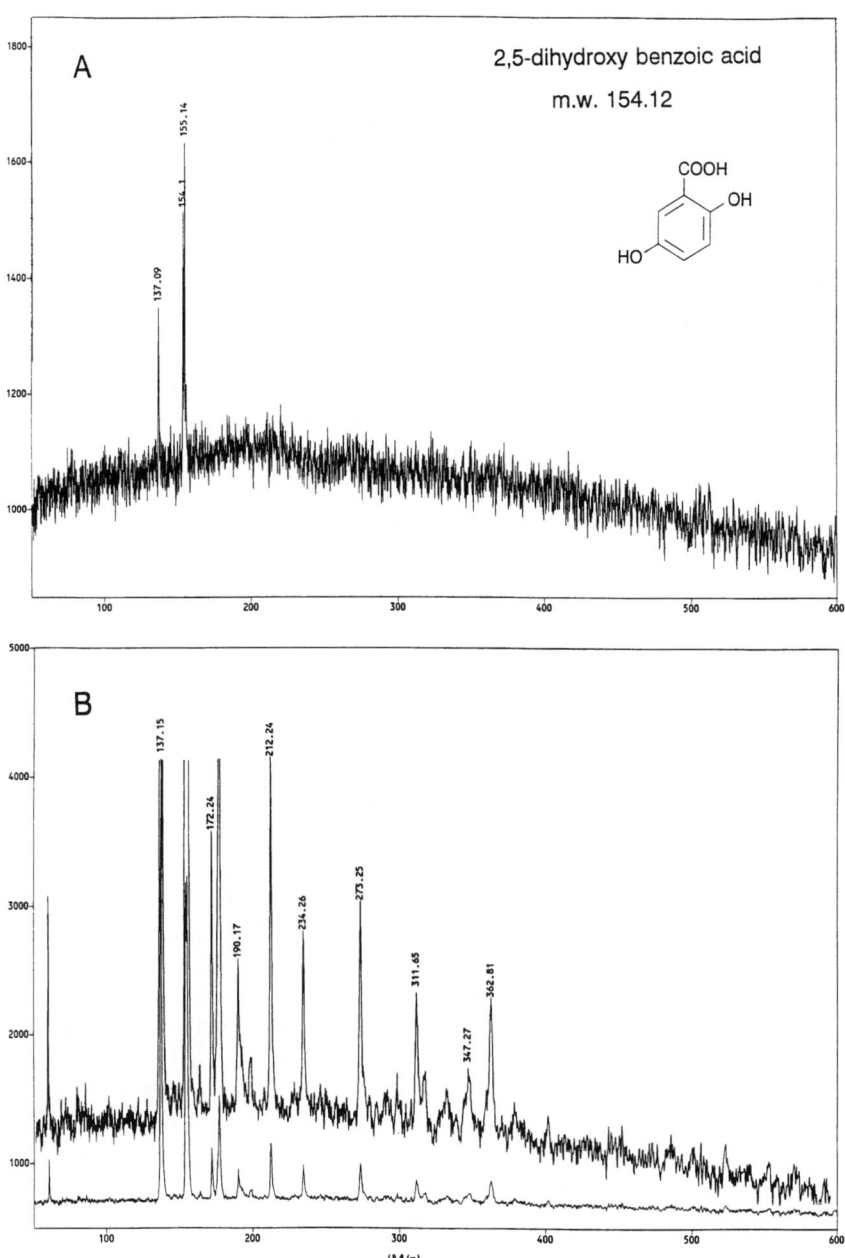

FIGURE 2 (continued)

Juhasz et al.[17] studied the utility of several common matrices with the β-chain of hemoglobin A (mol wt 15,127), horse myoglobin (mol wt 16,952), RNase B (mol wt 15,000), and bovine rod transducin α-chain (mol wt 44,000). For the small proteins around 15,000 Da, MALDI mass spectra, recorded using the matrices SA, CA, DHBA, and CCA, all showed 95 to 100% of the expected ions for the tryptic peptides and other proteolytic fragments. The matrix ATT was somewhat poorer than these, providing 49 to 100% of the expected fragments for a number of different protease digests. The larger protein, transducin α-chain, showed 82% of the expected tryptic fragments in the spectrum.

Billeci and Stults[16] reported results obtained by using several co-matrices with DHBA for the analysis of the tryptic digests of recombinant human growth hormone (rhGH) (mol wt 22,125) and recombinant tissue plasminogen activator (r-tPA) (mol wt 59,008). The r-tPA is also interesting because it contains several sites of glycosylation. Glucose, fructose, fucose, β-lactose, maltose, and sucrose were used in several molar ratios. For rhGH, a matrix consisting of fucose plus DHBA (1:1) produced an ion for 24 out of the 25 expected tryptic fragments at a level of 280 fmol of digested protein. These authors noted that slow drying of the matrix/sample mixture was important to achieve optimal results. Compared to a spectrum obtained from DHBA alone, the fucose/DHBA matrix gave a spectrum that had an increase in the absolute signal for many peptides, less intense ions derived from the matrix itself, improved resolution, and decreased adduct ion formation, e.g., $(M+Na)^+$. A similar study for a tryptic digest of r-tPA gave similar but less dramatic results when compared to the same digest analyzed in DHBA alone. Overall, 40 of the 51 expected r-tPA tryptic fragments were unambiguously identified, or 78% of the total. Nine of the other peptides were not resolved from others, or only partially resolved from neighboring peaks, making identification somewhat uncertain. Five of the six peptides that did not produce ions were di- and tripeptides.

Lecchi and Caprioli[18] examined the effect of several matrices for the analysis of the tryptic digest of *E. coli* recombinase A (recA) (mol wt 37,000). The same matrices as those cited above were used, including the fucose/DHBA matrix cited previously. In this experiment, only fragments between m/z 470 to 3500 were recorded. Out of the 29 expected tryptic fragments, 45% were seen with SA, 48% with FA, 90% with CCA, and 48% with DHBA/fucose. These results are summarized in Table 1. The best results were obtained from CCA for this digest, as shown in Figure 3, with the DHBA/fucose mixture being less effective. Although CCA gave the best results overall, the DHBA/fucose gave fewer and less intense background ions below m/z 500. No single matrix showed all of the expected fragments for this and several other protease digests analyzed, while each gave slightly different spectra. The recA digest is an important example from the point of view that FAB analyses also failed to record many peptides; only LC/MS

TABLE 1

Effectiveness of MALDI Matrices[a] in Identifying Fragments in a Single Sample of Rec A Digest

Fragment	Mol wt	CCA	SA	FA	ATT	DHBA + fucose
F4	479.6	X			X	
F25	487.6	X			X	
F33,F36	502.6	X				
F23	516.6	X		X		
F39	518.6					
F5	563.7	X		X	X	X
F6	589.6	X	X			X
F18	615.7	X			X	
F30	646.8	X		X	X	X
F1	689.9		X			
F22	742.8	X			X	X
F38	774.8					
F16	855.1	X		X	X	X
F37	875.0	X				
F34	946.0	X				
F3	1085.3	X	X	X	X	X
F26	1177.2	X	X			X
F8	1279.4	X		X	X	X
F9	1386.6	X	X	X	X	X
F19	1560.8	X	X			
F14	1742.0	X		X	X	X
F15	1757.3	X	X	X	X	X
F21	1808.3	X	X	X	X	X
F11	1907.1	X	X	X		
F7	2650.0	X	X	X		
F31	2704.0	X	X			
F13	2974.3	X	X	X		
F42	3043.1	X	X	X		

[a] Abbreviations defined in text.

using CF-FAB did a nearly complete analysis, identifying all but one of the fragments in this mass range.

4 SAMPLE PREPARATION

At the present (early) stage in the development of MALDI for mixture analysis, it is clear that much is yet to be done in the area of sample preparation. Current methods have been empirically devised and provide adequate methodology, but it is known that these are far from ideal. Typically, solutions of matrix compounds are prepared in approximately 0.1 M concentrations in a solvent containing 70% acetonitrile, 30% water, and

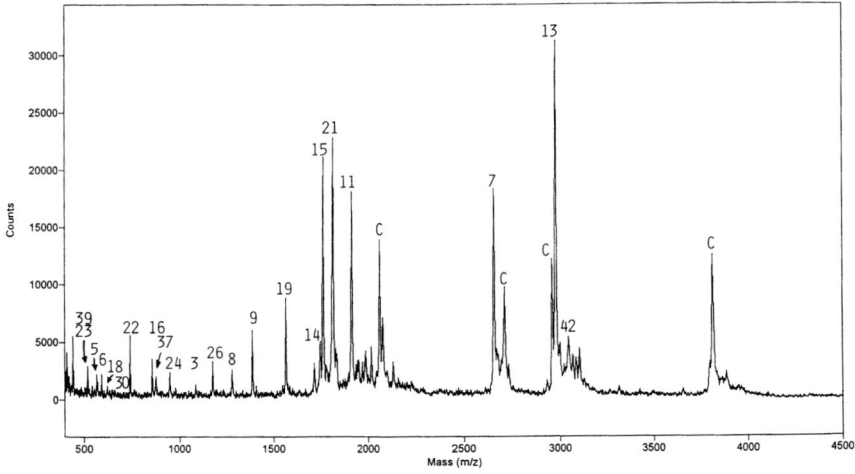

FIGURE 3
MALDI mass spectrum of the tryptic digest of *E. coli* recombinase A using CCA as the matrix: 1 pmol of digested protein was spotted on the target. The numbers refer to the tryptic fragments listed in Table 1, and "c" denotes a fragment produced by a chymotryptic or tryptic/chymotryptic cleavage.

0.1% trifluoroacetic acid. In the case of DHBA and fucose, and other more water-soluble compounds, a 98:2 water/acetonitrile solution containing 0.1% trifluoroacetic acid solution is used. Approximately 1 µL of matrix solution is mixed on the target with 0.5 to 1.0 µL of sample solution containing 0.5 to 1 pmol of peptide or protein and the sample allowed to slowly dry.

Air drying or using a gentle stream of dry clean nitrogen gas gives the best results, and this correlates with the production of a highly crystalline sample surface. Rapid drying such as that which would be achieved using a hot air gun gives a decidedly poorer spectrum. This process can be clearly seen in Figure 4 for the preparation of a sample of cytochrome C using sinapinic acid as a matrix. In the case of the slowly dried sample, the target surface appears highly crystalline. The spectrum obtained from this target shows an $(M+H)^+$ ion at m/z 12,360 with the actual molecular weight of the protein being 12,359. The less intense peak on the high mass side is the dehydrosinapinic acid adduct peak 206 mass units higher, and the doubly charged molecular species $(M+2H)^{+2}$ is recorded at m/z 6180. The resolution was measured to be approximately 200. In many spectra, particularly those of proteins, a dimer $(M_2+H)^+$ of low intensity is often recorded. These are generally laser-induced and are not usually biochemically significant. In the case of the incorrectly prepared sample (rapid heat drying), the surface appears amorphous and noncrystalline. Moreover, the best spectrum that could be obtained, shown in the figure, is significantly poorer in quality than that shown above. The molecular species gave a weak, broad and ill-defined peak which showed no fine structure, with a resolution of 10 or less. Although this spectrum was produced from a target that was deliberately dried with a hot air gun to create a bad target preparation, it shows the dependence of the technique on careful preparation procedures.

The amount of sample loaded on the target can also have an effect on the relative intensities of ions recorded in the spectrum. The MALDI analysis of proteins and peptides generally gives the best spectrum with sample loadings of 0.5 to 5 pmol in a 2 μL sample/matrix spot. Higher sample amounts can lead to ion suppression effects. The relative changes in intensities that can occur are shown in Figure 5 for a portion of the digest of recA when the spectra were prepared from two dilutions of a digest mixture containing 2.5 and 25 pmol, with all other experimental parameters remaining the same. The area of the spectrum around m/z 3000 is particularly noteworthy in that relative intensities change drastically, particularly for fragments T7, T13, and T42. The precise cause of this effect is not known in this case, but nevertheless points out the importance of sample concentration. In cases where measurements of relative abundances are important, it is advisable to run each digest at several dilutions, say, 0.3, 0.8, and 2 pmol, in order to determine whether this effect is significant in a given sample.

5 INSTRUMENT CALIBRATION

The flight time of an ion is converted to mass by calibration procedures that use a simple mathematical equation and known mass standards. Generally, two (or more) calibrants are needed to specify the time-to-mass conversion for a given analysis. Two basic techniques can be used: external and internal calibrations. For external calibration, the known mass standards are run as a separate sample, usually within a reasonably close time to that of the sample of interest. Internal calibration requires the specification of two or more known masses within a given spectrum; often this is achieved by adding two calibrants to the sample peptide mixture itself. The advantage of external calibration is that it does not require adulteration of the sample, further complicating an already complex protein digest mixture. On the other hand, since this is done at a different time than that of the sample of interest, slight changes in any of the instrument or analysis parameters can give increased mass measurement error. Typically, one can expect a mass measurement accuracy of about ±0.3 to 0.05% with protease digests with careful external calibration. Internal calibrations either require the analyst to accurately know the masses of two or more ions in a spectrum, or require the addition of internal calibrants. The first is problematic since a "known" mass peak may be slightly shifted due to an unknown and unresolved component, while the second requires adulteration of the sample, potentially leading to ion suppression effects and further complicating the mixture. However, internal calibration

FIGURE 4
(A) Target tips for (right) properly prepared sample and (left) poorly prepared sample of cytochome c, using sinapinic acid as the matrix. (B) Mass spectrum from the correctly prepared sample, and (C) that from the incorrectly prepared sample.

gives the best mass measurement accuracy, typically 0.05 to 0.01%. Other techniques meant to bridge the two, e.g., split sample targets where one side contains calibrants and the other sample, do not offer any particular advantage and give results roughly between the two.

If we consider the recA digest once again, an example of the mass measurement accuracy of one analysis using external standards is given in Table 2. In this measurement, potassium (m/z 39) and mellitin (m/z 2848.5) ions were used as mass standards, with the calibration spectra acquired just before the tryptic digest analysis. These data are particularly interesting because the calibrated masses all show a negative deviation from the actual masses, ranging from approximately −0.1 to −0.3%. It is clear that some important parameter was different between the calibration analysis and sample analysis, giving a different flight time of a given ion in each analysis. Internal standards obviate this problem. Table 2 also shows the results of the analysis of the same recA digest using internal standards, again using m/z 39 and m/z 2848.5 as calibrants, the latter being doped into the sample prior to analysis. Here, the magnitude of the error is approximately in the range ±0.05 to 0.01%, with both positive and negative variances. Clearly, internal standard calibration gives superior results and is preferable for obtaining the most accurate mass measurements.

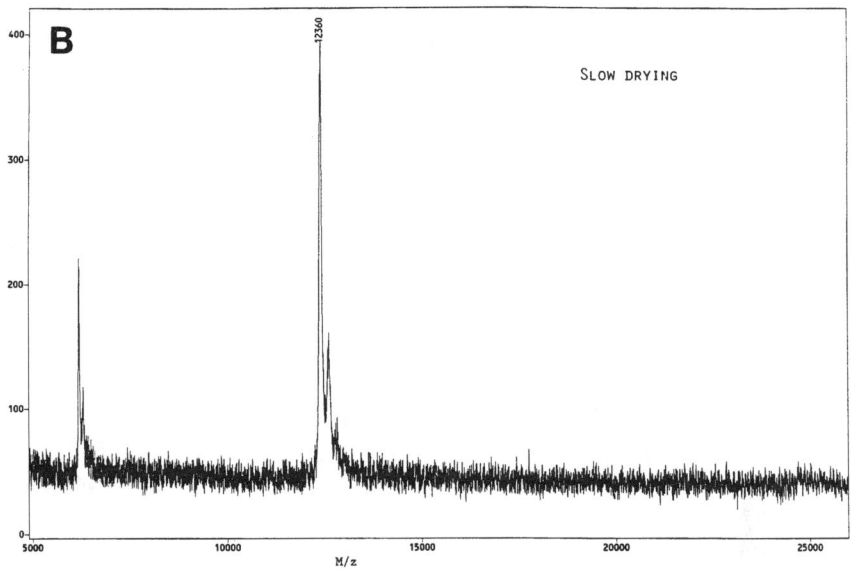

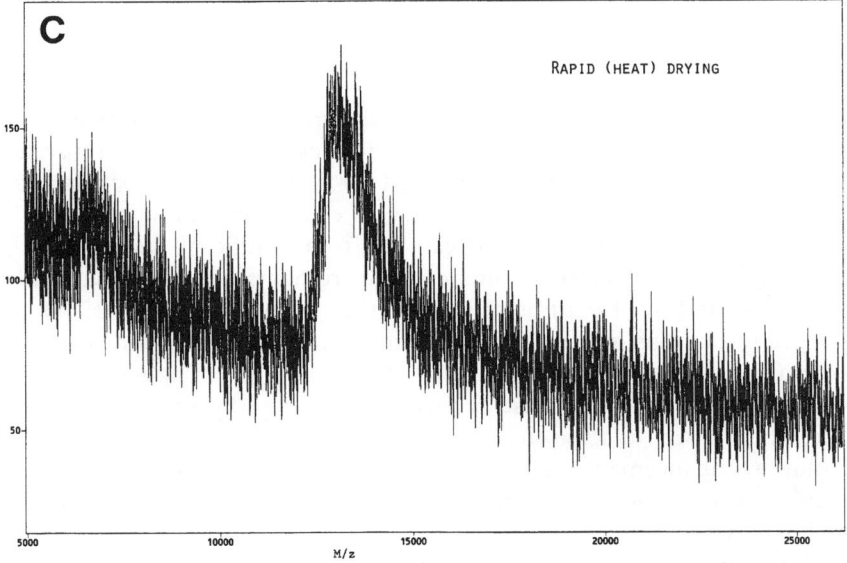

FIGURE 4 (continued)

6 RESOLUTION

The resolution of a mass spectrometer is its capability for the separation of two adjacent mass peaks, usually at some specified condition such as 10% valley, 50% valley, or half-width at full maximum (HWFM) for a given ion or ions. Generally, the equation R = M/ΔM, where R is the resolution, M the

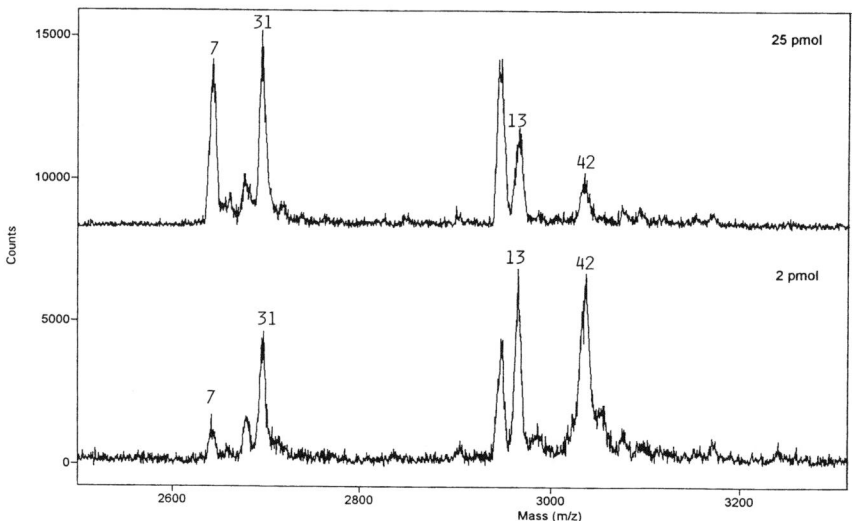

FIGURE 5
MALDI mass spectra of the tryptic digest of recombinase A using sinapinic acid. (A) Spotting of 25 pmol on target, and (B) 2.5 pmol on the target from 10 × dilution of that used in (A).

nominal mass, and ΔM the difference in mass between the two adjacent ions, describes this parameter (but in itself does not define the degree of separation achieved). The resolution usually quoted for MALDI/TOF instruments is obtained from measurements made under the very best conditions, and for a given analysis several conditions can lead to a significant decrease in this resolution. For many current commercial instruments, a linear design with a flight tube of 1 to 1.5 m at an accelerating voltage of about 25 kV can produce resolutions greater than 2000 or so under optimal conditions for a protein in the 10,000 Da range. Use of a reflectron, or 'ion mirror', will produce a considerable increase in resolution, perhaps by a factor of three or four. This device, at least those available today, does not provide such a resolution enhancement above about 10,000 Da, and in fact may lead to a significantly decreased signal intensity beyond this range when compared to the linear system.

Resolution, like sensitivity, is difficult to define in terms of how much is enough. Generally, one can say that for mixtures of peptides where most masses lie below m/z 2000 to 3000, a practical (routinely achievable) resolution of 2000 is good. However, this will not always be sufficient since there will be cases where two peptide fragments are close in mass, for example, the case where both have the same nominal mass but differ by their fractional mass. For optimal results in the analysis of protease digests, unit resolution (HWFM) up to about m/z 2000 to 3000 is highly desirable.

TABLE 2

Calibration[a] of MALDI MS Analysis of Rec A Tryptic Digest

Actual mass	% Mass measurement accuracy	
	External	Internal
479.6	−0.32	+0.03
516.6	−0.23	+0.04
589.6	−0.31	−0.06
742.8	−0.14	+0.12
855.1	−0.24	+0.06
1085.3	−0.29	+0.05
1177.2	−0.19	+0.02
1279.4	−0.27	+0.02
1386.6	−0.14	−0.02
1560.8	−0.13	−0.02
1742.0	−0.18	−0.01
1757.3	−0.14	+0.05
1808.3	−0.32	−0.06
1907.1	−0.16	+0.08
2650.0	−0.11	+0.03
2947.3	−0.14	+0.01

[a] Both internal and external calibration used m/z 39 and 2848.5 (melittin).

From sample to sample, the resolution of the mass peaks in a particular spectrum is dependent not only on the inherent design of the instrument, but also on specific analysis parameters, such as the method of sample preparation (see Figure 4) and laser intensity used to generate the spectrum. High laser power will cause a decrease in the resolution in a particular spectrum, although ion intensities may be increased. This energy spread in a given ion is lowest, i.e., the resolution is greatest, at the 'threshold' level of a given ion. This is seen in Figure 6 for the analysis of cytochrome c at both threshold and high laser power. In the former case, the resolution was calculated to be about 400, while in the latter, it was 150.

7 PEPTIDE MAPPING

The effectiveness of MALDI MS for peptide mapping has been demonstrated by several investigators for proteins ranging from about 15,000 to 50,000 Da. Relatively speaking, there is not yet a great deal published, and general statements and conclusions about the overall effectiveness for MALDI for the analysis of peptide digests are difficult to make. The chemical properties

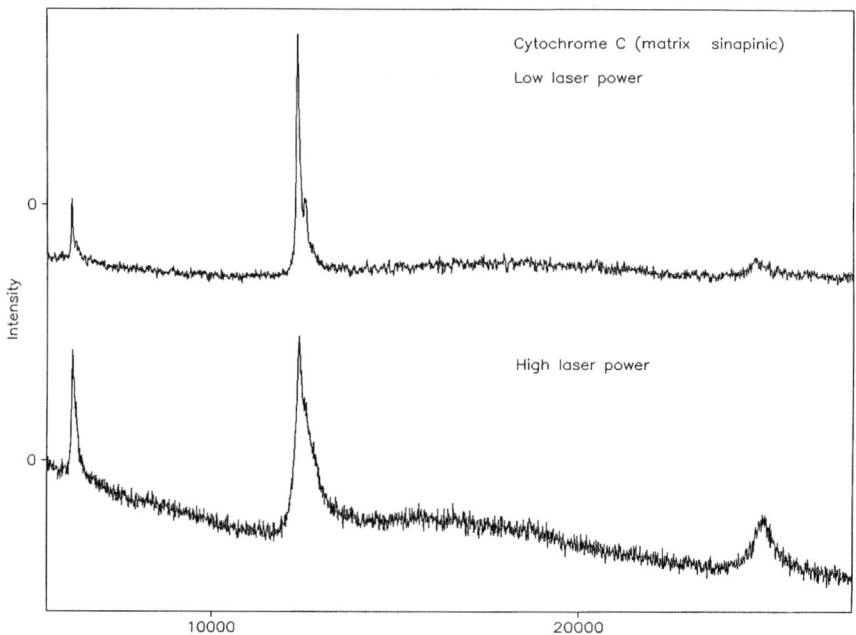

FIGURE 6
MALDI mass spectra of cytochrome c at different laser powers.

of fragment peptides, sample contaminants, sample preparation techniques, matrix effects, and even operator experience with such a new technology all can vary appreciably for a given sample. Nevertheless, the following paragraphs will attempt to summarize the state of knowledge based on what has been published and presented thus far.

It is not surprising that for relatively small proteins, say in the 10,000 to 15,000 mol wt range, MALDI appears to provide quite complete mapping information, usually identifying between 95 to 100% of the fragments. This has been shown by Juhasz et al.[17] for hemoglobin A β-chain with tryptic, endo-Asp-N, and endo-Glu-C digestions, horse myoglobin with endo-Asp-N digestion, and RNase B with tryptic digestion, and employing common matrices such as SA, CA, DHBA, and CCA. Annan et al.[19] identified 24 peptides in a partial tryptic digest of sperm whale apomyoglobin using MALDI on a magnetic MS equipped with an array detector. However, small proteins produce only a relatively small number of fragments and, for most, represent digests of purified proteins of known sequence and purity.

A few investigators have reported on the analysis of tryptic digests of large proteins by MALDI MS, i.e., proteins above 25,000 Da and, as expected, the situation becomes significantly more complicated. Billeci and Stults[16] have published the most extensive investigation thus far in which they considered various parameters using rhGH (mol wt 22,125) and r-tPA (mol wt 59,008).

As discussed earlier, these workers tested the effectiveness of MALDI using common matrices, including one they introduced that is a mixture of DHBA and fucose. Using the latter matrix, all but 1 of the 21 tryptic fragments of rhGH were observed from a sample containing 2.8 pmol of the digested protein. For the more complex digest of r-tPA, the analysis of 4.5 pmol of digested protein led to 40 of 51 peptides unambiguously identified, or 78%, while another 5 peptides were tentatively assigned, giving a total of 88% by this count. The mass measurement accuracy for these analyses were generally about ±1 Da or better, using internal calibrations where one of the tryptic fragments itself was the calibrant. Juhasz et al.[17] also analyzed the tryptic digest of bovine rod transducin β-chain (mol wt 44,000) and were able to identify 82% of the tryptic fragments. Lecchi and Caprioli[18] analyzed the tryptic digests of rhGH (mol wt 22,125) and *E. coli* recA (mol wt 36,700) and identified 87% and 90%, respectively, using CCA as the matrix. For recA, additional analyses with different matrices cumulatively led to the identification of 95% of the fragments. This conclusion, i.e., the use of several analyses with different matrices for a single sample mixture, was also made by others[16] and represents the most effective means to date to maximize the effectiveness of MALDI MS for complex mixture analysis.

The foregoing discussion has dealt with the analysis of whole protein digests, without removal of buffer salts and enzymes or fractionation of the mixture of peptides prior to MS analysis. It is clear that fractionation of the peptides, even into a relatively small number of samples, can greatly increase the effectiveness of the process. Peptides that were not detected in the MALDI analysis of the mixture, could be detected after the mixture was fractionated and purified by LC methods.[20] Beyond a doubt, LC separation is highly recommended when a rigorous identification is needed for all the peptides, although this comes at the expense of time and, often, requires increased amounts of sample.

8 SEQUENCE-ORDERED PEPTIDE MAPS

Often it is desirable to know the sequence position of a particular protease digest fragment within the protein chain. Normally, when a proteolytic (e.g., tryptic) digest of a protein is performed, the digest is allowed to proceed for many hours to achieve as complete a hydrolysis as possible before the fragment peptides are identified, for example, by FAB mapping techniques. Such an analysis generates a list of peptides identified by their molecular weights, but of course provides no information about how these were arranged in the original protein.

A time-course analysis of the digestion of a protein contains a great deal of information besides the identification of the final limit digest peptides. Most importantly, the digestion process is spread out in time: a consequence of steric hinderance in which not all of the peptide bonds susceptible to a given protease are available to that protease at the same time, and also because of different rates of hydrolysis of susceptible bonds. This produces intermediate fragment peptides which are themselves further degraded. Such precursors and products are simply related by mass; the molecular weights of two product peptides minus the mass of a water molecules equal that of the precursor.

Mass spectrometry can easily be used to identify these precursor/product relationships which can, by simple calculation, be used to determine a nearest-neighbor analysis of the proteolytic fragments. This was shown by Whaley and Caprioli[21] for the nearest-neighbor analysis of several peptides in the 3000 to 4000 Da range using trypsin digestion and CF-FAB to measure the molecular weights of the peptides. Substrate-to-enzyme ratios of 1:1000 were used to maximize the lifetime of precursor peptides for MS analysis. It is emphasized that in such an analysis, although time-course aliquots are removed and analyzed, the measurement of time itself is unnecessary; the time-course merely allows transient precursors to be measured. Analysis of time-course aliquots from proteolytic digests to allow identification of intermediates using mass spectrometry have been reported in several recent publications.[20-23]

MALDI MS has been used by Caprioli et al.[20] to generate a sequence-ordered proteolytic map of ribonuclease A. The reduced, pyridylethyl derivative of this protein was hydrolyzed with clostripain and samples removed at various time intervals, with the last being at 24 h. Approximately 8 pmol of the digest were then analyzed by MALDI MS using sinapinic acid as a matrix. Table 3 shows the masses of the fragments identified along with the mass measurement accuracy for the cumulative analysis of the time-course aliquots. Ten different intermediate peptide fragments could theoretically be produced during the digestion with an additional five fragments remaining at the limit digest of 24 h (see Table 3). Experimentally, seven intermediate peptides were identified by MALDI ranging in molecular weight from 3,384 to 13,403. The three intermediate peptides not seen in the spectra involved the hydrolysis of the arg-85 residue, adjacent to a pyridylethyl cysteine (position 84). It is presumed the bulky pyridylethyl group protected this bond from significant hydrolysis. From these data, 11 precursor/product relationships were easily made and these are given in Figure 7. It is possible that mathematically, given the error in determining the molecular weight of the peptides based on the mass measurement accuracy of 0.02 to 0.04%, an erroneous fit might be made. However, the redundancy of overlap information would make such an error obvious and, in fact, no such matching errors occurred in this case. Thus, the sequence order starting at the N-terminal position, was determined to be m/z 1151 (1 to 10), m/z 2658 (11 to 33), m/z 747 (34 to 39), m/z 5600 (40 to 85), and m/z 4454 (86 to 124). For

TABLE 3
Clostripain Fragments of Pyridylethylated Ribonuclease A

Predicted fragment	Calc (M+H)⁺	Measured (M+H)⁺	% Error
(1–10)[a]	1151.3	1151.1	–0.02
(11–33)[a]	2657.9	2657.9	0.00
(34–39)[a]	746.8	746.8	0.00
(40–85)[a]	5600.4	5600.1	–0.01
(86–124)[a]	4452.0	4453.7	0.04
(1–33)	3790.2	3790.2	0.00
(11–39)	3385.7	3386.3	0.02
(34–85)	6328.2	6325.2	–0.04
(40–124)	10033.4	10035.3	0.02
(1–39)	4518.0	4518.8	0.02
(11–85)	8967.1	—	—
(34–124)	10761.2	10763.3	0.02
(1–85)	10099.4	—	—
(11–124)	13400.1	13402.9	0.02
(1–124)	14532.4	14532.4	0.00

[a] Peptides remaining after complete digestion.

ribonuclease A, this verified the sequence by establishing the order of the limit digest peptides. In the case of an unknown, unless the N-terminal or C-terminal peptide were known, the reading direction of the ordered fragments would not be known, i.e., one would generate two possible sequence-ordered maps that would be the reverse order of each other. This problem is easily overcome even in the absence of any other sequence information, because the C-terminal peptide can be easily identified if the hydrolysis is carried out in oxygen-18-labeled water.[21]

Other investigators have also demonstrated the use of MALDI with the time-course analysis of protease reactions. Schar et al.[23] used mixtures of carboxypeptidases to digest parathyroid hormone (1 to 34) and identified 18 peptides that differed in sequentially missing the C-terminal residue. Although several intermediates were not identified because their ion intensities were too low, it is interesting to note that the procedure allowed sequencing beyond these missing fragments because of the mass specificity of the technique. Beavis and Chait[22] used MALDI MS and time-course sampling with incomplete cyanogen bromide digestion of human apolipoprotein AI, a protein containing 243 residues and generating 5 cyanogen bromide fragments on total hydrolysis. In this case, five overlapping fragment correlations were able to be made.

a) List of Observed Clostripain Fragments

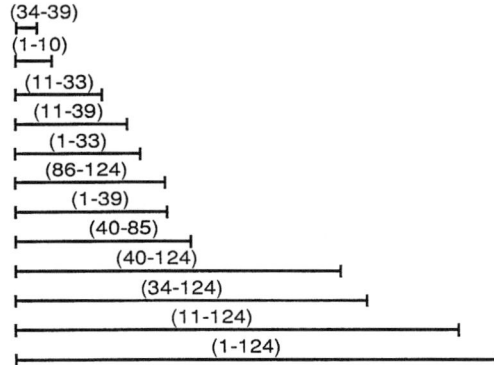

b) Precursor/Product Correlations

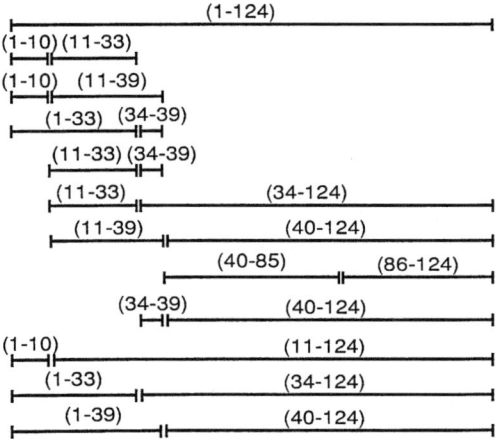

FIGURE 7
(a) Peptide fragments identified in the digestion of pyridylethylated ribonuclease A by clostripain throughout the time-course reaction, and (b) precursor/product correlations constructed from the data in (a).

The ability to construct sequence-ordered protease digest maps is a consequence of the inherent sequence overlap information present in a time-course digest. The ability of mass spectrometry to identify all peptide fragments generated during the reaction allows the ordered construct to be made with certainty. Such a construct provides unique information in checking the sequence of proteins using an analytical method that is very rapid, requires about a picomole of material, and involves simple sample preparation procedures.

9 SUMMARY AND OUTLOOK

The analysis of complex mixtures of peptides has been a long standing need in biological research, and it is clear that mass spectrometry is particularly well suited to this application. Contributions from technological innovations including FAB, PD, and electrospray ionization, LC/MS, tandem MS, and ion trapping instrumentation, as well as many others, continue to provide new methods to analyze mixtures. MALDI adds to this list, providing several advantages that are especially powerful, if not unique. However, as in all this specialized technology, the performance of the instrument and thus the quality of the spectra can be correlated with the effort and care used in the analysis.

With respect to the application of MALDI MS to the analysis of protease digests, one can consider two approaches:

1. A survey analysis, in which the analytical task is not critical in that not all peptide fragments are required to be identified and mass measurement accuracy is not critical. Such a case, for example, would be the checking of the presence of a peptide in a known system or following the extent of degradation or reaction of a protein. In this case, MALDI is indeed an extremely rapid and sensitive MS technique able to provide mass specific information in minutes. A single matrix compound can be used, e.g., CCA, using about 1 pmol of digested protein. External calibration is sufficient, typically giving ±0.3% mass assignment accuracy. One can expect to easily identify 50 to 75% of the peptides present in the mixture. The total time for such an analysis of a digest can be expected to be 15 to 30 min or less.

2. Optimal performance is achieved with significantly more diligence. The analytical task here is to identify all peptide fragments with the highest mass measurement accuracy possible. Often, high sensitivity is required. In this case, several matrices should be used individually, e.g., CCA, DHBA/fucose, SA, and others, to maximize the total number of fragments observed. The sample should be analyzed at several different concentrations (for example, 5, 1, and 0.5 pmol loaded) and data acquired at several different laser power settings. The use of internal standards is very important, and care must be taken to ensure that peaks from these do not overlap with any of those of the peptides in the digest. Under these conditions, one can expect 90 to 100% of the fragments to be identified, and typically with ±0.03% mass measurement accuracy. An analysis time of several hours can be expected.

MALDI MS brings to the biological research community a powerful analytical instrument with which to probe the molecular weights of compounds. Its ease of use and speed in obtaining data is most attractive and it is at a price that is in line with other high-technology instrumentation in biological research. At the moment, it has its greatest utility in the field of protein and peptide analysis, although one can expect significant advances in this methodology in other areas in the near future. It is clear that MALDI will be used extensively in biological research.

ACKNOWLEDGMENTS

The authors thank Drs. Piero Bianco and George Weinstock for the generous gift of *E. coli* recombinase A, and NIH (grant #GM43783-04) for partial support of this work.

REFERENCES

1. Barber, M., Bordoli, R.S., Sedgwick, R.D., and Tyler, A.N., *J. Chem. Soc. Chem. Commun.*, 325, 1981.
2. Barber, M., Bordoli, R.S., Elliott, G.J., Sedgwick, R.D., and Tyler, A.N., *Anal. Chem.*, 54, 645A, 1982.
3. Morris, H.R., Panico, M., and Taylor, G.W., *Biochem. Biophys. Res. Commun.*, 117, 299, 1983.
4. Caprioli, R.M., Fan, T., and Cottrell, J.S., *Anal. Chem.*, 58, 2949, 1986.
5. Caprioli, R.M., *Continuous-Flow Fast Atom Bombardment Mass Spectrometry,* Caprioli, R.M., Ed., John Wiley & Sons, Chichester, 1990, 1.
6. Caprioli, R.M., DaGue, B.B., and Wilson, K., *J. Chromatogr. Sci.*, 26, 640, 1988.
7. Torgerson, D.F., Skowronski, R.P., and MacFarlane, R.D., *Biophys. Res. Commun.*, 60, 616, 1974.
8. Jonsson, G.P., Hedin, A.B., Håkansson, P.L., Sundqvist, B.U.R., Göran, B., Säve, S., Nielsen, P.F., Roepstorff, P., Johansson, K.E., Kamensky, I., and Lindberg, M.S.L., *Anal. Chem.*, 58, 1084, 1986.
9. Dole, M., Mack, L.L., Hines, R.L., Mobley, R.C., Ferguson, L.D., and Alice, M.B., *J. Chem. Phys.*, 49, 2240, 1968.
10. Yamashita, M. and Fenn, J.B., *J. Phys. Chem.*, 88, 4451, 1984.
11. Karas, M. and Hillenkamp, F., *Anal. Chem.*, 60, 2299, 1988.
12. Beavis, R.C. and Chait, B.T., *Rapid Commun. Mass Spectrom.*, 3, 432, 1989.
13. Mamyrin, B.A., Karatajev, J.J., Shmikk, D.V., and Zagulin, V.A., *Sov. Phys. JETP*, 37, 45, 1973.
14. Beavis, R.C., Chaudhary, T., and Chait, B.T., *Org. Mass Spectrom.*, 27, 156, 1992.
15. Köster, C., Castoro, J.A., and Wilkins, C.L., *J. Am. Chem. Soc.*, 114, 7572, 1992.
16. Billeci, T.M. and Stults, J.T., *Anal. Chem.*, 65, 1707, 1993.
17. Juhasz, P., Papayannopoulos, I.A., Zeng, C., Papov, V., and Biemann, K., Proc. 40th ASMS Conf. Mass Spectrom. and Allied Topics, Washington, D.C., May 1992, 1913.
18. Lecchi, P. and Caprioli, R.M., Proc. 41st ASMS Conf. Mass Spectrom. and Allied Topics, San Francisco, CA, June, 1993.
19. Annan, R.S., Köchling, H.J., Hill, J.A., and Biemann, K., *Rapid Commun. Mass Spectrom.*, 6, 298, 1992.
20. Caprioli, R.M., Whaley, B., Mock, K.K., and Cottrell, J.S., *Techniques in Protein Chemistry,* Vol. II, 48, 497, 1991.
21. Whaley, B.S. and Caprioli, R.M., *Biol. Mass Spectrom.*, 20, 210, 1991.
22. Beavis, R.C. and Chait, B.T., *Proc. Natl. Acad. Sci. U.S.A.*, 87, 6873, 1990.
23. Schär, M., Bornsen, K.O., and Gassman, E., *Rapid Commun. Mass Spectrom.*, 5, 319, 1991.

INDEX

A

Accuracy
  computer simulation, 51, 53
  mass measurement, 222–223, 229–230
N-Acetylhexosamine, 204, 207
N-Acetyl-neuraminic acid, 204, 207
Activase, 85–87
Affinity chromatography, 129–133
Amino-terminal truncations, 11–16
Artifacts
  from electrospray, 209–211
  isolation, 11–15
  from peptide mapping, 23–24
Asparagine residues, 19–20
Aspartic acid side chains, 16–19
Automatch, 74, 75
6-Aza-2-thiothymine (ATT), 223

B

Buffer strength, 102

C

Caffeic acid, 223
Calibrations
  HPLC, 76–77
  MALDI, 229–231, 233
  wavelength, 64
Calmodulin, 33
Capillary diameter
  CE-MS, 171–175
  HPLC, 189
Capillary electrophoresis, 97–115
  and mass spectrometry, 143–176
Capillary HPLC, 122–123, 124, 189–190

Carbohydrate chemistry, 182, 191–194
  sugar residue differences, 203–207
Carboxyl-terminal truncations, 11–16
β-Casein, 111, 132–133
CF-FAB. *See* Continuous-flow FAB (CF-FAB)
Charge state, 126, 137, 197, 203
Chemical degradations, 15–16
Chemical modifications, 22–23
Chromatograms, 126, 128–129
Chymotrypsin, 109
Clostripain, 2
Collision-induced dissociation (CID), 131, 183, 197–201, 207–209
Columns
  capillary, 123
  characteristics, 122
  coating, 109
  effects from, 39, 41, 46, 80, 81–83
  temperature, 46–51
Complementary techniques, 105, 114
Computer simulation
  need for software, 94
  of peptide mapping, 33–54
  peptide mixture example, 36–37, 38–41
  potential problems, 51, 53
  rhGH, 52
  r-tPA, 36, 42–45, 47–50
Concanavalin A, 111
Confirmation of structure, 98
Continuous-flow FAB (CF-FAB), 120–121, 145, 146–147, 220, 227. *See also* Fast atom bombardment (FAB)
Contour plots, glycopeptides, 201–202
Correlation coefficient, 75
Critical resolution, 36, 37, 44, 52
Crossmatch, 74, 75
Cyanogen bromide, 109

α-Cyano-4-hydroxy cinnamic acid, 223
Cytochrome c, 114, 115, 158–161, 224–225, 234
  sample preparation, 228, 230–231

D

Deamidification, of asparagine residues, 19–20
Degradation
  chemical, 15–16
  Edman, 133
  products of, 105
Derivative spectra, 89–90, 109
Desalting, 137, 187
Descending paper chromatography, 2
Detection limits, 120, 121, 126, 168–171
Detection modes, 105, 108
DHBA, 223, 226, 227, 235
2,5-Dihydroxybenzoic acid (DHBA), 223, 226, 227, 235
Diode array detection, 60
Discriminator value (D-value), 74, 75, 79–80
Dissimilarity score, 62–64
  and noise, 65–71
  and peak identification, 77–79, 91–92
  and purity, 72
Disulfides, 3–9
DryLab, 34–45, 53
D-value, 74, 75, 79–80

E

E. coli recombinase A, 226–227, 230, 232, 235
Edman degradation, 3, 133
Efficiency, 99, 147, 170
Electric field strength, 102
Electroosmotic flow, 100, 104, 150
Electropherograms, 158, 159, 160, 161, 174
Electrophoresis, 2, 97–115
Electrospray ionization mass spectrometry (ESI-MS), viii, 60, 119–137, 145, 163, 168
  artifacts, 209–211
  carbohydrate identification, 182
  historical background, 183–184
  interfaces, 147–152
  ion sources, 183–187, 188

Endoproteinase Lys-C, 2
ENHANCE, 125
Equation R, 231–232
Erythrocytes, 166
ESI-MS. See Electrospray ionization mass spectrometry (ESI-MS)
External calibration, 229, 233

F

FAB. See Fast atom bombardment (FAB)
Fast atom bombardment (FAB), viii, 59, 120, 183, 220, 226. See also Continuous-flow FAB (CF-FAB)
Ferulic acid, 223
"Fingerprints" of proteins, 2, 3–24, 98
Fluorescamine, 109, 115
Fluorescence techniques, 111
Formic acid, 189
Forward library searching, 76
Fourier transform ion cyclotron resonance mass spectrometer (FT-ICR), 121, 164, 166, 167
Fucose, 191, 194, 204, 235

G

Gas chromatograph/mass spectrometer (GC/MS), quadrupole, 187–189
Glucagon, 111, 162
Glycoproteins
  α_1-acid, 111
  cleavage methods, 194–196
  map analysis of, 181–211
  recombinant HIV-1 envelope, 3, 19–20
  structure, 98
Glycosylated peptides, 112, 128
  contour plots, 201–202
  locating, 196–202
  predicting sites, 195–196
  series deductions, 206, 208
gp120, 3, 19–20
Gradient steepness, 35, 42, 46–51
Gradient time, 36–37

H

Hemoglobin, 2, 105, 108, 161, 166, 234
HexNAc+Hex, 204

Hexose, 204, 207
hGH. *See* Human growth hormone (hGH)
High-performance liquid chromatography (HPLC)
   capillary HPLC, 122–123, 124, 189–190
   HPLC-DAD, 60
   reversed phase. *See* RP-HPLC
High-voltage electrophoresis, 2
Historical background, vi–viii, 2, 183–184, 220
HIV-1 envelope glycoprotein, 3, 19–20
HPLC-DAD, 60
HSA, 109, 110, 163–164
Human AGP, 111
Human CD4, 3
Human growth hormone (hGH), 3, 4. *See also* Recombinant human growth hormone (rhGH)
   deamidation, 19
   isolation artifacts, 11–15
   methionone oxidation, 31
   shelf life, 15–16
Human serum albumin (HSA), 109, 110, 163–164

I

IGF-1. *See* Insulin-like growth factor (IGF-1)
Immobilized metal-ion affinity chromatography, 129, 132–133
Immobilized proteases, 124
Indirect fluorescence detection, 111
Information aspects
   library seaching, 76–77, 82
   multidisciplinary approach, 94
   peptide mass database, 134–136
   software, need for, 94
   spectral database, need for, 93–94
In situ digestion, 124, 136
Insulin, 2, 22
Insulin-like growth factor (IGF-1), 3, 4, 5–6, 9
   chemical modifications, 22–23
   RP-HPLC, 14
   shelf life, 15–16
   V_8 protease map, 14
Internal calibration, 229–230, 233
Internal cleavages, 9–11

Ionic strength, of buffer, 102, 104
Ion trap mass spectrometer (ITMS), 165–166
Isolation artifacts, 11–15
Isomerization, of aspartic acid and proline residues, 16–19
ITMS, 165–166

L

β-Lactoglobulin A, 109
Laser-induced fluorescence detection, 111
LC/MS. *See* Liquid chromatography-mass spectrometry (LC/MS)
Library searching, 76–77, 82
Linearity, 79
Liquid chromatography-mass spectrometry (LC/MS), viii, 60, 121, 190–191
   and capillary electrophoresis, 143–176
   complex series of ions, 203–207
   from glycoprotein digests, 196–211

M

MALDI. *See* Matrix-assisted laser desorption/ionization (MALDI)
Mannose, 191, 193–194, 203
Mass chromatogram, 126, 128–129
Mass resolution, 126
Mass spectrometry, 93, 120–121, 153–156, 183, 220. *See also individual methods*
Match factor (MF), 62
Match limit, 64, 73–76
Match threshold, 60
Matrix-assisted laser desorption/ionization (MALDI), viii, 120, 146, 195, 219–240
Matrix compounds, for peptide digests, 223–227
MCA, 88, 89
MECC, 114
Methionine, 20–22, 84, 109
MF, 62
MHC Class-I molecules, 133
Micellar electrokinetic capillary chromatography (MECC), 114
Mixtures, protein, 36–37, 38–41
Monosaccharides, 191–194, 204
MS/MS, 131, 133, 135, 158, 195, 207

Multicomponent analysis (MCA), 88, 89
Myoglobin, 109, 111

N

Natriuretic peptide receptor C, 129, 130
Nearest-neighbor analysis, 236
N-linked structures, 191–194
  predicting glycosylation sites, 195–196
Noise, from instrumentation, 64–71, 222
Normalization, of spectrum, 63

O

Oligosaccharides, 191, 193
O-linked monosaccharide structures, 194
Orthogonal separation, 98
Oxidation, of methionine residues, 20–22
Oxonium ions, 197–200

P

PD/MS, 146, 220
Peak correlation, 80
Peak identification, 60, 76–80, 97
  discriminator value, 79–80
  reduced dissimilarity score, 77–79
Peak purity, 60, 71–73
Peak shape, 101, 109
Peak tracking, 53, 80–83
Peptide mapping
  historical background, vii–viii, 183–184, 220
  LC-MS, 124–137
  MALDI, 233–238
  sequence-ordered, 235–238
  subtleties, 1–25
Peptide sequencing, 131, 133–136
pH
  capillary electrophoresis, 100
  disulfide exchanges, 5
  reversed phase chromatography, 189
  trypsin-catalyzed peptide bond synthesis, 8–9
Phosphopeptides, 128, 129
Phosphorylated proteins, 98, 111
Plasma desorption/mass spectrometry (PD/MS), 146, 220
Post-translationally modified proteins, 112, 126–131
Potency, 97
Precision, 100–101
Predictive accuracy, 51, 53
Problems and limitations
  capillary electrophoresis, 99
  capillary HPLC, 122, 124
  CE-MS, 145, 164–168
  computer simulation, 51, 53
  discriminator values, 79–80
  HPLC-DAD, 60
  off-line separation, 145
  sample purity, 137
  spectral matching, 64, 71, 74–76
  tryptic mapping, 59
ProDigest-LC, 34
Prolines, 197
Protease reactions, 237, 239
Protein identity, 97–98, 134–136
Protein mixtures, 36–37, 38–41, 136–137
Protropin, 84–85
Purity
  peak, 60, 71–73
  of protein sample, 32–33, 37, 72, 97
PVDF membrane, 135, 136

Q

Quadruple ion trap, 121
Quadrupole mass filter, 187–189
Quadrupole mass spectrometry, 121, 126, 156, 158, 197
Quality control, 97

R

Recombinant human erythropoietin (rHuEPO), 112, 113
Recombinant human glycoproteins, 181–211
Recombinant human growth hormone (rhGH), 10
  computer simulation, 52
  LC-MS, 125
  match limit example, 74
  peak analysis, 84–85
  tryptic digest, 37–41, 45, 46, 47–49, 226
  tryptic mapping, 59, 61, 68–70, 78, 90–91
Recombinant human tissue factor, 3

Recombinant tissue plasminogen activator
  (r-tPA), 10–11, 12, 17, 187
 Asp-N digests, 199
 computer simulation, 36, 51
 contour plots, 202
 tryptic digest, 33, 41–45, 47–50, 190,
   199, 205
 tryptic mapping, 85–87
Reduced elution speed, 168–171
Reflectron, 222, 232
Relaxin, 3–4, 5, 7–8, 9
 deamidation, 19, 20
Repeatability. *See* Reproducibility
Reproducibility, 99, 101, 109, 112, 150
Resolution
 CE, 100
 computer simulation, 35
 MALDI, 222, 228, 231–233
 rhGH, 44, 126
 r-tPA, 51
Reversed-phase high performance liquid
  chromatography. *See* RP-HPLC
Reversed-phase peptide maps, analysis of,
  181–211
Reverse library searching, 76–77, 78
rhGH. *See* Recombinant human growth
  hormone (rhGH)
rHuEPO, 112, 113
Robustness, 99, 100, 101, 150
RP-HPLC, 2–3
 detection limit of protein variants, vi–vii
 IGF-1, 14, 16
 with mass spectrometry, 119–137
 relaxin, 7, 8
 rhGH, 81
 r-tPA, 10–11
r-tPA. *See* Recombinant tissue plasminogen
  activator (r-tPA)
Ruggedness. *See* Robustness

S

Sample preparation
 capillary LC-MS, 123–124
 CE-MS, 145–146
 MALDI, 227–229
Sample stacking, 102
SDS-PAGE, 195, 196
Selected ion monitoring (SIM), 156
Selectivity, 2, 32–33, 37, 99, 100
 column temperature, 46–51
 gradient steepness, 35, 42, 46–51

Sensitivity, CE-MS, 168–171
Separation buffer, 102, 103
Separation optimization, 99,
  152–153
Series deductions, glycopeptides, 206,
  208
Sheath flow, 147, 148, 149, 150, 151
Shelf life, 15–16
Similarity profiles and curves, 70, 72, 73,
  78–79
Similarity score, 62–64, 72
Sinapinic acid, 223, 236
Software, need for, 94
Solvent splitting, 122
Specificity. *See* Selectivity
Spectral library, 76–77, 82
 database development, 93–94
Spectral matching, 60, 62–65, 90
Spectral similarity, 62
Spectral smoothing, 89–92
Spectral stripping, 83, 87–89
Staphylococcus aureus V_8, 2, 8, 9
Structural variants, vi–vii, 98. *See also*
  Variant identification
Substitutions, 22

T

Tandem mass spectrometry, 131, 144,
  165
TFA, 2–3, 32, 84, 189
Time-course analysis, 235, 237
Time-of-flight (TOF), viii, 121, 195,
  221–222
Total ion current (TIC), 124–125, 126,
  134, 137
Transpeptidation, 8–9
Trifluoroacetic acid (TFA), 2–3, 32, 84,
  189
Trypsin, 2, 8, 23, 106, 109, 195
Tryptic digest
 β-casein, 111, 132–133
 bovine somatotropin, 161
 E. coli recombinase A, 228, 230, 232
 glucagon, 162
 hGH, 127
 human serum albumin, 110
 rhGH, 37–41, 47–49, 125, 226,
   235
 rHuEPO, 112, 113
 r-tPA, 41–46, 47, 48, 85–87, 190, 199,
   205, 235

Tryptic mapping, vii, 5, 58–59
  analysis of, 181–211
  aspartic acid and proline residues, 17–18
  β-casein, 111
  calmodulin, 33
  deamidation, 19
  hemoglobin, 108, 161
  hGH, 21, 23
  human AGP, 111
  relaxin, 20, 21
  rhGH, 59, 61, 68–70, 78, 90–91
  RP-HPLC vs. CE, 105, 107
  r-tPA, 19, 32
  UV-spectral information, 57–94
Two-dimensional separation, 114, 134, 136, 201

U

UV electropherogram, 158, 159, 160, 161
UV-spectral information, 57–94

V

Variant identification
  amino acid substitutions, 22
  amino- and carboxyl-termini truncations, 11–16
  artifacts, 11–15, 23–24
  chemical modifications, 22–23
  deamidification, asparagine residues, 19–20
  internal cleavages, 9–11
  isomerization, aspartic acid and proline residues, 16–19
  oxidation, methionine residues, 20–22
V_8 protease maps
  IGF-1, 5, 6, 9, 14, 17, 24
  Staphylococcus aureus V_8, 2, 8, 9

W

Wavelength calibration, 64